Introduction to Multidimensional Nonstationary Radar Signal Analysis Using Fourier Transforms

This book bridges the gap between theoretical signal processing and practical radar applications. It presents cutting-edge Fourier-based methods for analyzing, decomposing, and extracting information from complex radar signals that change over time and space.

Introduction to Multidimensional Nonstationary Radar Signal Analysis Using Fourier Transforms provides the first comprehensive treatment of time-frequency and multidimensional signal analysis specifically tailored for radar applications. It features proprietary algorithms for one-, two-, and multidimensional signal decomposition using advanced techniques like the windowed two-dimensional Fourier transform (W2D-FT). This book includes numerous real-world radar examples, datasets, and visualization results that demonstrate practical implementation, while the accompanying MATLAB® toolbox enables immediate hands-on application. By combining classical Fourier methods with recent advances such as synchrosqueezing and adaptive decompositions, this book helps readers transform raw radar data into meaningful information about objects, motion, and environmental conditions. End-of-chapter exercises and questions are included to promote practical and theoretical comprehension.

This book is specifically designed for senior undergraduate and graduate students studying electrical engineering, signal processing, radar, sonar, or related fields, as well as practicing engineers, researchers, and specialists involved in radar system design, development, or analysis. This content is particularly valuable for professionals working in defense applications, civilian radar systems, and communication technologies who need to understand how mathematical concepts translate into real-world signal processing solutions. A solutions manual is available to adopting professors.

Introduction to Multidimensional Nonstationary Radar Signal Analysis Using Fourier Transforms

Karol Abratkiewicz

CRC Press
Taylor & Francis Group
Boca Raton London New York

CRC Press is an imprint of the
Taylor & Francis Group, an **informa** business

MATLAB® is a trademark of The MathWorks, Inc. and is used with permission. The MathWorks does not warrant the accuracy of the text or exercises in this book. This book's use or discussion of MATLAB® software or related products does not constitute endorsement or sponsorship by The MathWorks of a particular pedagogical approach or particular use of the MATLAB® software.

First edition published 2027
by CRC Press
2385 NW Executive Center Drive, Suite 320, Boca Raton FL 33431

and by CRC Press
4 Park Square, Milton Park, Abingdon, Oxon, OX14 4RN

CRC Press is an imprint of Taylor & Francis Group, LLC

ISBN: 978-1-041-28649-3 (hbk)
ISBN: 978-1-041-28650-9 (pbk)
ISBN: 978-1-003-76578-3 (ebk)

DOI: 10.1201/9781003765783

Typeset in Times
by codeMantra

Dedicated to my father, who inspired my path,
and to my mother, whose strength has carried me along it.

Contents

Preface ... x
Author ... xiii

Chapter 1 Radar Concept ... 1

1.1 Radio Detection and Ranging ... 1
1.2 Main Radar Configurations ... 3
1.2.1 Principle of Operation ... 4
1.2.2 Geometry ... 4
1.2.3 Applications ... 4
1.2.4 Frequency Bands ... 6
1.2.5 Waveform Types ... 6
1.3 Basic Formulas ... 7
1.3.1 Radar Range Equation ... 7
1.3.2 Target Range Estimation ... 8
1.3.3 Range Resolution ... 9
1.3.4 Target Velocity Estimation ... 10
1.3.5 Velocity Resolution ... 10
1.3.6 Correlation ... 11
1.3.7 Matched Filter ... 12
1.3.8 Ambiguity Function ... 12
1.4 Summary ... 15
1.5 Chapter in a Nutshell ... 15
1.6 Exercises ... 16
Notes ... 17

Chapter 2 Analysis of One-Dimensional Radar Signals ... 18

2.1 Introduction ... 18
2.2 One-Dimensional Fourier Transform ... 19
2.2.1 Continuous Fourier Transform ... 20
2.2.2 Discrete-Time Fourier Transform ... 21
2.2.3 Discrete Fourier Transform (DFT) ... 22
2.2.4 Fast Fourier Transform ... 24
2.2.5 Fourier Transform Properties ... 26
2.2.6 Examples ... 33
2.3 Stationarity ... 35
2.4 Short-Time Fourier Transform ... 36
2.4.1 Inverse Short-Time Fourier Transform ... 38
2.4.2 Spectrogram ... 39

2.4.3 Examples 40
2.4.4 Window Width Choice 41
2.5 Adaptive Short-Time Fourier Transform 46
2.6 Short-Time Chirplet Transform 49
2.7 Time-Frequency Reassignment 58
2.8 Multitaper STFT and Reassignment 68
2.9 Time-Frequency Synchrosqueezing 76
2.9.1 Vertical Synchrosqueezing 77
2.9.2 Horizontal Synchrosqueezing 92
2.10 Signal Parameter Estimation 97
2.11 STFT and Spectrogram Morphology 106
2.12 Summary 111
2.13 Chapter in a Nutshell 112
2.14 Exercises 113
Notes 114

Chapter 3 Analysis of Two- and Multidimensional Radar Signals 115

3.1 Introduction 115
3.2 Spatial Micro-Doppler Signature 117
3.3 ISAR Image Filtering 124
3.4 Two-Dimensional Fourier transform 131
3.4.1 Continuous Two-Dimensional Fourier Transform 132
3.4.2 Two-Dimensional Discrete-Time Fourier Transform 133
3.4.3 Two-Dimensional Discrete Fourier Transform 133
3.4.4 Two-Dimensional Fast Fourier Transform 133
3.4.5 Two-Dimensional Fourier Transform Properties ... 134
3.4.6 Example 140
3.5 From One-Dimensional Signal to Bivariate Spectrum 141
3.6 STFT for Two-Dimensional Signal – Windowed Two-Dimensional Fourier Transform 148
3.7 Improving W2D-FT Readability 153
3.7.1 Adaptive W2D-FT 153
3.7.2 Adaptive Windowed Two-Dimensional Chirplet Transform 159
3.7.3 Real-Life Example 162
3.8 W2D-FT Concentration 166
3.8.1 Four-Dimensional Reassignment 166
3.8.2 Generalization 171
3.8.3 Dimension-Selective Reassignment of W2D-FT... 173
3.9 Signal Retrieval from the Concentrated W2D-FT 190

3.10 Multitaper Reassignment in W2D-FT 198
3.10.1 Multitaper Dimension-Selective W2D-FT Reassignment: ISAR Image Noise Flattening 198
3.10.2 Multitaper W2D-FT Reassignment in Four Dimensions: Range-Velocity Map Case Study 209
3.11 Summary 215
3.12 Chapter in a Nutshell 217
3.13 Exercises 218
Notes 219

Chapter 4 Target Kinematic Parameter Estimation in Radars 220

4.1 Introduction 220
4.2 Range and Velocity Estimation in Radars 221
4.3 Problem of Range and Velocity Migration 225
4.4 Acceleration Estimation 231
4.4.1 Extended Cross-Ambiguity Function 232
4.4.2 Generalized Radon-Fourier Transform for Long-Time Coherent Integration 235
4.4.3 Zak Transform-Based Target Acceleration Estimation 241
4.5 High-Order Target Kinematic Parameter Estimation 252
4.6 Multitaper Spectrum Estimation in Radars – Efficient Way to Noise Flattening and Simultaneous Echo Sharpening 261
4.7 Summary 268
4.8 Chapter in a Nutshell 269
4.9 Exercises 270
Notes 272

Chapter 5 Conclusions, Challenges, Prospects 273

References 275

Index 289

Preface

129,864,880—according to a Google Books analysis, this is the estimated number of distinct book titles that had been cataloged by the year 2010 [1]. Since then, it is believed that up to four million new titles are published annually, bringing us to approximately 190 million books in existence. Among them are the most influential and timeless works of human civilization: *The Bible*, *The Quran*, *The Dialogues of Plato*, Darwin's *On the Origin of Species*, and other classics that have shaped our understanding of ethics, nature, and knowledge. Literature has flourished in countless directions, from the theatrical brilliance of Shakespeare, the dystopian foresight of Orwell, the mythic landscapes of Tolkien, to the philosophical depths of Dostoyevsky and the intricate narratives of Tokarczuk. Add to that the vast sea of scientific publications, textbooks, essays, and technical manuals that guide entire disciplines. Out of this immense ocean of human thought and expression, you have chosen this book—a text on the specialized subject of radar signal processing. Perhaps by curiosity, perhaps by necessity, you are here. And that, in itself, is a rare and humbling distinction. For one among millions, your attention has settled here, and it is my privilege to speak to you through these pages.

The story behind this book began in 2014, when I joined the student research group at Gdańsk University of Technology. Under the guidance of my future friend Marek Płotka, I took my first steps into the world of radar. Later, as a master's student between 2012 and 2017, I continued this journey and, in 2016, worked with my colleagues on an ambitious student project: building a small-scale synthetic aperture radar system from scratch. This initiative was part of one of our advanced coursework assignments in the later semesters of our degree program. Despite our enthusiasm and the technical effort we invested, the project was met with skepticism. As a result, the radar never received the institutional support we had hoped for. Still, the idea stayed with me. That early attempt, though unofficial and academically disregarded, sparked an interest that has continued to grow ever since. What was once dismissed as a side curiosity became the seed of a long-term passion and, eventually, the foundation of this book.

In 2018, I left my position at the Gdańsk University of Technology and moved to Germany for a three-month internship at the Fraunhofer Institute for High Frequency Physics and Radar Techniques (FHR). There, under the guidance of Dr. Diego Cristallini, Dr. Daniel O'Hagan, and Martin Ummenhofer, I began my professional journey in radar signal processing and was introduced to the field of passive radar systems. Following that experience, I commenced my PhD studies at Warsaw University of Technology under the supervision of Prof. Piotr Samczyński. In 2022, I successfully defended my doctoral dissertation entitled "Methods for analysis and synthesis of radar signals using chirp rate estimation in the time-frequency domain." Currently, I am pursuing a habilitation degree at the same university and remain actively engaged in numerous research and development projects related to radar technology and digital signal processing, as a member of the Radar Techniques Research Lab at the Institute of Electronic Systems.

This book would never have been written without the team I am fortunate to be a part of. First and foremost, I would like to express my deepest gratitude to my supervisors: Dr. Krzysztof Czarnecki, who introduced me to the beauty of time-frequency signal processing during my master's studies, and Prof. Piotr Samczyński, who opened the door to the world of radar. I cannot imagine my career path without their guidance and support. Their encouragement, insightful discussions, and contributions, many of which are reflected in the results and analyses presented in this book, have been instrumental in shaping its content. I would also like to extend my sincere thanks to my colleagues from the Warsaw University of Technology: Prof. Mateusz Malanowski, Dr. Zbigniew Gajo, Dr. Gustaw Mazurek, Dr. Marcin Bączyk, Dr. Marcin Żywek, Dr. Sandhi Wangiyana, Dr. Jędrzej Drozdowicz, Marek Płotka, Jacek Gambrych, and Bartosz Falęcki, whose collaboration, critical feedback, and contributions of data significantly enriched this work.

I would also like to express my heartfelt gratitude to my family and loved ones, whose unwavering support has accompanied me throughout my entire educational journey. Thank you for unconditionally standing by me, even when the purpose of my work may not have always been clear. I am especially grateful to my mother and to my beloved Gosia for their constant encouragement and presence. This book is dedicated to my parents, with a special remembrance of my father, who first introduced me to the world of electrical engineering. May he rest in peace.

I wrote this book as the one I wish I had when I first began my journey with time-frequency signal analysis. Although the field is vast and built upon decades of extensive research, I found a lack of a single, comprehensive resource that addressed the specific challenges I encountered throughout nearly ten years of experience. The theoretical foundations are scattered across numerous articles, reports, and books, often presenting differing definitions, explanations, and notations. This inconsistency frequently led to confusion, especially when trying to develop a coherent understanding of the underlying principles of signal processing. With this book, I aim to clarify and unify these scattered concepts within a single, cohesive framework tailored to radar signal processing. It presents a general signal-processing perspective applied to radar, supported by a solid theoretical foundation in multidimensional and nonstationary signal analysis. My goal is to extend generic signal-processing methods with rigorous mathematical formulations and numerical experiments, offering a comprehensive and practical reference for researchers and practitioners alike.

Most of the MATLAB functions implementing the techniques described in this book are publicly available on my GitHub profile:

`https://github.com/kabratkiewicz`

Readers are encouraged to use, extend, and contribute to this toolbox. The implementations are designed to be generic and versatile, allowing the processing of any signal. However, specific signals (especially those originating from military systems) cannot be freely shared. The same limitation applies to certain radar-related implementations, as they are subject to export license regulations. Consequently, the methods described in Chapter 4 are not publicly available due to legal restrictions.

This book is structured as follows: Chapter 1 introduces the general concept of radar systems. It covers fundamental principles, commonly used terminology, and the basics of radar operation.

In Chapter 2, we explore one-dimensional signal-processing methods with a focus on frequency and time-frequency analysis of radar signals. Topics include the Fourier transform and its variants, signal energy representation using spectrograms, and signal reconstruction and denoising techniques for more accurate interpretation.

Chapter 3 extends this framework to two-dimensional (and, more broadly, multidimensional) signals, addressing practical challenges in radar data analysis. We extract signal features that support accurate representation and parameter estimation through two-dimensional Fourier analysis. Particular attention is given to range-velocity processing and radar imaging, including enhancement and filtering techniques for multidimensional radar data.

Chapter 4 applies the techniques discussed in earlier chapters to estimate target kinematic parameters in both active and passive radar scenarios. While the previous chapters focus on general signal-processing techniques with selected radar applications, this chapter specifically addresses the analysis of target motion, particularly variations in velocity.

Finally, Chapter 5 presents concluding remarks, highlights current challenges, discusses future research directions, and outlines opportunities for further development in the field of radar signal processing.

Author

Karol Abratkiewicz was born in Brodnica, Poland, in 1993. He received his B.Sc. (with honors) and M.Sc. in Electronics and Telecommunications from the Gdańsk University of Technology (GUT), Gdańsk, Poland, in 2016 and 2017, respectively. He earned his Ph.D. (with honors) in Telecommunications from the Warsaw University of Technology (WUT), Warsaw, Poland, in 2022.

From 2015 to 2018, he was affiliated with GUT, participating in international research and development projects focused on providing wireless and embedded infrastructure for the high-tech industry. In the summer of 2018, he completed a research internship at Fraunhofer Institute for High Frequency Physics and Radar Techniques (Fraunhofer FHR) in Wachtberg, Germany, with the Department of Passive Radar and Antijamming Techniques.

Since 2018, he has been a researcher with the Radar Research Group at WUT and has held the position of Associate Professor since 2023. He is actively involved in national and international radar research projects, working on advanced radar signal processing, including time-frequency analysis, signal parameter estimation, and range-Doppler processing.

He has published extensively in leading journals in the field, including *IEEE Transactions on Geoscience and Remote Sensing*, *IEEE Transactions on Image Processing*, *IEEE Signal Processing Letters*, *IET Radar, Sonar & Navigation*, *IEEE Sensors Journal*, and *IEEE Journal of Selected Topics in Applied Earth Observations and Remote Sensing*. He also serves as a reviewer for numerous high-impact journals in the areas of radar, remote sensing, and signal processing.

He has been recognized for his contributions with several prestigious awards, including the IEEE Michael C. Wicks Radar Student Travel Grant at the 2022 IEEE Radar Conference, the Sensors 2022 Travel Award in Remote Sensors and Radar Sensors, the Ministerial Scholarship for Outstanding Young Scientists, and the START Scholarship for Exceptional Young Researchers from the Foundation for Polish Science.

He is an active member of several professional organizations, including the IEEE Aerospace and Electronic Systems Society, the IEEE Signal Processing Society, and the IEEE Geoscience and Remote Sensing Society. He currently serves as the Secretary of the Polish Joint Chapter of Antennas and Propagation, Aerospace and Electronic Systems, and Microwave Theory and Techniques, as well as the Treasurer of the Polish Chapter of the IEEE Signal Processing Society.

His research contributions combine theoretical developments with practical implementations, aiming to advance radar-sensing capabilities and signal-processing techniques for both civilian and defense applications.

1 Radar Concept

1.1 RADIO DETECTION AND RANGING

Radar, an acronym for **Ra**dio **D**etection **a**nd **R**anging, is a technology that uses electromagnetic waves to detect, locate, and track objects. Initially developed for military applications in the early 20th century, radar systems have become indispensable in various fields, including aviation, maritime navigation, weather forecasting, and automotive safety [2, 3]. The origins of radar technology are deeply connected to early discoveries in electromagnetic waves and radio communication [4,5]. Russian physicist Alexander Popov, a pioneer in wireless communication, conducted experiments in 1897 in the Baltic Sea that hinted at the potential for detecting objects using radio waves. In his report, Popov describes how Lieutenant Il'in's ship interrupted the radio communication link between two other vessels, Europe and Africa. This observation demonstrated how physical objects could influence radio waves, laying an essential conceptual foundation for radar technology.

A few years later, in 1904, German engineer Christian Hülsmeyer developed and patented the Telemobiloskop, a device designed to detect ships using reflected electromagnetic waves [6]. Though it could not measure distances, it represented a significant step toward practical radar systems, even if it received little attention then.

The significant breakthroughs came in the 1930s and 1940s, with radar systems being developed independently in various countries, including the United Kingdom, the United States, and Germany. British scientist Sir Robert Watson-Watt, for example, demonstrated a radar system in 1935 that could detect aircraft, a technology that became crucial during World War II [5]. Early radar systems, though large and reliant on vacuum tubes, revolutionized military operations and paved the way for a wide range of civilian applications in the years to follow. By analyzing the time delay and frequency changes of reflected signals, radars can provide detailed information about an object's distance, speed, and direction. This chapter explores the fundamental principles of radar technology, its components, and its diverse applications in modern industries.

A simplified radar architecture[1] is presented in Figure 1.1. The signal is generated by the oscillator. Next, the modulator modifies the signal parameters like phase, frequency, amplitude, and so on to obtain an assumed waveform at its output. The signal is amplified and passed through the duplexer, whose function is to direct the signal into the transmitting antenna, which emits the wave into space. The signal received by the antenna is passed through the duplexer to the low-noise amplifier. The received signal is mixed with a reference signal (or local oscillator signal derived from the transmitter), and subsequently processed using matched filtering or correlation-based processing to extract echoes.

From the signal-processing perspective, radars' operation principle is trivial, and the high-level concept is shown in Figure 1.2. The radar system transmits a pulse (of

DOI: 10.1201/9781003765783-1

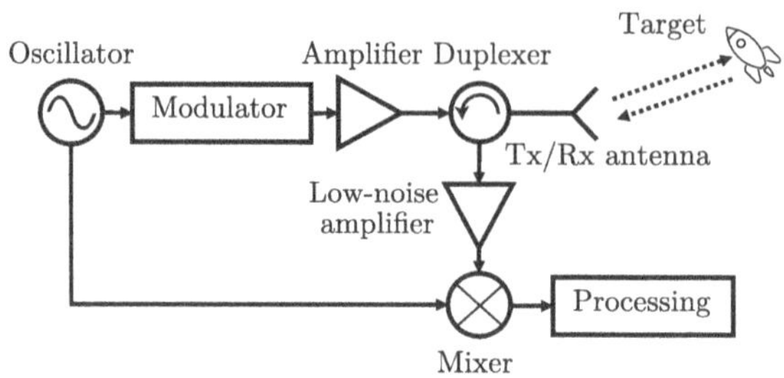

Figure 1.1 Simplified radar architecture.

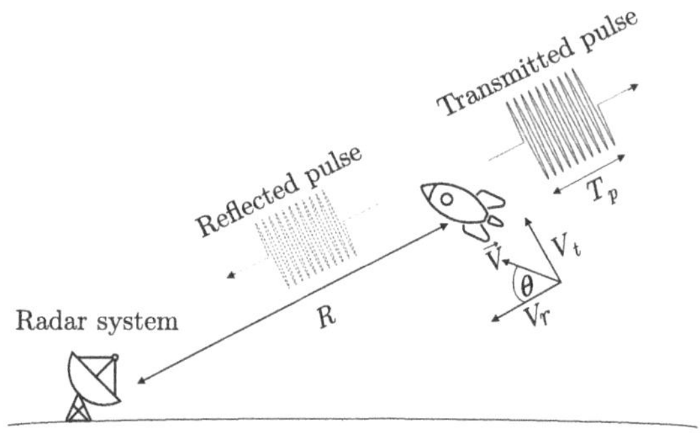

Figure 1.2 The radar operation principle.

a duration of T_p), propagating in space and reflecting from obstacles. At the moment when the pulse impinges the target (graphically represented by the rocket in Figure 1.2), the distance between it and the radar system is R. Thus, the reflected signal reaches the radar receiver after

$$\Delta t = \frac{2R}{c}, \tag{1.1}$$

where c is the propagation speed (in this book, the propagation speed is assumed to be equal to the speed of light, which in vacuum is $c = 299792458$ m/s). Since the pulse travels forth and back (from radar to the target and back to the radar), the range in (1.1) is multiplied by 2. Having Δt, one can compute the target range.

As shown in Figure 1.2, the target moves toward the radar with the velocity $\vec{V}$, which can be decomposed into two counterparts – radial V_r and transversal V_t components.[2] Based on the Doppler phenomenon, radar systems allow the radial components to be estimated. The velocity is calculated indirectly as the frequency shift between the transmitted and the received pulses. The relationship describing the velocity and the Doppler frequency is given as

$$f_D = -\frac{2V}{\lambda}\cos(\theta), \tag{1.2}$$

where f_D is the Doppler frequency, V is the target velocity, λ denotes the wavelength ($\lambda = c/f_c$ with the pulse carrier frequency f_c), and θ expresses the angle between the propagation direction of the transmitted/reflected signal and the target movement direction.

The presented radar principles are relatively straightforward and do not delve into details related to software and hardware concepts. The next section gives an overview of the most common radar configurations with a brief description of each of them.

1.2 MAIN RADAR CONFIGURATIONS

Although the first radar experiment was conducted in a bistatic geometry [6, 7], research and development of later radar systems were oriented to provide monostatic active sensing, which resulted in, e.g., Chain Home – the British early warning system [3]. Over 120 years (counting from Hülsmeyer's patent in 1904), radar technology has evolved significantly, yielding flourishing radar systems in various configurations and applications. Nowadays, radar systems cannot be easily categorized as their principle of operation, geometry, frequency band, applications, and waveform types differ. Figure 1.3 delineates the general overview of radar systems. Note that the presented division is not definitive. One can certainly add other groups

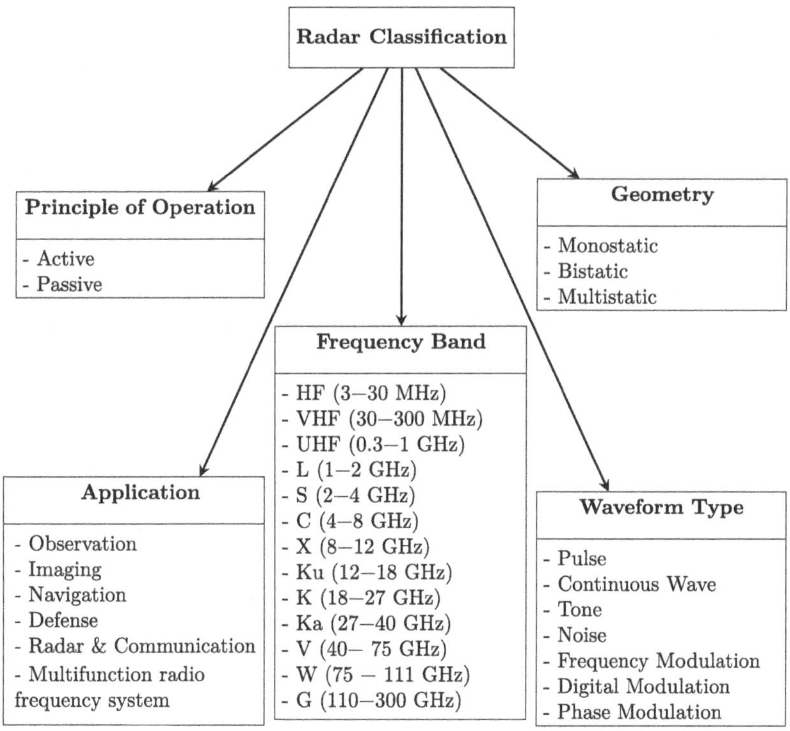

Figure 1.3 Radar classification considering the main system features.

or their members, e.g., **Scanning Method** with members like *mechanical* or *electronic scanning*, or **Deployment** with members like *ground*, *airborne*, *shipborne*, etc. This chart differentiates between the main radar configurations.

1.2.1 PRINCIPLE OF OPERATION

First, recent radar systems can be considered in active and passive configurations. In short, an active radar system emits a signal and receives echoes reflected from obstacles [8]. The system possesses information about the waveform parameters and transmission timestamps. In turn, passive radars do not have their own signal source [9, 10]. Instead, they utilize so-called signals of opportunity to detect moving targets. For example, passive radars typically use telecommunications signals like radio or television broadcasting for situational awareness. Their detection and jamming are hampered by the cost of more complex processing and the lack of waveform control.

1.2.2 GEOMETRY

A crucial aspect related to the radar principle of operation is its geometry. Monostatic radars are typically active systems where the transmitter and receiver are collocated. We call the system bistatic when the transmitter and receiver are spatially separated. Such a configuration can usually be found in passive radars, where the transmitter (e.g., radio transmitter) is far from the radar receiver. Note that such a configuration is generally called a non-cooperative bistatic configuration, which means both system parts work independently. An extension of a bistatic configuration is the multistatic case. While in the bistatic configuration, one can have one transmitter-receiver pair; in the multistatic configuration, several pairs exist. An example is a single television broadcast transmitter and three passive radar receivers. Bi- and multistatic configurations can be found in both active and passive radars. There is also another class, often neglected in such juxtapositions. Active radars can sometimes work with two separate antennas, but are located in close vicinity. According to the definition, their antennas are separated, so they should be assigned to bistatic radars. However, if the distance between antennas is negligibly small compared to the radar range, such a configuration is called quasi-monostatic. All mentioned geometries are shown in Figure 1.4.

1.2.3 APPLICATIONS

Itemizing all officially available applications of radar systems is extremely tough, if possible at all. Radars are used in many applications, for example, in air traffic control to control the situation in civil aviation [11], air defense for military purposes [12], Earth surface imaging to ice or flooding effect monitoring [13–15]. Modern cars are equipped with anti-collision systems based on radars [16]. Also, the maritime domain is full of radar technology, e.g., marine radars [14, 17] used to detect ships or land hazards. Nowadays, due to a

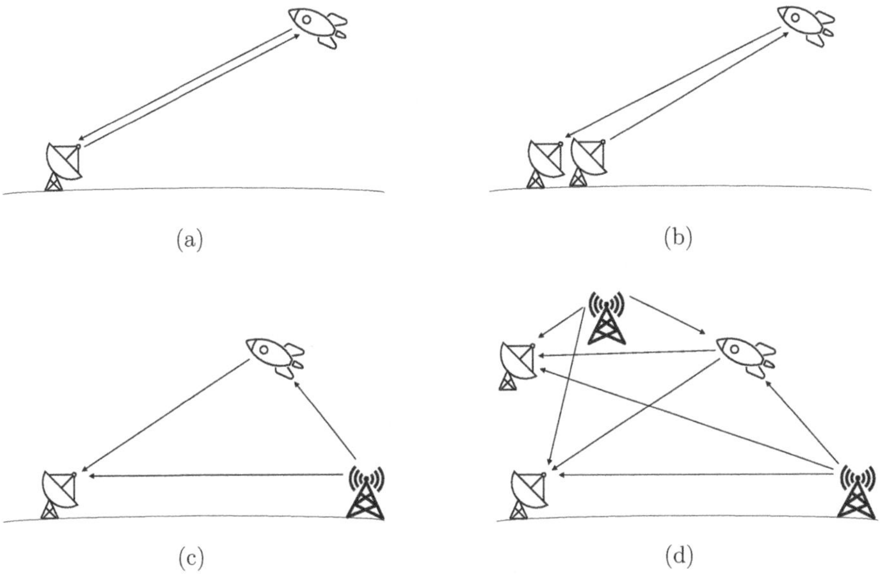

Figure 1.4 An illustration of different radar geometries. (a) Monostatic configuration – radar transmitter and receiver are in the same place. (b) Quasi-monostatic geometry – radar transmitter and receiver are separated but close to each other. (c) Bistatic geometry – radar receiver is far from the transmitter, the example of which is the telecommunication broadcasting station as in passive radar, but in general, it can be a dedicated radar transmitter. (d) Multistatic configuration – there are several spatially dislocated transmitter-receiver pairs.

highly dense spectrum allocation, radar and communication subsystems are integrated into a single and joint system, delivering extended functionalities in contrast to those subsystems working separately [18]. The mentioned applications are only a tiny scratch of the available areas where radar systems are employed. Certainly worth noting are weather radars [19], over-the-horizon radars [20], Internet of Things and industrial applications [21], including those for presence detection and activation using a simple Doppler radar, or intelligent transport systems for adaptive cruise control, collision avoidance, and traffic monitoring. More about radar applications can be found in the extensive literature [22, 23].

In addition to conventional operation, more advanced effects and configurations also play an important role. One example is the micro-Doppler phenomenon, where small-scale target motions, such as vibrations or rotations, create additional spectral features that can be used for target classification [24]. Another example is the forward scattering radar configuration, which exploits scattering in the forward direction to achieve high sensitivity for target detection [25]. Both of these topics will be discussed in greater detail in later chapters.

1.2.4 FREQUENCY BANDS

The choice of radar carrier frequency depends on several factors. First of all, the system's range is taken into account. Usually, the lower the frequency is, the higher the range (assuming the same emitted power). Another factor dictating the radar frequency is the available maximum power. Legal standards restrict this aspect, limiting the upper value of the radiated power. With the carrier frequency, there is also a related signal bandwidth. Excluding wideband radars, one can coarsely assume that the maximum relative bandwidth equals 10% of the carrier frequency (this value is assumed as a practical compromise between range resolution, signal energy, hardware complexity, and regulatory spectrum limitations). Therefore, the higher the carrier frequency, the wider the available bandwidth. For instance, 10% for high-frequency (HF) radar, e.g., over-the-horizon radar working at the frequency of 10 MHz is 1 MHz, while in practice, this value is $10 - 100\times$ smaller due to the hardware limitation. In turn, radars working in the W band, e.g., at the frequency of 100 GHz, may have a bandwidth of approximately 10 GHz. As shown in the next section, both carrier frequency and bandwidth play a key role in the system's performance. Increasing the system carrier frequency leads to hardware miniaturization but affects other radar system parameters.

Another class constitutes wideband and ultrawideband radars, whose bandwidth is much more significant than the typical 10% of the carrier frequency [26]. They often exceed 500 MHz or a relative bandwidth greater than 20%, enabling exceptional range resolution – the ability to distinguish between closely spaced targets. Ultrawideband radars are particularly effective in environments with clutter or when penetrating through obstacles like walls or foliage, making them ideal for applications such as ground-penetrating radar, imaging, and short-range communications. However, the design and implementation of ultrawideband systems present challenges, including higher power requirements, antenna design, and compliance with regulatory limits on spectral emissions. Despite these challenges, such technology plays a critical role in modern radar systems, offering capabilities that narrowband radars cannot achieve.

1.2.5 WAVEFORM TYPES

The last main class the radar can be assigned to considers its waveform. Regardless of the waveform, it can be continuous (the system emits the signal all the time) or interrupted. The latter case is usually referred to as a pulse radar. Beginning from the very straightforward example when the radar transmits a monochromatic wave, ending with digitally modulated signals carrying telecommunication information, radar waveforms differ primarily in their auto-ambiguity function. In short, the auto-ambiguity function informs about the autocorrelation (self-similarity) of the signal with different frequency shifts. Consideration of the Doppler shift results from radar's main functionality, i.e., detection of delayed and moving targets, whose velocity affects the frequency of the radar signal. Therefore, the better the signal correlates with itself, keeping sidelobes at the lowest possible level, the better the waveform for radar purposes. More about the auto-ambiguity function can be found in the next section.

1.3 BASIC FORMULAS

Among a variety of radar classes and applications, gathering all available mathematical formulas used to calculate, estimate, or scale parameters existing in those systems deserves a dedicated book (although there are some good quality handbooks [2], the technology rapidly grows, requiring updates). Let us focus on the most important ones, which are exploited in the remaining part of this book. They are fundamental to our further considerations and constitute basic knowledge for radar engineers.

1.3.1 RADAR RANGE EQUATION

One of the most important formulas defines the received signal power for assumed system parameters. The radar equation takes several factors into account. Firstly, let us look into the monostatic case, describing the signal power, which takes on the form

$$P_r = \frac{P_t G_t G_r \lambda^2 \sigma}{(4\pi)^3 R^4 L}, \tag{1.3}$$

where:

- P_r – Received power [W],
- P_t – Transmitted power [W],
- G_t – Transmitter antenna gain [unitless],
- G_r – Receiver antenna gain [unitless] (equal to G_t in monostatic radars),
- λ – Wavelength of the transmitted signal [m],
- σ – Radar cross-section (RCS) of the target [m^2],
- R – Range to the target [m],
- L – System losses (due to hardware, atmosphere, etc.) [unitless].

The monostatic radar equation describes the received power P_r of a radar signal reflected from a target. It assumes the same antenna is used to transmit and receive the signal. The received power decreases with the fourth power of the range R, making radar systems susceptible to the distance to the target. In many practical systems, especially those working in pulse mode, $G_t = G_r$ as the signal is emitted and received by the same element. From the above, the range equation can be written as

$$R = \sqrt[4]{\frac{P_t G_t G_r \lambda^2 \sigma}{(4\pi)^3 P_r L}}. \tag{1.4}$$

The situation is a bit different for the bistatic case, as the transmitted and reflected signals travel in different ways. Therefore, the power received by the bistatic system takes on the following form

$$P_r = \frac{P_t G_t G_r \lambda^2 \sigma}{(4\pi)^3 R_1^2 R_2^2 L}, \tag{1.5}$$

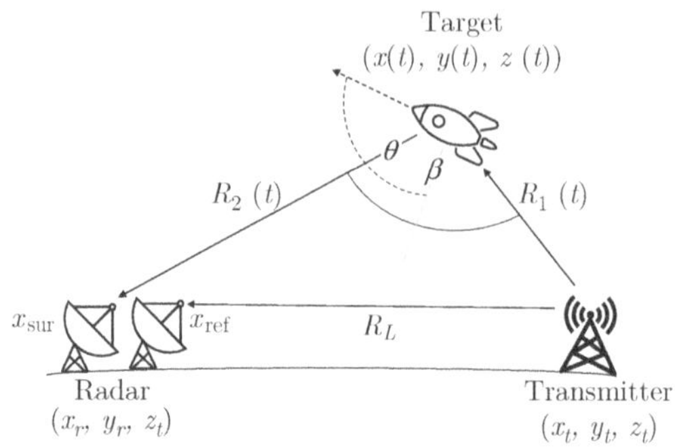

Figure 1.5 Passive bistatic geometry with transmitter, receiver, and target coordinates.

where R_1 and R_2 are the distances from the transmitter to the target and from the target to the receiver, respectively (see Figure 1.5). Both are expressed in meters. Consequently, the bistatic range equation can be written as

$$\sqrt{R_1 R_2} = \sqrt{\frac{P_t G_t G_r \lambda^2 \sigma}{(4\pi)^3 L}}. \tag{1.6}$$

1.3.2 TARGET RANGE ESTIMATION

The basic radar function is target range estimation. The indirect measurement utilizes the time difference between the transmitted and received signals. As already defined for the monostatic (and quasi-monostatic) case (see Figure 1.2), for the target at the range R, this time difference reads as

$$\Delta t = \frac{2R}{c}. \tag{1.7}$$

In practice, R is our unknown variable, but all remaining factors (including the propagation speed as the speed of light c) are given. R can thus be estimated as

$$R = \frac{\Delta t c}{2}. \tag{1.8}$$

The problem of range estimation is more challenging in bistatic radars. To explain this nontrivial issue, let us explain the bistatic geometry shown in Figure 1.5, which actually presents the geometry of the bistatic passive radar. The signal is transmitted by the stationary antenna located in (x_t, y_t, z_t). The signal goes to the radar receiver (also stationary), composed of two antennas: reference x_{ref} collecting the signal pattern, and surveillance x_{sur} for acquiring echoes from obstacles. Let us assume that both radar antennas are co-located at (x_r, y_r, z_r). From these coordinates, we have the first range, called the baseline, which is given as

$$R_L = ||(x_t, y_t, z_t) - (x_r, y_r, z_t)||_2 = \sqrt{(x_t - x_r)^2 + (y_t - y_r)^2 + (z_t - z_r)^2}. \tag{1.9}$$

$R(1)$ and $R(2)$ can be written as

$$R_1 = \sqrt{(x-x_t)^2 + (y-y_t)^2 + (z-z_t)^2}, \tag{1.10}$$

$$R_2 = \sqrt{(x-x_r)^2 + (y-y_r)^2 + (z-z_r)^2}. \tag{1.11}$$

A bistatic radar can measure the bistatic range, defined as

$$R_b = R_1 + R_2 - R_L. \tag{1.12}$$

Therefore, we know the relative range comprising the baseline in bistatic radars, but this relation is less intuitive than that in monostatic systems. Similarly, as in the monostatic case, the range-time relation considers the propagation time and the speed of light, using the well-known formula

$$R_b = (\Delta t_{R_1} + \Delta t_{R_2} - \Delta t_{R_L})c, \tag{1.13}$$

where $\Delta t_{R_1}, \Delta t_{R_2}, \Delta t_{R_L}$ is the time that it takes the signal to propagate from the transmitter to the target, from the target to the receiver, and from the transmitter to the receiver, respectively.

In multistatic radars, the same operations as in a bistatic configuration are performed for each transmitter-receiver pair.

1.3.3 RANGE RESOLUTION

Range resolution is one of the key radar parameters. It defines the smallest distance, allowing the system to differentiate between two closely located targets. Again, we identify two cases related to radar geometry. For the (quasi) monostatic radars, the range resolution is described as

$$\Delta R = \frac{c}{2B}, \tag{1.14}$$

where B is the signal bandwidth.[3] It is worth noting that this parameter is not directly associated with the carrier frequency. Notwithstanding, usually, the bandwidth increases with the increase of the carrier frequency (as delineated in Section 1.2.4). Equation (1.14) shows that one has 1 km of range resolution in monostatic radar when the bandwidth is approximately $B = 300$ kHz. One-meter resolution is achievable for $B = 300$ MHz, while 1 cm resolution is available for $B = 30$ GHz.

In the bistatic case, there is a subtle modification of (1.14), and the bistatic range resolution takes on the following form

$$\Delta R_b = \frac{c}{B}, \tag{1.15}$$

with all symbols as established previously. Bistatic radars, when operating with the same bandwidth as monostatic systems, exhibit a range resolution that is typically worse—by up to a factor of two—compared to their monostatic counterparts.

1.3.4 TARGET VELOCITY ESTIMATION

The velocity estimation process is based on assessing how much the received signal frequency differs from the frequency of the transmitted wave. Therefore, this measurement is indirect, akin to the target range estimation.

In the (quasi) monostatic case, the formula describing the Doppler frequency shift and the target velocity was already defined by (1.2). The target velocity is then calculated as

$$V = -\frac{f_D \lambda}{2} = -\frac{c f_D}{2 f_c}. \tag{1.16}$$

The formula for bistatic target velocity is similar to (1.16) with a remark on vanishing the factor 2 in the denominator. In general, the bistatic Doppler frequency equals

$$f_D = -\frac{2V}{\lambda} \cos(\theta) \cos(\beta/2), \tag{1.17}$$

where β is the so-called bistatic angle, and θ is the angle between the velocity vector and the bistatic angle bisector (see Figure 1.5). Of course, the angles β and θ are unknown, and the bistatic velocity is, in practice, estimated as

$$V_b = -f_D \lambda = -\frac{c f_D}{f_c}. \tag{1.18}$$

Remark 1.1. *In the literature, (1.16) and (1.18) sometimes have a positive sign. In this book, the sign is negative, which originates from the convention that a positive Doppler frequency shift ($f_D > 0$) corresponds to a target moving toward the radar, while a negative Doppler frequency shift ($f_D < 0$) corresponds to a target moving away. This convention aligns with the classical Doppler effect, where the frequency of a wave increases as the source and observer approach each other and decreases as they move apart. The negative sign ensures that the velocity V has the correct sign: positive for approaching targets and negative for receding ones.*

1.3.5 VELOCITY RESOLUTION

Target detection and tracking are typically based on the target's range and velocity. The signal bandwidth determines the ability to distinguish between two closely spaced targets. However, targets may be at similar ranges and have nearly identical velocities. In such cases, velocity resolution depends on the observation (integration) time T_i and the wavelength λ. The longer the observation time or the shorter the wavelength, the higher the velocity resolution. Thus, we have the following relationship

$$\Delta V = \frac{\lambda}{M T_i}, \tag{1.19}$$

where $M = 2$ for the monostatic case and 1 for the bistatic configuration.

1.3.6 CORRELATION

Correlation is a fundamental mathematical operation that measures similarity between two signals or the signal itself. Correlation in radar systems is a technique used to compare received signals with known reference waveforms. It helps detect targets by identifying patterns in noisy data and estimating parameters such as time delay, which relates to distance, or frequency shift, which indicates velocity. Radar systems can improve detection sensitivity and measurement accuracy by maximizing the correlation function, especially in cluttered or noisy environments. The correlation of two signals $x(t)$ and $y(t)$ is defined as

$$\chi(t) = \int_{-\infty}^{\infty} x(\tau)y^*(\tau - t)\mathrm{d}\tau = \int_{-\infty}^{\infty} x(\tau + t)y^*(\tau)\mathrm{d}\tau, \tag{1.20}$$

where $(\cdot)^*$ denotes the complex conjugate. The correlation $\chi(t)$ measures the similarity between two signals (in this case, $x(t)$ and $y(t)$) as a function of the time shift t. It evaluates how well the signal $x(t)$ matches the time-shifted signal $y(t)$. Correlation is widely used in applications such as radar systems to determine the delay of an echo signal, which in turn allows for calculating the object's distance.

Remark 1.2. *In the literature, one can find two variants of correlation – autocorrelation and cross-correlation. Autocorrelation refers to the situation when $x(t) = y(t)$ and we measure the auto-similarity of a given signal. Cross-correlation is defined when $x(t) \neq y(t)$. In the remainder part of this book, correlation means cross-correlation unless otherwise stated. If the signal is correlated with itself, this is explicitly written as autocorrelation.*

Figure 1.6 shows an illustrative example of the correlation result. Two random signals $x(t)$ and $y(t)$ are correlated using (1.20). The latter one, $y(t)$, is delayed by 0.2 s with respect to $x(t)$. This imitates the situation when $y(t)$ is the echo of the target illuminated[4] by the radar with a noise waveform. The maximum $\chi(t)$ value shows the estimated delay and the related target range. The correlation function, as a fundamental radar tool, is intensely used in the further part of this book.

Remark 1.3. *As observed in (1.20), there are two equivalent formulations of the cross-correlation operation. In the first variant, the signal $y(t)$ is shifted toward positive values of t, and for each time shift, the scalar (inner) product is computed between the two signals, with one of them being conjugated. In the alternative formulation, $y(t)$ remains fixed while $x(t)$ is shifted in the negative time direction. This conceptual difference can be illustrated using the simulated example presented in Figure 1.6. To obtain the final outcome shown in Figure 1.6c, one may fix $y(t)$ and shift $x(t)$ toward positive time. Alternatively, an equivalent result can be achieved by keeping $x(t)$ fixed and shifting $y(t)$ toward negative infinity.*

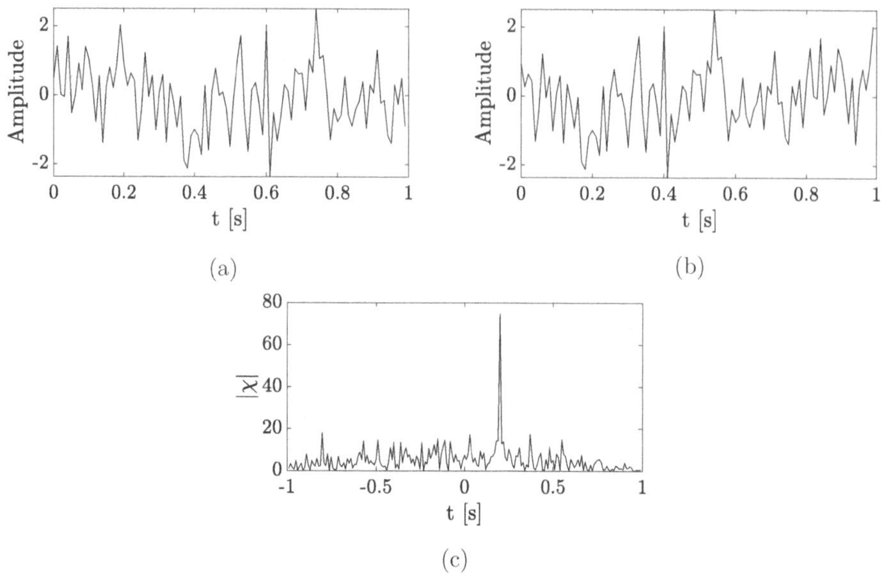

Figure 1.6 An example of correlation. (a) $y(t)$; (b) $x(t)$; (c) cross-correlation function.

1.3.7 MATCHED FILTER

The matched filter is a special type of linear filter that maximizes the signal-to-noise ratio (SNR) at its output [27]. This technique is popularly used in radars and, therefore, is a natural consequence of typical methods in the problem of signal parameter estimation and target detection. Generally, the matched filtering is the convolution of a signal and the filter response, which is a conjugated time-reversed reference signal pattern. In applications such as active radar, after the matched filtering of the transmitted and received signal, the reflected copies of the outgoing pulse are distinguished from the received signal with maximized SNR. The matched filtering of the signal $x(t)$ is defined as follows [28]:

$$\xi(t) = \int_{-\infty}^{\infty} x(\tau)g(t-\tau)\mathrm{d}\tau, \tag{1.21}$$

where $g(t)$ is the matched filter impulse response given by $g(t) = y^*(-t)$, where $y(t)$ denotes the reference signal (template) to be detected in the received signal $x(t)$.

1.3.8 AMBIGUITY FUNCTION

The correlation described in the previous subsection is routinely used for delay (range) estimation. Radars' utility, however, lies in their ability to estimate range and velocity. The latter one cannot be estimated from the simple correlation formula (1.20). The problem of simultaneous delay and frequency shift estimation between two signals can be considered using the extended correlation called the ambiguity function. Knowing the already defined correlation, the introduction of the ambiguity function comes down to adding the second parameter describing the Doppler shift as follows:

$$\Psi(t,\omega) = \int_{-\infty}^{\infty} x(\tau)y^*(\tau - t)e^{j\omega\tau}\mathrm{d}\tau = \int_{-\infty}^{\infty} x(\tau + t)y^*(\tau)e^{j\omega(\tau+t)}\mathrm{d}\tau, \qquad (1.22)$$

where $\omega = 2\pi f$ is the angular frequency and $j^2 = -1$ is the imaginary unit. Also, we differentiate between the two cases of the ambiguity function in this case. When $x(t) = y(t)$, we call (1.22) an auto-ambiguity function. Otherwise, we have a cross-ambiguity function. In general, (1.22) is a complex-valued two-dimensional (2D) distribution. It describes how the signal correlates with another one (or with itself for auto-ambiguity function), including a frequency shift of one of the signals. This makes this function a perfect tool for radar signal analysis. Firstly, it allows echo extraction from the received radar signal (using the cross-ambiguity function). Secondly, it allows for signal analysis from the perspective of their utility for target detection. Using the auto-ambiguity distribution, one can easily notice how the waveform behaves regarding the peak width, peak-to-sidelobe ratio, integrated sidelobe ratio, etc. The mentioned parameters describe the waveform in terms of its usage for radar purposes. Let us briefly describe the waveform parameters:

- **Peak Width**: The peak width refers to the width of the main lobe of the ambiguity function, typically measured at a specific threshold (e.g., half-maximum). It represents the system's resolution capability in terms of delay and Doppler shifts.
- **Peak-to-Sidelobe Ratio**: The peak-to-sidelobe ratio is the ratio of the maximum value of the main lobe to the highest sidelobe level in the ambiguity function. It indicates the system's ability to distinguish the main signal from interference caused by sidelobes.
- **Integrated Sidelobe Level**: The integrated sidelobe level is the measure of the energy contained in all the sidelobes in the ambiguity function. It provides descriptive information on the overall sidelobe contamination in the analyzed signal.

The example of the three-dimensional auto-ambiguity function for a fifth-generation (5G) mobile network synchronization signal, which was used for passive radar, is shown in Figure 1.7 [29]. The main peak of the ambiguity function occurs at zero delay ($t = 0$) and zero Doppler shift ($f_D = 0$). This peak represents the maximum correlation of the transmitted signal with itself, corresponding to a perfect alignment in both time and frequency. A strong main peak is desirable because it ensures high detectability of a target under ideal conditions.

However, real-life signals also exhibit sidelobes, which are smaller but noticeable features that appear at offsets from the main peak in either delay or Doppler. Sidelobes arise due to the finite bandwidth and duration of the signal, as no practical waveform can be perfectly concentrated in both time and frequency. Sidelobes can lead to ambiguity, where false detections occur because an echo is misinterpreted as

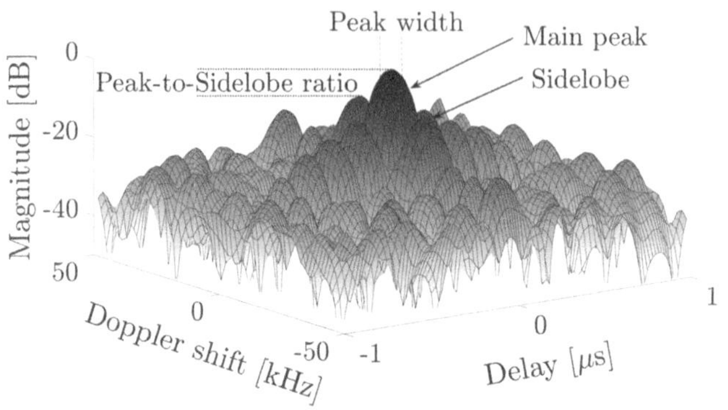

Figure 1.7 3D ambiguity function with crucial metrics.

a target. Minimizing sidelobes is crucial to reduce interference between targets and improve the radar's ability to resolve closely spaced objects.

The peak-to-sidelobe ratio is an essential metric for quantifying the strength of the highest sidelobe relative to the main peak. It is defined as the ratio of the maximum sidelobe magnitude to the magnitude of the main peak. This ratio is often expressed in decibels (dB). A higher peak-to-sidelobe ratio indicates better suppression of sidelobes, leading to improved target detection and reduced false alarms.

Together, these parameters, the main peak, sidelobes, and peak-to-sidelobe ratio, define the quality of a radar signal in terms of its resolution, ambiguity, and reliability in detecting and distinguishing targets. Selected real-life signals' ambiguity functions in the 2D plane (a projection of the three-dimensional (3D) function on a flat surface) are shown in Figure 1.8.

Figure 1.8a presents the auto-ambiguity function of the linear chirp, usually called a linear frequency-modulated signal. The ambiguity function of a chirp signal appears as a tilted line because the signal exhibits a linear frequency modulation, resulting in a strong correlation along a range of delay and Doppler shift pairs that match the chirp's instantaneous frequency trajectory. An almost perfect shape is obtainable for band-limited white Gaussian-distributed noise, which is shown in Figure 1.8b. Such a signal is characterized by a strong and relatively narrow main peak and low sidelobes, leading to perfect target localization. 13-bit Barker code, whose auto-ambiguity function is delineated in Figure 1.8c, yields a constant sidelobe level and a characteristic X-shape distribution. The last presented example is shown in Figure 1.8d and presents the auto-ambiguity function for the FM radio. This kind of signal is typically used in passive radars where illuminators of opportunity are commercial broadcasting systems. It has to be highlighted that in FM radio, the shape of the auto-ambiguity function can vary dynamically depending on the transmitted signal. The higher the dynamic of the transmitted signal (e.g., rock music), the better the radar properties. For calm content (classical music), the ambiguity function becomes less useful for radar operation.

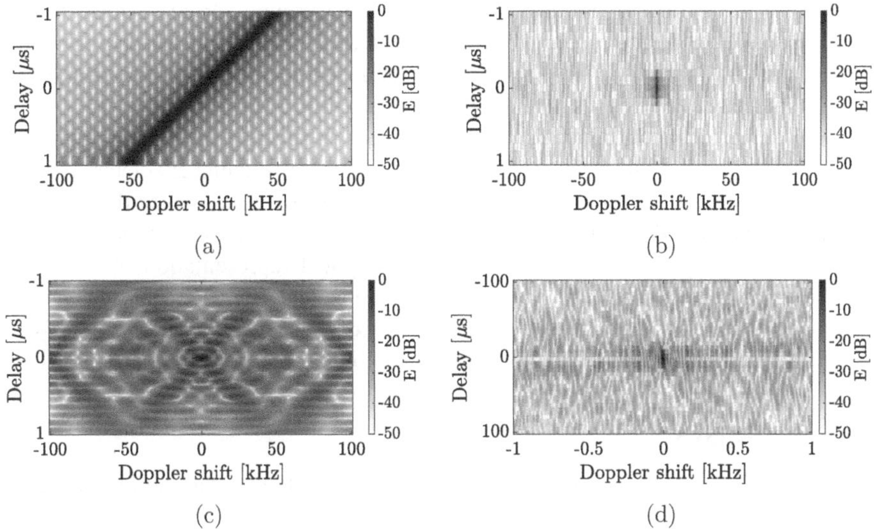

Figure 1.8 Four examples of auto-ambiguity function. (a) Linear frequency-modulated chirp; (b) noise signal; (c) Barker code (13-bit); (d) frequency-modulated radio signal.

1.4 SUMMARY

This chapter has provided a concise introduction to the fundamental principles of radar systems. While the presented content does not cover all aspects of radar signal analysis, it offers a solid foundation for understanding the concepts that will be expanded upon in subsequent chapters. Readers interested in a more comprehensive treatment of radar theory are encouraged to consult the standard references in the field, which explore topics ranging from hardware implementations to signal processing and system design [2,6,7,9,30].

The purpose of this chapter is not to offer an exhaustive overview but rather to establish a working understanding of radar concepts that are essential for the material covered later in this book. In subsequent chapters, various theoretical constructs, equations, and radar processing methods will be introduced and discussed. For this reason, it is highly recommended that the reader retain at least a basic familiarity with the topics introduced here, as doing so will facilitate a smoother comprehension of the advanced techniques to come.

1.5 CHAPTER IN A NUTSHELL

The key takeaways from this chapter can be summarized as follows:

- Radar systems are capable of indirectly measuring the range and radial velocity of objects by analyzing the reflection or scattering of electromagnetic waves. These waves may be emitted by the radar itself (active radar) or originate from external sources of illumination (passive radar).

- A wide variety of radar configurations exist, ranging from systems designed for high-resolution ground or target imaging to passive early warning systems. These systems differ in geometry, waveform design, operational objectives, and application domains.
- The performance of a radar system depends on several crucial parameters, including the carrier frequency, signal bandwidth, antenna characteristics, and transmitted power. These parameters collectively influence detection capability and are quantitatively described by the radar range equation.
- The ambiguity function provides an analytical representation of how a given radar waveform correlates with time- and frequency-shifted versions of itself. It is a fundamental tool used to evaluate the resolution and performance trade-offs of radar signals in both time and Doppler domains.

1.6 EXERCISES

1. Derive the formula for range resolution in monostatic radar and in bistatic radar.
2. Derive the formula for velocity resolution in monostatic radar and in bistatic radar.
3. Using any programming environment (e.g., MATLAB), generate a baseband harmonic pulse (without amplitude or phase modulation). Compute and plot the magnitude of its autocorrelation function. Vary the duration of this pulse (e.g., from 1 to 100 μs) and compare the results.
4. Using any programming environment (e.g., MATLAB), generate a baseband pulse with linear frequency modulation. Plot the magnitude of its autocorrelation function. Vary the frequency sweep bandwidth (e.g., between 1 MHz and 1 GHz) and compare the results. Compare them also with the non-FM pulse.
5. Simulate the operation of a simple radar. Generate any radar signal analyzed in previous exercises. Add to this signal its delayed version with amplitude equal to 10% of the original, delayed by some value T. Vary T and observe its influence on the magnitude of the correlation function. Repeat the exercise for different signals.
6. Plot how the Doppler frequency changes as a function of the angle θ (see formula (1.2)). Assume a constant velocity V and wavelength λ (e.g., $V = 50\,\mathrm{m/s}$, $\lambda = 10\,\mathrm{cm}$). Vary θ from $-\pi/4$ to $\pi/4$ and observe how θ affects the Doppler frequency variation.
7. Implement a radar range calculator that estimates radar range for user-defined parameters (formula (1.4)). Use initial values such as $P_t = 1\,\mathrm{W}$, $G_t = G_r = 10$, $\lambda = 10\,\mathrm{cm}$, $P_r = 0.01\,\mathrm{W}$, $L = 0.1$. Plot how the radar range changes for different radar cross-sections σ in the interval 0.1–$100\,\mathrm{m}^2$. Also observe the influence of other parameters, especially the system loss L, the wavelength, and antenna gain.
8. Implement a procedure that computes the cross-ambiguity function of two signals (see formula (1.22)). Implement it in the simplest way: multiply

one signal by a factor corresponding to a frequency shift (this factor should be complex-valued if negative frequencies are analyzed) and compute the correlation for each frequency shift.

9. Plot the magnitude of the cross-ambiguity function for a harmonic signal. Vary the pulse duration and compare the results.
10. Plot the magnitude of the cross-ambiguity function for a linearly FM signal. Vary the pulse duration and bandwidth and compare the results.

NOTES

1 We explain later in this chapter that this is an active monostatic radar architecture.

2 For the sake of simplicity, the term *velocity* instead of the *radial velocity* is used in the sequel of this book unless otherwise stated.

3 For simple monochromatic pulse, the bandwidth is defined as $B = 1/T_p$, where T_p is the pulse duration (see Figure 1.2).

4 *Illumination* is a jargon borrowed from optics, albeit widely used by the radar community.

2 Analysis of One-Dimensional Radar Signals

2.1 INTRODUCTION

Chapter 1 introduced general concepts of radar systems. Now, let us focus on more detailed aspects of signal processing in radars – one-dimensional signals. Their abundance in radars makes analysis challenging because, as shown in this chapter, the signals exhibit significantly different characteristics.

Selected examples of one-dimensional signals are presented in this chapter with particular emphasis on practical aspects. They are also gathered from real-life radar systems, thus leaving no doubts about the usability of the presented signal-processing techniques.

First of all, let us introduce what a one-dimensional signal actually is. A one-dimensional signal is a mathematical function that describes the variation of a certain quantity (e.g., amplitude, intensity, voltage, etc.) as a function of a single independent variable, most often time. In the context of radar signal processing, a one-dimensional signal is typically represented as $x(t)$ for continuous-time signals or $x[n]$ for discrete-time signals, where t is the continuous-time variable, and $x(t)$ is the value of the signal at a given instant. In turn, for discrete-time signals, n is the discrete-time variable (usually an integer), and $x[n]$ is the value of the signal at a given time step.

There are plenty of one-dimensional signals in radars on which we can focus our attention. Here are some selected examples of radar signals analyzed in the remainder of this chapter. They can be found in active and passive radars and radio frequency systems receiving different signals, including those originating from other radar systems.

- **Radar Pulses**: we recognize several types of radar pulses, like linear and nonlinear frequency-modulated chirps, digitally modulated pulses, etc. Generally, each waveform transmitted by radars can be deemed a one-dimensional signal. In other words, the transmitter and the received signals can be analyzed as one-dimensional signals.
- **Micro-Doppler Signatures**: each moving object reflects the radar signal with a frequency shift related to its velocity (see (1.2)). However, some targets may have several moving parts, each with a slightly different velocity. An example of such a target can be a drone, a running person, or a

DOI: 10.1201/9781003765783-2

helicopter. Rotating blades or waving human limbs reflect the radio signal with a Doppler shift proportional to its velocity.

- **Range-Compressed Signal**: this is the effect of cross-correlation (or matched filtering) of the transmitted and received signals. The outcome informs about reflections at different distances and is one-dimensional, as illustrated in Chapter 1 in Figure 1.6c.
- **Jamming Signals**: radars, especially military radars, are precious equipment during conflicts as they are used for early warning systems and short-range target detection. Enemies intend to disrupt these systems so that they do not fulfill their function, i.e., detect threats. Sending noise or signals similar to those emitted by radar is the basis of electronic warfare aimed at obstructing enemy radar systems. Such signals can also be considered as one-dimensional signals.

The above examples do not exhaust the topic, albeit they constitute some of the foundations that will be analyzed in this chapter. In the following, we focus on frequency and time-frequency analysis of one-dimensional radar signals with selected applications. Let us start with the frequency analysis, namely, the Fourier transform.

This chapter first delves into the one-dimensional Fourier transform and its properties. In Section 2.2, we discuss signal decomposition onto harmonic components and the possibilities the Fourier transform enables. Section 2.2.5 thoroughly discusses the Fourier transform properties. Next, we focus on multicomponent and nonstationary signals, whose Fourier spectrum sometimes suffers from insufficient time and frequency relations representation. Considerations about the short-time Fourier transform are described in Section 2.4, and further improvements of this technique, primarily based on adaptive readability improvement, are consequently presented in Sections 2.5 and 2.6. Next, in Section 2.7, we move to more advanced techniques for energy concentration in the time-frequency domain (so-called time-frequency reassignment) and, in Section 2.8, its combination with the multi-window technique for noise flattening. Further description focuses on different variants of reassignment, allowing for time-frequency representation sharpening but with preserved ability to retrieve the signal to its time- domain form. So-called time-frequency synchrosqueezing is then described in Section 2.9. Possibilities of signal parameter estimation in the time-frequency plane are delineated in Section 2.10. Prior to conclusions, this chapter reports some fascinating characteristics of the time-frequency energy distribution and insightful consequences of its geometry, allowing for signal extraction and filtering from noisy time-frequency representations.

2.2 ONE-DIMENSIONAL FOURIER TRANSFORM

The Fourier Transform is a fundamental tool in radars and generally in mathematics and engineering, providing a method to analyze signals in the frequency domain. Manifold and valuable books, chapters, articles, and reports have been developed about the Fourier transform so far [31–40]. This section briefly condenses all the information and introduces the Fourier transform's theoretical foundations and practical applications. In the following, we focus on the most critical aspects, remembering that the Fourier analysis is the topic to which a dozen books were dedicated.

The Fourier transform owes its name and development to Joseph Fourier, a French mathematician and physicist who introduced the concept in the early 19th century. Fourier's work was initially motivated by his studies on heat conduction, culminating in his seminal publication [41]. In his work, Fourier proposed that any periodic function could be expressed as a sum of sines and cosines, forming the basis of what is now called the Fourier series. This groundbreaking idea extended beyond periodic functions to non-periodic functions through the concept of the Fourier transform, enabling the analysis of signals in terms of their frequency components.

Fourier's insights laid the foundation for the field of harmonic analysis and have had profound implications across various domains. In mathematics, Fourier's ideas inspired the development of functional analysis and the theory of integral transforms. In physics, the Fourier transform became essential for understanding wave phenomena and quantum mechanics. In engineering, it revolutionized signal processing, communications, and control systems. In this chapter, we focus primarily on its utility in the radar field.

Over the decades, the Fourier transform has been refined and extended, leading to discrete versions such as the discrete Fourier transform (DFT) and computational algorithms like the fast Fourier transform (FFT). Today, it remains one of the most versatile and widely used tools in science and engineering. Let us start with some definitions.

2.2.1 CONTINUOUS FOURIER TRANSFORM

The Fourier transform converts a time-domain signal into its frequency-domain representation. It is defined for a continuous signal $x(t)$ as

$$F_x(\omega) = \mathscr{F}\{x(t)\} = \int_{-\infty}^{\infty} x(t)e^{-j\omega t}\mathrm{d}t, \tag{2.1}$$

where $F_x(\omega)$ is the Fourier transform of $x(t)$, ω represents the angular frequency, and j is the imaginary unit $j^2 = -1$.

The inverse Fourier transform allows the reconstruction of the original signal from its frequency components

$$x(t) = \mathscr{F}^{-1}\{F_x(\omega)\} = \frac{1}{2\pi}\int_{-\infty}^{\infty} F_x(\omega)e^{j\omega t}\mathrm{d}\omega. \tag{2.2}$$

At this point, we need to clarify why the exponent function e and imaginary unit play key roles in the Fourier transform.[1] The Fourier transform heavily relies on the mathematical properties of Euler's number e and the imaginary unit j, both essential for its formulation and practical applications.

Euler's number $e \approx 2.718$ is a fundamental mathematical constant that arises naturally in exponential growth, decay, and many other phenomena. In the context of the Fourier transform, e plays a pivotal role due to its relationship with complex exponentials:

$$e^{j\theta} = \cos\theta + j\sin\theta, \tag{2.3}$$

This identity, known as Euler's formula, allows for the compact representation of oscillatory functions, such as sine and cosine, which form the basis of Fourier analysis. Euler's number simplifies mathematical expressions: Complex exponentials like $e^{-j\omega t}$ combine trigonometric oscillations into a single term. It also enables efficient computation. The exponential function is well-suited for numerical and analytical manipulation in signal analysis. Moreover, it represents continuous and smooth transitions between frequencies in the Fourier domain.

The imaginary unit j, defined as $j^2 = -1$, is equally critical in the Fourier transform. It facilitates the transition from the time and real domain to the complex frequency domain. Thus, it encodes phase information. The imaginary part of the Fourier transform represents phase shifts, which are crucial for reconstructing the original signal. The complex nature of the imaginary unit ensures that frequency components are symmetrically represented, accounting for both positive and negative frequencies. It also simplifies operations using j, and the differentiation and integration of oscillatory functions become straightforward in the frequency domain.

Together, Euler's number e and the imaginary unit j provide the mathematical foundation that makes the Fourier transform a powerful and elegant tool for analyzing signals in both time and frequency domains.

Knowing where e and j are from, let us recall the Fourier transform formula (2.1)

$$F_x(\omega) = \int_{-\infty}^{\infty} x(t)e^{-j\omega t}\,\mathrm{d}t. \tag{2.4}$$

Substituting (2.3) into (2.4), we have

$$F_x(\omega) = \int_{-\infty}^{\infty} x(t)\left(\cos(\omega t) - j\sin(\omega t)\right)\mathrm{d}t. \tag{2.5}$$

As can be seen, the Fourier transform can be interpreted as a projection of the signal $x(t)$ onto complex exponential functions.

2.2.2 DISCRETE-TIME FOURIER TRANSFORM

The Fourier transform is used for analytical considerations, theoretical analysis, and derivations of new signal parameter estimators. Practical problems, especially those where noise plays a key role, can be extremely challenging to solve analytically. In the era of computing machines and discrete signals (signals acquired using sensors and represented numerically using analog-to-digital converters), the Fourier transform is usually computed from the sequence of the unknown signal samples to retrieve spectral characteristics and describe this signal. Such an analysis allows a profound insight into the signal source. For instance, we aim to protect a given perimeter, e.g., the embassy area, above which we encountered a drone. To neutralize the threat, we can shoot down the target or disturb its communication to make it unable to continue its mission. Analysis of the radio frequency spectrum allows the determination of the occupied bandwidth and its jamming. The process is possible owing to the radio receiver converting the microwave signal into its discrete samples. One can compute the discrete-time Fourier transform (DTFT) based on them.

The DTFT extends the Fourier transform to discrete-time signals that return a continuous-frequency representation. It takes discrete-time signal samples, equally spaced with an interval $T_s = 1/f_s$ (where f_s is the sampling rate), and produces a continuous spectrum (generally overlapped or spaced of f_s). In other words, the DTFT gives a continuous-frequency representation of a discrete signal. The DTFT is defined as

$$F_{x,D}(\omega) = \sum_{n=-\infty}^{\infty} x[n]e^{-j\omega n}, \tag{2.6}$$

while its inverse transform (IDTFT) reads as

$$x[n] = \frac{1}{2\pi}\int_{-\pi}^{\pi} F_{x,D}(\omega)e^{j\omega n}\mathrm{d}\omega. \tag{2.7}$$

2.2.3 DISCRETE FOURIER TRANSFORM (DFT)

The DFT is an operation that transforms a finite sequence of equally spaced samples of a function into another sequence of the same length, representing equally spaced samples of the DTFT. The DFT is a complex-valued function of frequency, and the spacing between its samples in the frequency domain corresponds to the reciprocal of the input sequence's duration. The DFT is defined as

$$F_x[k] = \sum_{n=0}^{N-1} x[n]e^{-j2\pi kn/N}, \tag{2.8}$$

where $x[n]$ is the discrete signal in the time domain, $F_x[k]$ is the DFT of the signal and N is the number of samples. Akin to the previously described Fourier transforms, the DFT is invertible

$$x[n] = \frac{1}{N}\sum_{k=0}^{N-1} F_x[k]e^{j2\pi kn/N}. \tag{2.9}$$

Figure 2.1 compares the continuous and discrete Fourier transforms. The spectra (S) were illustrated in a linear scale as modulus of the frequency representation (FR) so that

$$\mathrm{S}(\omega) = |\mathrm{FR}(\omega)|. \tag{2.10}$$

However, it has to be stressed that any Fourier transform gives complex-valued results with the amplitude and phase content. For the convenience of illustration, the results shown in the analyzed example are the magnitudes of spectra.

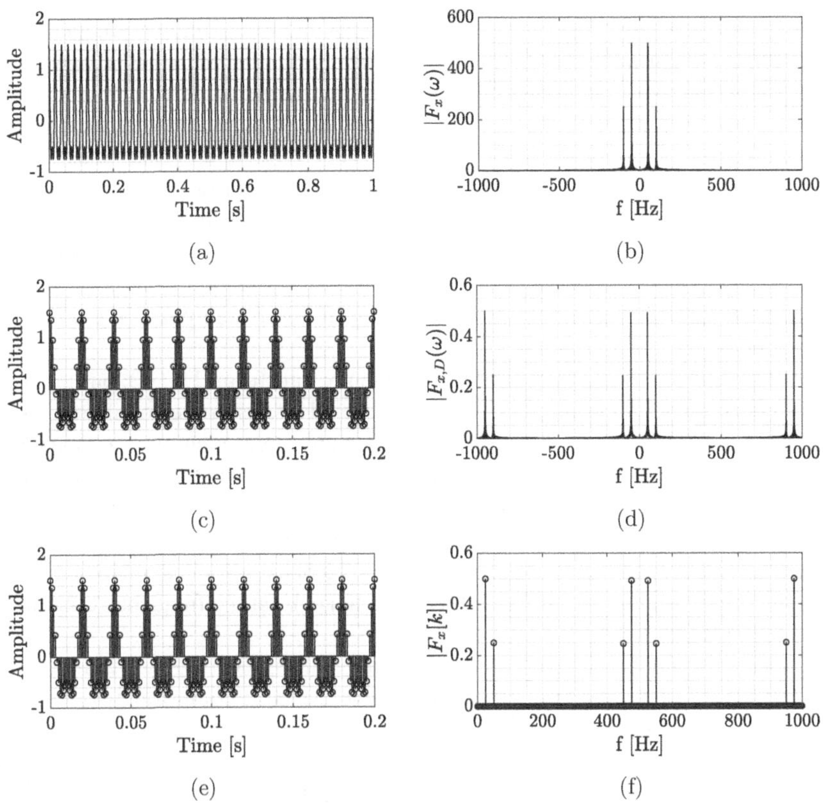

Figure 2.1 Depiction of continuous- and discrete-time signals and continuous- and discrete-frequency Fourier transforms. (a) Continuous-time signal; (b) continuous-frequency spectrum; (c) discrete-time signal (part); (d) DTFT; (e) discrete-time signal (part); (f) DFT.

The left column (Figure 2.1a, c and e) shows the time-domain signals, and their frequency domain counterparts are shown on the right-hand side (Figure 2.1b, d, and f). The example assumes a signal model

$$x(t) = \cos(2\pi f t) + 0.5 \cdot \cos(2 \cdot 2\pi f t), \tag{2.11}$$

where $f = 50$ Hz and the signal length is $T = 1$ s. This continuous-time signal is shown in Figure 2.1a. The signal in the discrete-time domain, sampled with a rate of $f_s = 1$ kHz, can be written as

$$x[n] = \cos(2\pi f n) + 0.5 \cdot \cos(2 \cdot 2\pi f n), \quad n = 0/f_s, 1/f_s, ..., (L-1)/f_s, \tag{2.12}$$

where $L = T/f_s$ is the signal length in samples. The discrete-time graphical representation of this signal is shown in Figure 2.1c and e (for clarity, only the signal part is shown to highlight its discrete nature).

First, let us consider the continuous-time signal and its continuous-frequency representation, shown in Figure 2.1a and b, respectively. The signal is a sum of two harmonic terms, with the first component with the frequency of $f = 50$ Hz and the second twice as weak with the frequency of 100 Hz. The spectral representation of the signal exhibits four peaks at ± 100 and ± 50 Hz with an appropriately scaled magnitude. The analyzed signal was real-valued. Thus, the spectrum is symmetrical, and we observe two identical peaks on each semi-axis. This occurs because real-valued signals satisfy the property of conjugate symmetry in the Fourier domain, meaning that each positive-frequency component has a corresponding complex-conjugate counterpart at the negative frequency, resulting in a mirrored peak on the opposite semi-axis.

For the discrete-time and continuous-frequency transform (DTFT), our input signal is represented by a vector composed of sampled values of the continuous signal, as shown in Figure 2.1c. Its continuous-frequency representation is delineated in Figure 2.1d. The signal peaks are periodic, with a span equal to the sampling rate f_s. Nevertheless, the spectrum is continuous.

Discrete-time and discrete-frequency signal representations are shown in Figure 2.1e and d, respectively. The input signal in the discrete-time domain is the same as in the previous case. Its DFT exhibits discrete peaks at the frequency coordinates equal to the signal frequencies.

The same signal and its DFT (for the positive half-axis) are shown in Figure 2.2. The left-hand side waveform is the original one. Since it comprises two harmonic terms, the Fourier transform allows for their decomposition. The first one, with the frequency of $f_1 = 50$ Hz, has a higher amplitude. The second mode, with the frequency of $f_2 = 100$ Hz, has an amplitude as weaker as twice. After the Fourier decompositions, these modes are represented as peaks with accordingly scaled amplitude in the frequency domain.

2.2.4 FAST FOURIER TRANSFORM

The FFT provides a highly efficient method for computing the DFT, described in the previous subsection. This algorithm is foundational to various practical

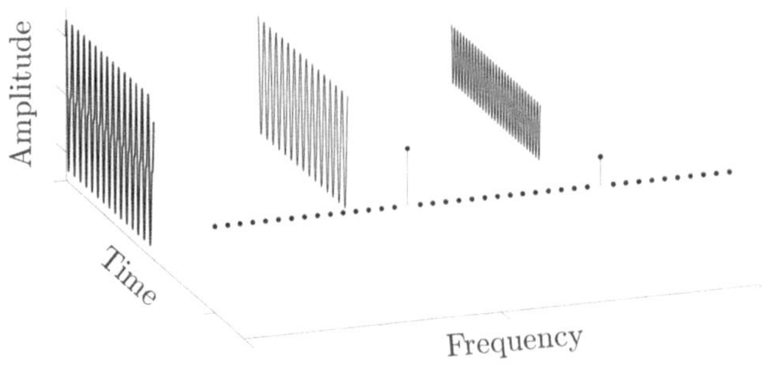

Figure 2.2 Discrete Fourier transform visualization.

applications, including signal processing, image analysis, data compression, numerical simulations, and radars.

Although the DFT emerged in the mid-20th century as a discrete counterpart to Fourier's continuous analysis, its computational complexity hindered widespread practical use. Calculating the DFT directly for N data points requires $O(N^2)$ operations, which is computationally expensive for large datasets [42]. This challenge led to the development of the FFT, which reduces the computational cost to $O(N \log_2 N)$, making it feasible for real-time and large-scale applications.

James Cooley and John Tukey popularized the modern formulation of the FFT in their landmark paper from 1965 [43]. However, historical evidence reveals that earlier versions of the algorithm were independently discovered by Carl Friedrich Gauss in 1805 [44] and later rediscovered by other mathematicians. Cooley and Tukey's contribution was formalizing and generalizing the algorithm using a divide-and-conquer strategy, breaking the DFT into smaller sub-problems that could be solved recursively. The FFT tangibly reduces computation time, making it indispensable in signal processing, image processing, and other fields.

The diagram provided in Figure 2.3 illustrates the Cooley-Tukey decimation-in-time (DIT) algorithm (so-called the 2-radix algorithm), which is the most commonly used FFT variant [43]. For the sake of clarity, the example focuses on the eight-point DFT and the signal of the same length, but it can be easily extended to longer sequences.

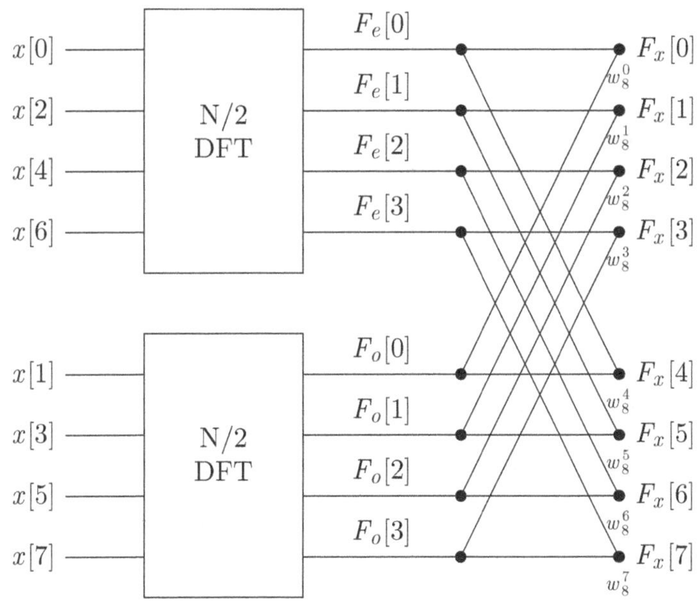

Figure 2.3 FFT computation using the 2-radix algorithm.

The diagram can be explained as follows:

1. **Input Splitting**: The input sequence $x[n]$ of length N is divided into two subsequences:
   - The even-indexed elements, $x[0], x[2], \dots, x[N-2]$.
   - The odd-indexed elements, $x[1], x[3], \dots, x[N-1]$.

   Each subsequence has a length of $N/2$.
2. **Recursive DFT Computation**: Separate $N/2$-point DFTs are computed for both $F_e[k]$ and $F_o[k]$. This recursion continues until the input sequences are of size 2, at which point the DFTs are computed directly.
3. **Combination Step**: The results of the two $N/2$-point DFTs are combined using the following equation:

$$F_x[k] = F_e[k] + w_N^k \cdot F_o[k], \quad k = 0, 1, \dots, \frac{N}{2} - 1, \tag{2.13}$$

$$F_x\left[k + \frac{N}{2}\right] = F_e[k] - w_N^k F_o[k], \quad k = 0, 1, \dots, \frac{N}{2} - 1, \tag{2.14}$$

   where $w_N^k = e^{-j\frac{2\pi k}{N}}$ is the kth twiddle factor.
   For $k = 0, 1, \dots, \frac{N}{2} - 1$, the FFT computes

$$F_x[k] = F_e[k] + w_N^k \cdot F_o[k]. \tag{2.15}$$

   For $k = \frac{N}{2}, \frac{N}{2} + 1, \dots, N-1$, the FFT computes

$$F_x[k] = F_e\left[k - \frac{N}{2}\right] - w_N^{k-\frac{N}{2}} \cdot F_o\left[k - \frac{N}{2}\right]. \tag{2.16}$$

4. **Output**: The algorithm outputs N frequency-domain components $F_x[k]$, representing the input sequence's Fourier transform.

Thanks to its versatility and low computational complexity, FFT is used in many modern systems like radars. This algorithm is also notably used in the remaining part of this book, especially in the published MATLAB toolbox with the implementation of the presented methods.

2.2.5 FOURIER TRANSFORM PROPERTIES

The Fourier Transform has several pivotal properties, making it a versatile signal analysis tool. These properties are summarized in Table 2.1.

Let us advocate these properties with proofs.

- **Linearity**: The Fourier transform of a linear combination of signals is the same linear combination of their Fourier transforms:

$$\mathscr{F}\{ax_1(t) + bx_2(t)\} = aF_{x_1}(\omega) + bF_{x_2}(\omega). \tag{2.17}$$

Table 2.1
Properties of the Fourier Transform

Property	Mathematical Expression
Linearity	$\mathscr{F}\{ax_1(t)+bx_2(t)\} = aF_{x_1}(\omega)+bF_{x_2}(\omega)$
Time shifting	$\mathscr{F}\{x(t-t_0)\} = F_x(\omega)e^{-j\omega t_0}$
Frequency shifting	$\mathscr{F}\{x(t)e^{j\omega_0 t}\} = F_x(\omega-\omega_0)$
Scaling in time	$\mathscr{F}\{x(at)\} = \frac{1}{\lvert a\rvert}F_x\left(\frac{\omega}{a}\right)$
Duality	$\mathscr{F}\{F_x(t)\} = 2\pi x(-\omega)$
Parseval's theorem	$\int_{-\infty}^{\infty} \lvert x(t)\rvert^2 \mathrm{d}t = \frac{1}{2\pi}\int_{-\infty}^{\infty} \lvert F_x(\omega)\rvert^2 d\omega$
Convolution theorem	$\mathscr{F}\{x_1(t)*x_2(t)\} = F_{x_1}(\omega)F_{x_2}(\omega)$
Differentiation in time	$\mathscr{F}\left\{\frac{\mathrm{d}^n x(t)}{\mathrm{d}t^n}\right\} = (j\omega)^n F_x(\omega)$
Time reversal	$\mathscr{F}\{x(-t)\} = F_x(-\omega)$
Multiplication in time domain	$\mathscr{F}\{x_1(t)x_2(t)\} = \frac{1}{2\pi}\left[F_{x_1}(\omega)*F_{x_2}(\omega)\right]$

Proof: By definition of the Fourier transform

$$\mathscr{F}\{x(t)\} = \int_{-\infty}^{\infty} x(t)e^{-j\omega t}\mathrm{d}t. \tag{2.18}$$

For $x(t) = ax_1(t)+bx_2(t)$, we have

$$\mathscr{F}\{ax_1(t)+bx_2(t)\} = \int_{-\infty}^{\infty} [ax_1(t)+bx_2(t)]\, e^{-j\omega t}\mathrm{d}t. \tag{2.19}$$

Using the linearity of the integral, we obtain

$$\mathscr{F}\{ax_1(t)+bx_2(t)\} = a\int_{-\infty}^{\infty} x_1(t)e^{-j\omega t}\mathrm{d}t + b\int_{-\infty}^{\infty} x_2(t)e^{-j\omega t}\mathrm{d}t. \tag{2.20}$$

Recognizing these as $F_{x_1}(\omega)$ and $F_{x_2}(\omega)$, one has

$$\mathscr{F}\{ax_1(t)+bx_2(t)\} = aF_{x_1}(\omega)+bF_{x_2}(\omega). \tag{2.21}$$

- **Time Shifting**: A time shift in the signal corresponds to a phase shift in the frequency domain:

$$\mathscr{F}\{x(t-t_0)\} = F_x(\omega)e^{-j\omega t_0}. \tag{2.22}$$

Proof: By definition

$$\mathscr{F}\{x(t-t_0)\} = \int_{-\infty}^{\infty} x(t-t_0)e^{-j\omega t}\mathrm{d}t. \tag{2.23}$$

Using a substitution $\tau = t - t_0$, so $t = \tau + t_0$ and $dt = d\tau$, we have

$$\mathscr{F}\{x(t-t_0)\} = \int_{-\infty}^{\infty} x(\tau)e^{-j\omega(\tau+t_0)}d\tau. \tag{2.24}$$

Separating the exponential yields

$$\mathscr{F}\{x(t-t_0)\} = e^{-j\omega t_0}\int_{-\infty}^{\infty} x(\tau)e^{-j\omega\tau}d\tau. \tag{2.25}$$

The integral is $F_x(\omega)$, so

$$\mathscr{F}\{x(t-t_0)\} = F_x(\omega)e^{-j\omega t_0}. \tag{2.26}$$

- **Frequency Shifting**: Multiplying a signal by complex exponential shifts its Fourier transform:

$$\mathscr{F}\{x(t)e^{j\omega_0 t}\} = F_x(\omega - \omega_0). \tag{2.27}$$

Proof: From the definition, we have

$$\mathscr{F}\{x(t)e^{j\omega_0 t}\} = \int_{-\infty}^{\infty} x(t)e^{j\omega_0 t}e^{-j\omega t}dt. \tag{2.28}$$

Combining the exponentials leads to

$$\mathscr{F}\{x(t)e^{j\omega_0 t}\} = \int_{-\infty}^{\infty} x(t)e^{-j(\omega-\omega_0)t}dt. \tag{2.29}$$

Using the definition of the Fourier transform gives

$$\mathscr{F}\{x(t)e^{j\omega_0 t}\} = F_x(\omega - \omega_0). \tag{2.30}$$

- **Scaling in Time**: Compressing or expanding a signal in time affects the frequency domain as follows:

$$\mathscr{F}\{x(at)\} = \frac{1}{|a|}F_x\left(\frac{\omega}{a}\right). \tag{2.31}$$

Proof: We have

$$\mathscr{F}\{x(at)\} = \int_{-\infty}^{\infty} x(at)e^{-j\omega t}dt. \tag{2.32}$$

Substituting $\tau = at$, so $t = \frac{\tau}{a}$ and $dt = \frac{d\tau}{a}$, leads to

$$\mathscr{F}\{x(at)\} = \int_{-\infty}^{\infty} x(\tau)e^{-j\omega\frac{\tau}{a}}\frac{d\tau}{a}. \tag{2.33}$$

Simplifying the equation gives

$$\mathscr{F}\{x(at)\} = \frac{1}{|a|}\int_{-\infty}^{\infty} x(\tau)e^{-j\frac{\omega}{a}\tau}\mathrm{d}\tau. \tag{2.34}$$

Recognizing the integral as $F_x\left(\frac{\omega}{a}\right)$ ends the proof

$$\mathscr{F}\{x(at)\} = \frac{1}{|a|}F_x\left(\frac{\omega}{a}\right). \tag{2.35}$$

For $a < 0$, the limits of integration reverse, introducing an additional sign that leads to the factor $\frac{1}{|a|}$.

- **Duality**: The Fourier transform of a Fourier transform exhibits symmetry:

$$\mathscr{F}\{F_x(t)\} = 2\pi x(-\omega). \tag{2.36}$$

Proof: Starting from the definition of the Fourier transform, we have

$$\mathscr{F}\{F_x(t)\} = \int_{-\infty}^{\infty} F_x(t)e^{-j\omega t}\mathrm{d}t. \tag{2.37}$$

Substituting the definition of $F_x(t)$ yields

$$F_x(t) = \int_{-\infty}^{\infty} x(\tau)e^{-j\tau t}\mathrm{d}\tau, \tag{2.38}$$

which gives

$$\mathscr{F}\{F_x(t)\} = \int_{-\infty}^{\infty}\left[\int_{-\infty}^{\infty} x(\tau)e^{-j\tau t}\mathrm{d}\tau\right]e^{-j\omega t}\mathrm{d}t. \tag{2.39}$$

Changing the order of integration leads to

$$\mathscr{F}\{F_x(t)\} = \int_{-\infty}^{\infty} x(\tau)\left[\int_{-\infty}^{\infty} e^{-jt(\tau+\omega)}\mathrm{d}t\right]\mathrm{d}\tau. \tag{2.40}$$

Using the identity

$$\int_{-\infty}^{\infty} e^{-j\alpha t}\mathrm{d}t = 2\pi\delta(\alpha), \tag{2.41}$$

we obtain

$$\int_{-\infty}^{\infty} e^{-jt(\tau+\omega)}\mathrm{d}t = 2\pi\delta(\tau+\omega). \tag{2.42}$$

Thus,

$$\mathscr{F}\{F_x(t)\} = 2\pi\int_{-\infty}^{\infty} x(\tau)\delta(\tau+\omega)d\tau = 2\pi x(-\omega). \tag{2.43}$$

- **Parseval's Theorem**: The total energy of a signal is equal in both time and frequency domains:

$$\int_{-\infty}^{\infty} |x(t)|^2 \mathrm{d}t = \frac{1}{2\pi} \int_{-\infty}^{\infty} |F_x(\omega)|^2 \mathrm{d}\omega. \tag{2.44}$$

Proof: Starting from the inverse Fourier transform

$$x(t) = \frac{1}{2\pi} \int_{-\infty}^{\infty} F_x(\omega) e^{j\omega t} \mathrm{d}\omega, \tag{2.45}$$

we consider the energy

$$\int_{-\infty}^{\infty} |x(t)|^2 \mathrm{d}t = \int_{-\infty}^{\infty} x(t) x^*(t) \mathrm{d}t. \tag{2.46}$$

Substituting both $x(t)$ and its conjugate gives

$$x^*(t) = \frac{1}{2\pi} \int_{-\infty}^{\infty} F_x^*(\nu) e^{-j\nu t} \mathrm{d}\nu, \tag{2.47}$$

thus

$$|x(t)|^2 = \frac{1}{(2\pi)^2} \int_{-\infty}^{\infty} \int_{-\infty}^{\infty} F_x(\omega) F_x^*(\nu) e^{j(\omega-\nu)t} \mathrm{d}\omega \mathrm{d}\nu. \tag{2.48}$$

Integrating over t yields

$$\int_{-\infty}^{\infty} e^{j(\omega-\nu)t} \mathrm{d}t = 2\pi\delta(\omega-\nu), \tag{2.49}$$

so

$$\int_{-\infty}^{\infty} |x(t)|^2 \mathrm{d}t = \frac{1}{(2\pi)^2} \int_{-\infty}^{\infty} \int_{-\infty}^{\infty} F_x(\omega) F_x^*(\nu) 2\pi\delta(\omega-\nu) \mathrm{d}\omega \mathrm{d}\nu. \tag{2.50}$$

Evaluating the delta function leads to

$$\int_{-\infty}^{\infty} |x(t)|^2 \mathrm{d}t = \frac{1}{2\pi} \int_{-\infty}^{\infty} |F_x(\omega)|^2 \mathrm{d}\omega. \tag{2.51}$$

- **Convolution Theorem**: Convolution in time corresponds to multiplication in frequency:

$$\mathscr{F}\{x_1(t) * x_2(t)\} = F_{x_1}(\omega) F_{x_2}(\omega). \tag{2.52}$$

Proof: The convolution in the time domain is defined as

$$x_1(t) * x_2(t) = \int_{-\infty}^{\infty} x_1(\tau)x_2(t-\tau)\mathrm{d}\tau. \tag{2.53}$$

Taking the Fourier transform yields

$$\mathscr{F}\{x_1 * x_2\} = \int_{-\infty}^{\infty} \left[\int_{-\infty}^{\infty} x_1(\tau)x_2(t-\tau)\mathrm{d}\tau \right] e^{-j\omega t}\mathrm{d}t. \tag{2.54}$$

Changing the order of integration gives

$$\mathscr{F}\{x_1 * x_2\} = \int_{-\infty}^{\infty} x_1(\tau) \left[\int_{-\infty}^{\infty} x_2(t-\tau)e^{-j\omega t}\mathrm{d}t \right] \mathrm{d}\tau. \tag{2.55}$$

Using substitution $u = t - \tau$ leads to

$$\int_{-\infty}^{\infty} x_2(t-\tau)e^{-j\omega t}\mathrm{d}t = e^{-j\omega\tau} \int_{-\infty}^{\infty} x_2(u)e^{-j\omega u}\mathrm{d}u. \tag{2.56}$$

Thus,

$$\mathscr{F}\{x_1 * x_2\} = \int_{-\infty}^{\infty} x_1(\tau)e^{-j\omega\tau}\mathrm{d}\tau \cdot F_{x_2}(\omega), \tag{2.57}$$

which yields

$$\mathscr{F}\{x_1 * x_2\} = F_{x_1}(\omega)F_{x_2}(\omega). \tag{2.58}$$

- **Differentiation in Time**: Differentiating a signal in time corresponds to multiplication by $j\omega$ in frequency:

$$\mathscr{F}\left\{\frac{\mathrm{d}^n x(t)}{\mathrm{d}t^n}\right\} = (j\omega)^n F_x(\omega). \tag{2.59}$$

Proof: By definition of the Fourier transform

$$\mathscr{F}\left\{\frac{\mathrm{d}^n x(t)}{\mathrm{d}t^n}\right\} = \int_{-\infty}^{\infty} \frac{\mathrm{d}^n x(t)}{\mathrm{d}t^n} e^{-j\omega t}\mathrm{d}t. \tag{2.60}$$

Using integration by parts (for the case $n = 1$), let

$$u = e^{-j\omega t}, \quad \mathrm{d}v = \frac{\mathrm{d}x(t)}{\mathrm{d}t}\mathrm{d}t. \tag{2.61}$$

Then, we obtain

$$\int \frac{\mathrm{d}x(t)}{\mathrm{d}t} e^{-j\omega t}\mathrm{d}t = x(t)e^{-j\omega t}\Big|_{-\infty}^{\infty} - \int x(t)\frac{\mathrm{d}}{\mathrm{d}t}\left(e^{-j\omega t}\right)\mathrm{d}t. \tag{2.62}$$

The boundary terms vanish when $x(t) \to 0$ as $t \to \pm\infty$. Simplifying the derivative yields

$$\frac{\mathrm{d}}{\mathrm{d}t}\left(e^{-j\omega t}\right) = -j\omega e^{-j\omega t}. \tag{2.63}$$

Thus

$$\int \frac{\mathrm{d}x(t)}{\mathrm{d}t} e^{-j\omega t}\,\mathrm{d}t = j\omega \int x(t)e^{-j\omega t}\,\mathrm{d}t. \tag{2.64}$$

Recognizing the integral as $\mathscr{F}\{x(t)\} = F_x(\omega)$ simplifies the equation to the following form:

$$\mathscr{F}\left\{\frac{\mathrm{d}x(t)}{\mathrm{d}t}\right\} = j\omega F_x(\omega). \tag{2.65}$$

For higher-order derivatives ($n > 1$), the process can be repeated iteratively, yielding

$$\mathscr{F}\left\{\frac{\mathrm{d}^n x(t)}{\mathrm{d}t^n}\right\} = (j\omega)^n F_x(\omega). \tag{2.66}$$

- **Time Reversal**: Reversing a signal in time reverses its Fourier transform in frequency:

$$\mathscr{F}\{x(-t)\} = F_x(-\omega). \tag{2.67}$$

Proof: Recalling the Fourier transform definition

$$\mathscr{F}\{x(-t)\} = \int_{-\infty}^{\infty} x(-t)e^{-j\omega t}\,\mathrm{d}t. \tag{2.68}$$

Using a substitution $\tau = -t$, so $t = -\tau$ and $\mathrm{d}t = -\mathrm{d}\tau$ leads to

$$\mathscr{F}\{x(-t)\} = \int_{-\infty}^{\infty} x(\tau)e^{-j\omega(-\tau)}(-\mathrm{d}\tau). \tag{2.69}$$

Simplifying the equation yields

$$\mathscr{F}\{x(-t)\} = \int_{-\infty}^{\infty} x(\tau)e^{j\omega\tau}\mathrm{d}\tau. \tag{2.70}$$

Recognizing this as the Fourier transform of $x(t)$ evaluated at $-\omega$ ends the proof

$$\mathscr{F}\{x(-t)\} = F_x(-\omega). \tag{2.71}$$

- **Multiplication in Time Domain**: The multiplication property in the time domain for the one-dimensional Fourier transform is given by

$$\mathscr{F}\{x_1(t)x_2(t)\} = \frac{1}{2\pi}F_{x_1}(\omega) * F_{x_2}(\omega), \tag{2.72}$$

where $*$ denotes convolution.

Proof: By the definition of the Fourier transform, the transform of the product $x_1(t)x_2(t)$ can be written as

$$F_{x_1x_2}(\omega) = \int_{-\infty}^{\infty} x_1(t)x_2(t)e^{-j\omega t}\mathrm{d}t. \tag{2.73}$$

Expressing $x_2(t)$ using its inverse Fourier transform leads to

$$x_2(t) = \frac{1}{2\pi}\int_{-\infty}^{\infty} F_{x_2}(\xi)e^{j\xi t}\mathrm{d}\xi, \tag{2.74}$$

substituting this into the integral

$$F_{x_1x_2}(\omega) = \int_{-\infty}^{\infty} x_1(t)\left[\frac{1}{2\pi}\int_{-\infty}^{\infty} F_{x_2}(\xi)e^{j\xi t}\mathrm{d}\xi\right]e^{-j\omega t}\mathrm{d}t. \tag{2.75}$$

Rearranging the order of integration yields

$$F_{x_1x_2}(\omega) = \frac{1}{2\pi}\int_{-\infty}^{\infty} F_{x_2}(\xi)\left[\int_{-\infty}^{\infty} x_1(t)e^{-j(\omega-\xi)t}\mathrm{d}t\right]\mathrm{d}\xi. \tag{2.76}$$

Recognizing that the inner integral is the Fourier transform of $x_1(t)$, we obtain

$$F_{x_1x_2}(\omega) = \frac{1}{2\pi}\int_{-\infty}^{\infty} F_{x_1}(\omega-\xi)F_{x_2}(\xi)\mathrm{d}\xi. \tag{2.77}$$

This is the convolution of $F_{x_1}(\omega)$ and $F_{x_2}(\omega)$ in the frequency domain, completing the proof.

2.2.6 EXAMPLES

Let us analyze two different radar pulses of unknown parameters. The signals were recorded during the trials; we deem their characteristics unknown. The time and frequency distributions of the signals are delineated in Figures 2.4 and 2.5. They present the time-domain real and imaginary signal parts[2] and the corresponding Fourier spectra obtained using the FFT method. From the plots, we can deduce the pulse duration (assessing the first and the last pulse samples), modulation type (from the shape of the waveform), and bandwidth (establishing the upper and lower frequency range from the spectrum). Especially it can be easily done for the second case in Figure 2.5c, where the noise level is much lower. Although these examples are pretty informative, we still do not have complete knowledge of these waveforms. For instance, we are unsure about the modulation type, frequency modulation rate, and possible interference in the analyzed band. The aforementioned signal features can be regarded as nonstationary. Although this term was already mentioned in this book, let us discuss this concept in more detail.

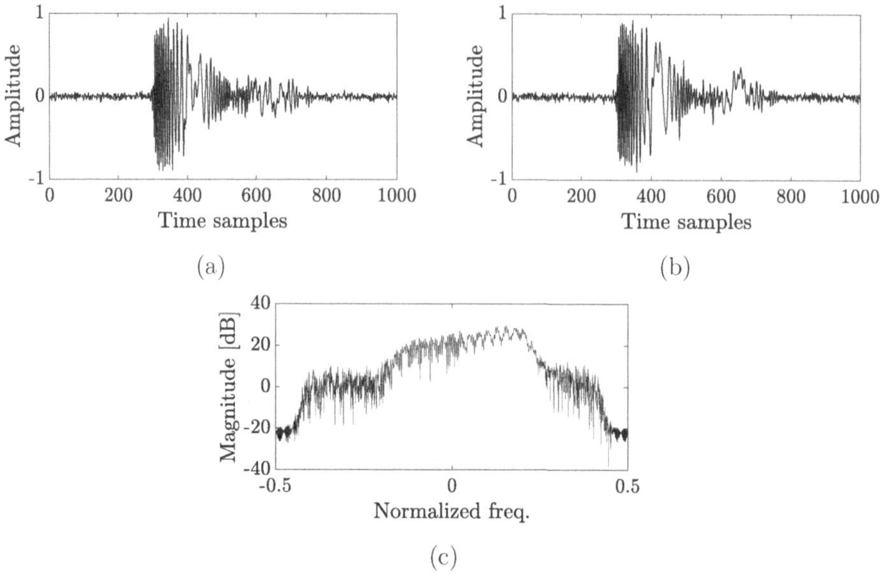

Figure 2.4 Time and frequency distributions of the first radar pulse. (a) Real ($\Re(x(t))$) and (b) imaginary ($\Im(x(t))$) parts of the real-life radar pulse; (c) spectrum of the pulse.

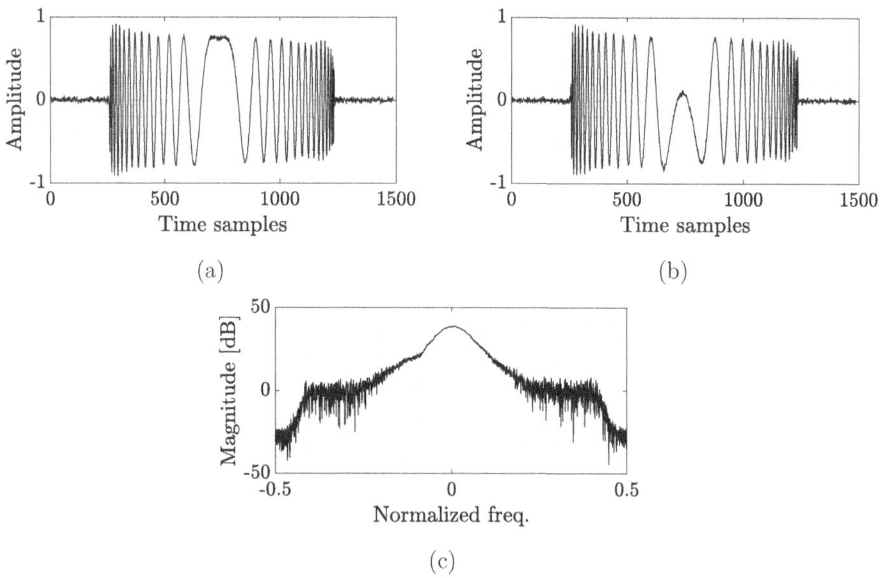

Figure 2.5 Time and frequency distributions of the second radar pulse. (a) Real ($\Re(x(t))$) and (b) imaginary ($\Im(x(t))$) parts of the real-life radar pulse; (c) Spectrum of the pulse.

2.3 STATIONARITY

Before explaining the matter of time-frequency signal analysis, it would be handy to establish why nonstationary signals perfectly fit this kind of analysis and what stationarity actually means. From the mathematical setting, a stationary signal obeys two conditions [45, 46]:

- The mean value is a time-invariant constant.
- The covariance function depends solely on the time difference, not absolute time.

Both of them are hardly ever met in practice, mainly in two respects [46]. Stationarity is a notion referring to stochastic signals, which practically encompasses both stochastic and deterministic. Secondly, translational invariance is typically simplified to local invariance, thereby introducing terms called local stationarity [47]. Those remarks permit verification (testing) of stationarity in a given observation scale [48, 49].

From the abovementioned observations, we infer the statement of relativity when contemplating practical aspects of stationarity. A given signal may be classified as stationary in a given time frame, but wide observation breaks down this condition. This claim is supported by the air traffic control signal displayed in Figure 2.6 (for simplicity, only the real part of the signal is displayed).

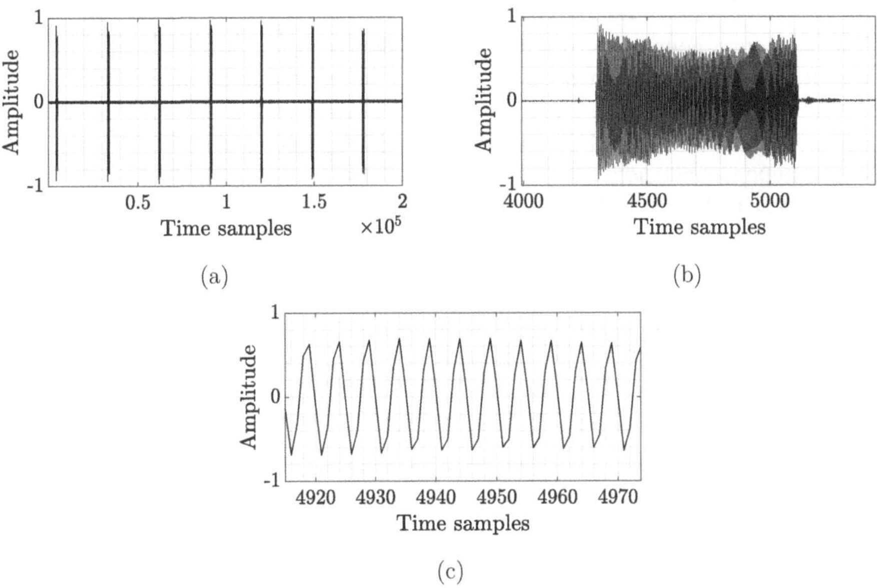

Figure 2.6 Air traffic control pulse train as an example of relative stationarity. (a) A pulse train containing several pulses; (b) a single pulse; (c) close-up on the selected pulse fragment.

Firstly, we see a waveform with seven pulses. Their regular pattern leads to it being commonly agreed as stationary. A periodic structure with regularly distributed pulses is a perfect example of what we loosely call stationary. Next, a single pulse exhibits drastic changes with the occurrence of envelope variation. Such an observation scale makes our pulse nonstationary as its time-varying amplitude does not meet any criteria of stationarity. However, a close-up of the short-lasting pulse part again permits us to infer the signal stationarity.

Vast radar signals are nonstationary in a broad sense, thus including their amplitude, frequency, and phase deviation. Nonstationarity, in the loose meaning, is, therefore, everything occurring between stationarity, e.g., radar pulses between the stationary 'silence'. The same observation comes from the signal spectrum observation, and following Flandrin, we may deduce [46]

> "A realization of a signal observed over a given observation scale will be said to be stationary relative to this scale if its local spectra undergo no time evolution."

This crucial observation imposes the need to analyze nonstationary signals using a dynamic spectrum considering their time evolution. Local stationarity can be analyzed using a short-time spectrum from a local signal part. Stationarity, in a global sense, can, therefore, be deduced from comparing subsequent spectra, leading to a deep insight into the signal character. This possibility is obtained using the short-time Fourier transform defined in the next section.

2.4 SHORT-TIME FOURIER TRANSFORM

The short-time Fourier transform (STFT) is the extension of the classical Fourier transform. While the latter provides insight into the frequency information for an entire signal, the former offers a way to examine how the frequency content of a signal evolves over time. This feature is particularly interesting when dealing with multicomponent and nonstationary signals in radars.

The STFT of the signal $x(t)$ is defined as

$$F_x^h(t,\omega) = \int_{-\infty}^{\infty} x(\tau)h^*(\tau-t)e^{-j\omega(\tau-t)}\mathrm{d}\tau = \int_{-\infty}^{\infty} x(\tau+t)h^*(\tau)e^{-j\omega\tau}\mathrm{d}\tau, \qquad (2.78)$$

where $x(t)$ is the signal to be analyzed, $h(t)$ is a window function (usually a bell-shaped function), $e^{-j\omega t}$ represents the Fourier kernel, and the subscript and superscript in $F_x^h(t,\omega)$ denote the analyzed signal and the window, respectively.

Remark 2.1. *In general, the STFT can be formulated in another, albeit equivalent, way. This perspective interprets the STFT as a convolution between the signal under analysis, $x(t)$, and the complex-conjugated impulse response of a band-pass filter centered at an angular frequency ω, denoted as $g(t,\omega) = h^*(t)e^{j\omega t}$. In this case, for any filter $g(t)$, the STFT is given by the inner product as follows:*

$$F_x^g(t,\omega) = \int_{-\infty}^{\infty} x(t-\tau)g(t,\omega)\mathrm{d}\tau. \qquad (2.79)$$

This representation can be visualized as the result of filtering the signal $x(t)$ through a bank of band-pass filters, each tuned to a specific frequency. The output of each filter, associated with an angular frequency ω, provides a localized measure of the signal's frequency content at a given time t. Therefore, the STFT reveals how the signal's spectral content evolves over time and within the neighborhood of ω.

Equation (2.78) can be explained twofold:

- We fix the position of the signal $x(t)$ in time, and shift the window $h(t)$ toward the right-hand side ($+\infty$). We compute the Fourier transform from the window-signal product in each window position. As a result, we have a set of Fourier spectra for each window position.
- We fix the window position in $t = 0$ and shift the signal to $-\infty$ (this is why we add t in the signal argument in the right-hand side equation). Akin to the first case, we multiply the window by the signal and compute its Fourier transform for each signal position.

A graphical interpretation of the two equipollent definitions (2.78), in particular, the way how the window and signal are shifted with respect to each other, is delineated in Figure 2.7.

Remark 2.2. *The definition (2.78) assumes a general case when $h(t) \in \mathbb{C}$. For the sake of simplicity, we use the notation without the conjugation symbol whenever the window is real-valued. Unless otherwise stated or explicitly highlighted, the window is implicitly assumed to be real $h(t) \in \mathbb{R}$.*

The window function $h(t)$ plays a key role in the STFT representation. It isolates a portion of the signal around the time instant t, enabling the Fourier transform to analyze only that segment. The result is a function $F_x^h(t,\omega)$ that describes the signal's frequency content as it changes over time and is the complex-value two-dimensional distribution so that $F_x^h(t,\omega) \in \mathbb{C}$. The window's width determines the trade-off between time and frequency resolution. A narrower window offers better time resolution but worse frequency resolution, while a wider window does the opposite. In the further part of this book, we use the simplified term *narrow window* for these narrow in time (but wide in the frequency) and *wide window* for windows expansive in the time domain (but narrow in frequency). It is essential to differentiate between time and frequency dependencies to define further considerations clearly. The influence of window types is profoundly discussed in the latter part of this chapter. The most common windows used in the STFT are as follows:

- **Rectangular Window**: A simple window with constant amplitude within a fixed duration.
- **Hamming Window**: A smoother window that reduces spectral leakage.
- **Gaussian Window**: Provides excellent time-frequency localization.

Their examples in the time and frequency domains are shown in Figure 2.8. In short, the narrowest peak is the feature of the rectangular window. Nevertheless, the sidelobes in the frequency domain are relatively high, especially the first sidelobe at

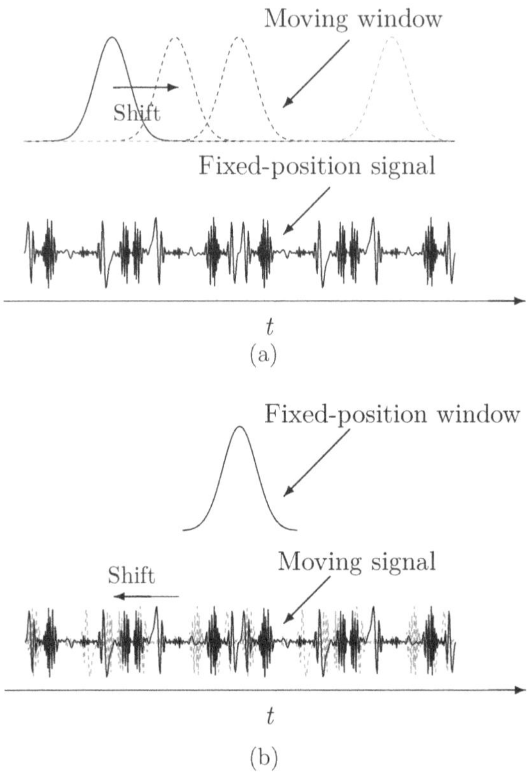

Figure 2.7 Depiction of two equivalent STFT computation ways according to (2.78). (a) Left-hand side definition interpretation. (b) Right-hand side definition interpretation.

the level of ~ -13 dB with respect to the main lobe. In turn, the Gaussian window has a wide main peak, but the spectral leakage and the level of sidelobes are small. Typically, any bell-shaped function can make the time-frequency localization more precise than the rectangular one for which the high sidelobes can cover weaker signal components.

2.4.1 INVERSE SHORT-TIME FOURIER TRANSFORM

Now, let us focus on the inversion formula. The STFT is invertible, allowing the signal to be reconstructed in its time-domain form. By the definition of the STFT, we have

$$x(t) = \int_{-\infty}^{\infty}\int_{-\infty}^{\infty} F_x^h(t,\omega)h(\tau-t)e^{j\omega(\tau-t)}\mathrm{d}t\mathrm{d}\omega = \frac{1}{2\pi h(0)}\int_{-\infty}^{\infty} F_x^h(t,\omega)\mathrm{d}\omega, \quad (2.80)$$

where $h(0)$ is the central (maximal) window value. From (2.80), we can see that the inverse STFT (ISTFT) is the STFT integrated over the frequency and normalized by the window in $t=0$.

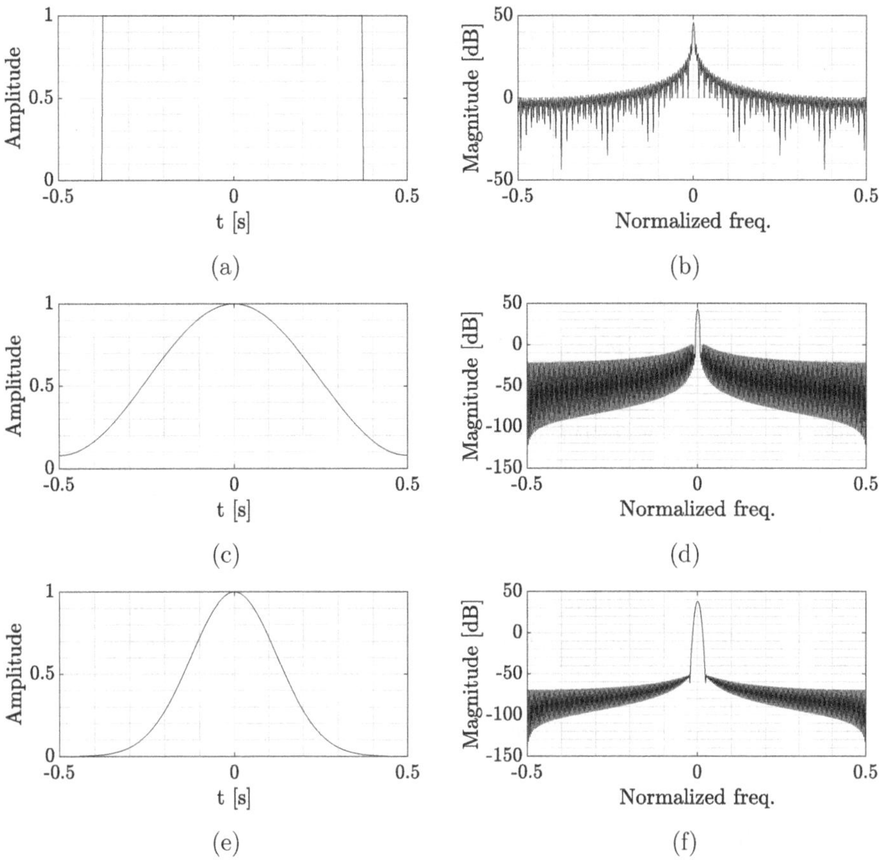

Figure 2.8 Time and frequency representations of the most commonly used windows. (a) Rectangular window in the time domain; (b) rectangular window in the frequency domain; (c) Hamming window in the time domain; (d) Hamming window in the frequency domain; (e) Gauss window in the time domain; (f) Gauss window in the frequency domain.

2.4.2 SPECTROGRAM

As mentioned, the STFT is the complex-value distribution. The spectrogram, usually referred to as the energy distribution, is typically used to both graphically visualize the STFT and present the energetic contribution of particular signal modes in the entire signal content. The spectrogram is defined as the squared modulus of the STFT

$$E_x^h(t,\omega) = \left|F_x^h(t,\omega)\right|^2. \tag{2.81}$$

In practice, the STFT (and spectrogram) is usually computed using the FFT algorithm for each window position. The computation diagram is shown in Figure 2.9. Our signal under consideration is multiplied by the window of the width W. The

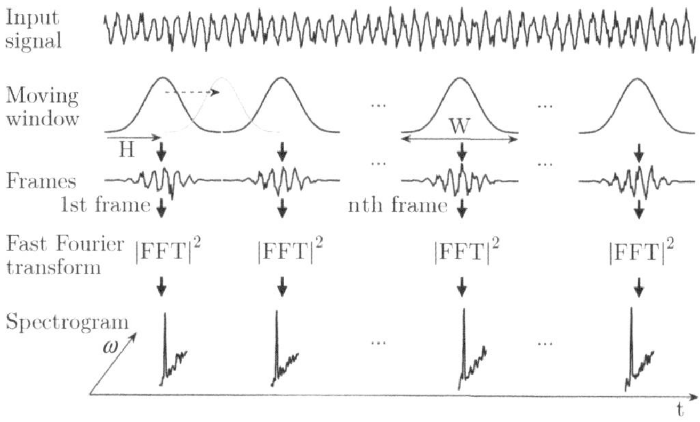

Figure 2.9 Spectrogram computation pipeline. A window of length W is shifted with a stride of H. At each position, the product of the signal and the window forms the nth frame, which is then transformed by the FFT. The squared magnitude of the FFT output yields a spectrogram column.

window moves along the time axis, shifting H samples in each iteration. For each window position, we compute the FFT from the frame. Its squared absolute value gives the spectrogram column. The computational complexity of the spectrogram computation using the FFT algorithm equals $\mathcal{O}(N \log_2 N)$, where N is the FFT size. One can notice that the computational complexity of the STFT in the Landau notation equals the complexity of the FFT.

2.4.3 EXAMPLES

The real-life examples presented radar pulses were shown in the time and frequency domains in Figures 2.4 and 2.5. Now, let us analyze them using the STFT to stress out its advantages in nonstationary and multicomponent signal analysis. The spectrograms of these signals are shown in Figure 2.10.

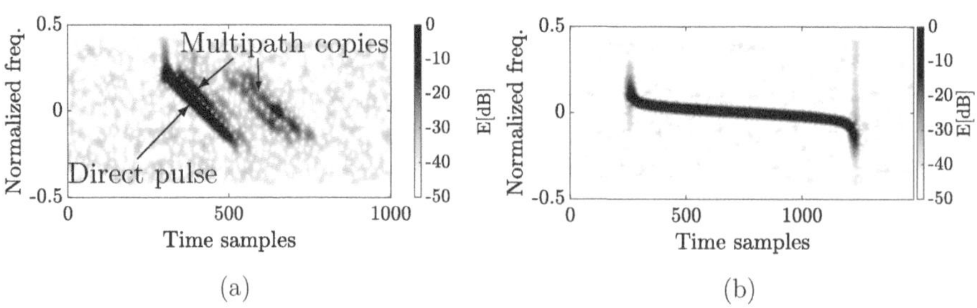

Figure 2.10 Spectrograms of the real-life radar pulses. (a) Spectrogram of the first pulse – linear frequency-modulated signal with apparent multipath propagation effect. (b) Spectrogram of the second pulse – nonlinear frequency-modulated signal.

The difference between the joint time-frequency analysis and the analysis only in the time or frequency domain is tremendous. Firstly, the spectrogram allows for the deduction of the signal character and interference, as shown for the linear chirp from Figure 2.10a. The waveform is disturbed by multipath propagation, manifesting itself by the delayed pulse copies at the same frequency. They cannot be seen in the spectrum analysis as they lie at an identical frequency as the original pulse. We can deduce the number of reflected signal copies and their energy. In turn, we can identify the modulation type for the nonlinear chirp, as shown in Figure 2.10b. The spectrogram shows that the pulse has a nonlinear FM, which was not noticeable in the time or frequency domain analysis. These examples show the superiority of the STFT over classical Fourier analysis when working with such signals.

2.4.4 WINDOW WIDTH CHOICE

The choice of the window is crucial for the STFT (and spectrogram) readability. This parameter directly influences the resolution of the resulting time-frequency representation. This effect is often characterized as a trade-off between time and frequency resolution due to the Heisenberg-Gabor uncertainty principle [50]. Let us consider these metrics separately.

The time resolution Δ_t is approximately proportional to the effective duration of the window function $h(t)$. A narrower window provides better localization in time but inferiorizes the frequency resolution. This can be expressed as

$$\Delta_t \propto \sqrt{\int_{-\infty}^{\infty} t^2 \left|h(t)\right|^2 \mathrm{d}t}. \tag{2.82}$$

The frequency resolution Δ_ω is inversely proportional to the bandwidth of the window $h(t)$ in the frequency domain. The wider the window in the time domain, the narrower the window in the frequency domain, and the frequency resolution is improved according to the formula

$$\Delta_\omega \propto \sqrt{\int_{-\infty}^{\infty} \omega^2 \left|H(\omega)\right|^2 \mathrm{d}\omega}, \tag{2.83}$$

where $H(\omega)$ is the Fourier transform of $h(t)$.

The time-frequency resolution of the STFT is fundamentally limited by the uncertainty principle, which states

$$\Delta_t \Delta_\omega \geq \frac{1}{4\pi}. \tag{2.84}$$

This inequality shows that improving time resolution comes at the cost of worsening frequency resolution and vice versa. In other words, the choice of the $h(t)$ (in particular, its width) determines how well features like transients or oscillatory components are captured. A narrower window is preferable for analyzing rapid transients, while a wider window is better suited for identifying steady-state sinusoidal components.

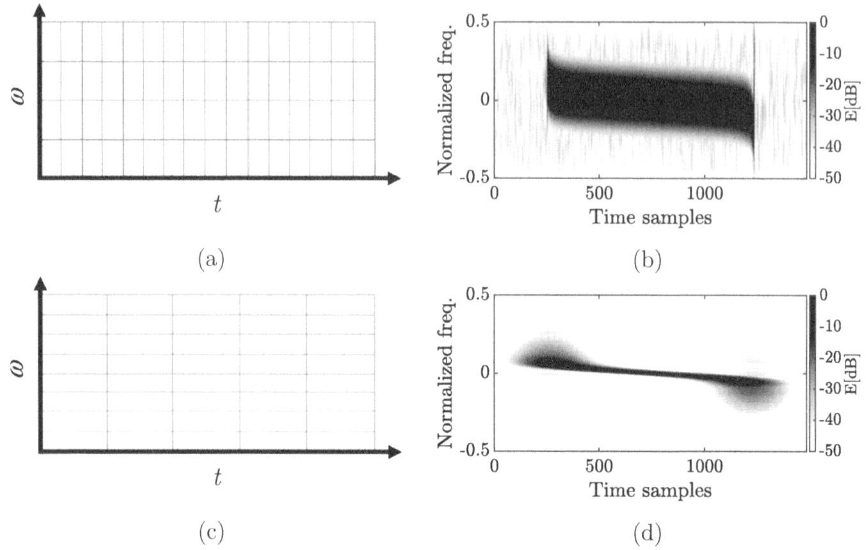

Figure 2.11 Depiction of the influence of the window width on the spectrogram resolution with two distinctive examples using a nonlinear frequency-modulated pulse. (a) Graphical interpretation of the spectrogram Heisenberg-Gabor uncertainty principle for the narrow window. (b) Nonlinear frequency-modulated pulse spectrogram for the narrow window. (c) Graphical interpretation of the spectrogram Heisenberg-Gabor uncertainty principle for the wide window. (d) Nonlinear frequency-modulated pulse spectrogram for the wide window.

The problem of the influence of the window width on the spectrogram resolution is graphically presented in Figure 2.11 and supported by a real-life example. A narrow window (see Figure 2.11a and b) allows for achieving a good time resolution. This feature is handy for assessing discontinuities, transients, or signal duration. However, the frequency resolution is poor, leading to biased information on the signal bandwidth and instantaneous frequency.

Wide window, in turn, yields better frequency resolution, thanks to which one can estimate the signal's instantaneous frequency, especially for the linear part of the chirp. Notwithstanding, the outcome is disturbed by the deficient time resolution disabling pulse duration estimation.

Two methods allow for spectrogram optimization, and the simple one is described first. For a single-component signal with pure and linear FM, Cohen [51] derived a relationship between the FM factor and the window width. As such, we can write

$$\text{Optimal window width} \sim \frac{1}{2\left|\phi''(t)\right|}, \tag{2.85}$$

where $\phi''(t)$ is the second-order derivative of the signal phase. For the linear chirp (with a unitary amplitude), one has

$$x(t) = \exp\left(j(\phi_0 + \omega_0 t + \alpha t^2/2)\right) = \exp\left(j\phi(t)\right), \tag{2.86}$$

where ϕ_0 is the initial phase, ω_0 is the carrier angular frequency, and α is the FM factor (chirp rate). The signal phase and its first- and second-order derivatives read as, respectively:

$$\phi(t) = \phi_0 + \omega_0 t + \alpha t^2/2, \tag{2.87}$$

$$\phi'(t) = \frac{\mathrm{d}\phi(t)}{\mathrm{d}t} = \omega_0 + \alpha t, \tag{2.88}$$

$$\phi''(t) = \frac{\mathrm{d}^2\phi(t)}{\mathrm{d}t^2} = \alpha. \tag{2.89}$$

Equation (2.88) is usually called the *instantaneous frequency*. As follows from (2.89), for linear frequency-modulated signals, the second-order phase derivative equals the frequency rate.

The relationship defined by (2.85) does not explicitly define how the window width should be selected but describes the standard deviation of the modified signal and window at time t. For the Gaussian window with a time spread σ

$$h(t) = \frac{1}{\sqrt{2\pi}\sigma} e^{-\frac{t^2}{2\sigma^2}}, \tag{2.90}$$

we can derive a direct formula for the optimal σ (σ_o) when analyzing frequency-modulated signal [52], which can be defined as

$$\sigma_o = \sqrt{\frac{1}{|\alpha|}}, \quad \alpha \neq 0. \tag{2.91}$$

Note that in (2.91), α is expressed in rad/s^2.

The considerations presented so far were conducted under the prerequisite of linear FM. Although such signals certainly exist, especially in radars, many practical considerations go beyond this narrow case. Now, attention is paid to the more rigorous method based on the Rényi entropy [53,54], which is a spectrogram information content measure. The γ-order Rényi entropy is defined as

$$H_R^\gamma\left(E_x^h(t,\omega)\right) = \frac{1}{1-\gamma}\log_2\left(\frac{\int\limits_{-\infty}^{\infty}\int\limits_{-\infty}^{\infty}\left(E_x^h(t,\omega)\right)^\gamma \mathrm{d}t\mathrm{d}\omega}{\int\limits_{-\infty}^{\infty}\int\limits_{-\infty}^{\infty}\left(E_x^h(t,\omega)\right)\mathrm{d}t\mathrm{d}\omega}\right), \tag{2.92}$$

and is expressed in bits. The lower the value of (2.92), the better the spectrogram concentration (components are more focused on the instantaneous frequency ridges). This approach is beneficial when considering multicomponent signals or components with other modulation types than linear FM. As such, manipulating the window width can obtain the spectrogram with the lowest entropy that guarantees the optimal resolution. Of course, the situation is more complicated in practice when multiple components of different natures can coexist. Thus, the optimal window width cannot simultaneously offer the best resolution for each of them.

The selection of the order of Rényi entropy is a critical issue related to the interpretation of results and the efficiency of the analysis. Rényi entropy is a measure of uncertainty and dispersion of information, which is applied in time-frequency signal analysis to assess the concentration of a signal's energy in the time-frequency distribution. The order of Rényi entropy, denoted as γ, determines how weights are assigned to different values of the time-frequency distribution. The parameter γ controls the measure's sensitivity to various distribution aspects. For $\gamma = 2$, Rényi entropy is particularly sensitive to regions with high energy concentration, making it useful for analyses where capturing dominant components of the signal is essential. For $\gamma \to 1$, Rényi entropy approaches Shannon entropy, which evenly considers all distribution elements. On the other hand, for larger values of γ, the entropy becomes increasingly focused on the peaks of the distribution.

Choosing an appropriate value of γ poses a challenge because its selection must align with the characteristics of the signal and the analysis objective. For signals with highly dispersed energy in the time-frequency domain, a large γ value may lead to the loss of information about weaker components. Conversely, a small γ value may fail to adequately emphasize distinct and significant signal components, resulting in blurred analysis outcomes. In practice, the value of γ is often selected empirically, leading to issues stemming from the lack of universal selection criteria. Adaptive methodologies are frequently employed to adjust γ based on the analyzed signal dynamically, or comparisons of results across different entropy orders are used to identify the most representative values. Ultimately, the choice of the Rényi entropy order represents a trade-off between the level of detail and the stability of the analysis. Optimizing this parameter requires careful consideration of the signal's properties and the research objectives, making it a significant topic in developing time-frequency analysis methods.

The literature [53, 55–58] shows that $\gamma = 3$ ensures stable results and is the best choice for information analysis. This value is used in the study in this book unless otherwise stated.

To present the process of window width optimization using the Rényi entropy, let us consider three signals with the following models:

$$x_1(t) = \exp\left(j2\pi\left(\frac{1000t^2}{2}\right)\right), \tag{2.93}$$

$$x_2(t) = \exp\left(j2\pi\left(200t + \frac{500t^2}{2}\right)\right), \tag{2.94}$$

$$x_3(t) = x_1(t) + x_2(t). \tag{2.95}$$

The signals were sampled with the rate of $f_s = 1000$ Sa/s and disturbed by noise, giving the signal-to-noise ratio (SNR) of 20 dB. The spectrogram was computed for each signal with different σ parameters in the Gaussian window. Next, the third-order Rényi entropy was calculated for each case. Figure 2.12 depicts the outcomes of the optimization process and the optimized spectrograms.

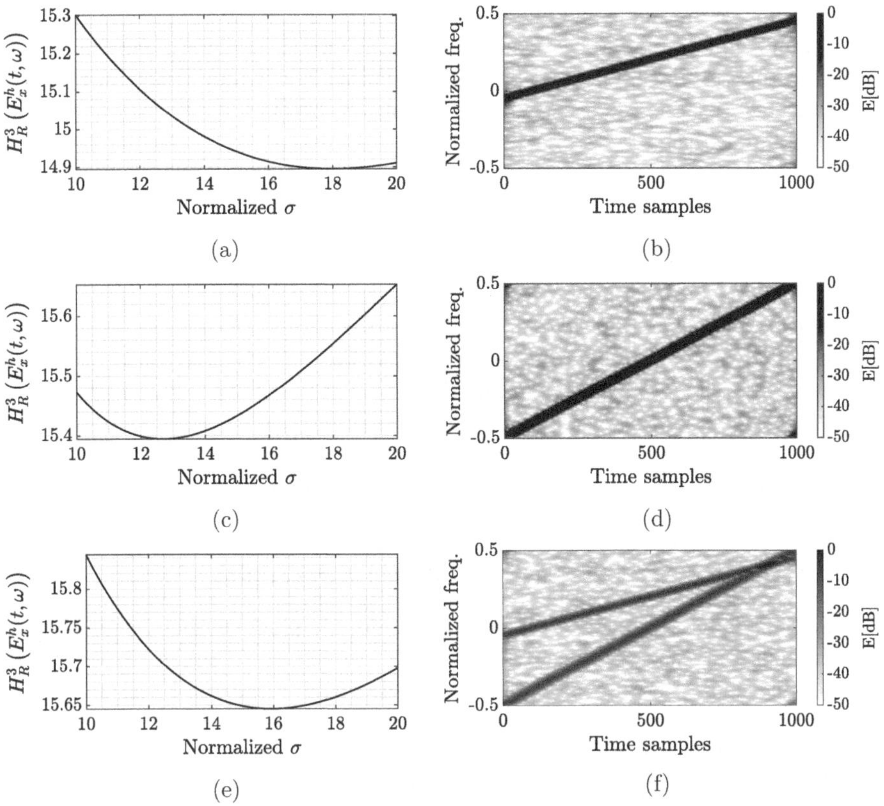

Figure 2.12 Visualization of the window width optimization process for different linear chirps. (a) Rényi entropy for different Gaussian window widths for the slow chirp. (b) Spectrogram of the slow chirp with the optimized window. (c) Rényi entropy for different Gaussian window widths for the fast chirp. (d) Spectrogram of the fast chirp with the optimized window. (e) Rényi entropy for different Gaussian window widths for the sum of slow and fast chirps. (f) Spectrogram of the sum of two chirps with the optimized window.

The Rényi entropies for each analyzed case presented in Figure 2.12a, c, and e in principle illustrate relation (2.91), saying that the optimal Gaussian window-time spread parameter is inversely proportional to the signal chirp rate. We need a narrower window to capture fast-changing FM when the chirp rate increases (fast chirp case). In turn, when we analyze a slow chirp, its frequency deviation is progressing slowly; thus, a wider window is required. A combination of these two cases leads to a trade-off. In the analyzed case, it happened that the optimal window width for the sum of two chirps is approximately the mean of the two cases, but it is not a rule. The problem is more complex and detailed in [54].

Concluding, the design of the window $h(t)$ is a crucial step in STFT analysis. The trade-off between time and frequency resolution requires careful consideration based on the nature of the signal and the specific application. The use of adaptive windowing or multi-window methods can aid in alleviating these constraints. One can use the Rényi entropy for linear chirps to find the optimal window width. Also, knowing the FM rate in chirps, one can use a simple relationship (2.91) for the Gaussian window. Worth noting is that (2.91) works directly for the Gaussian window, while the Rényi entropy can be used for any window (Hann, Blackman-Harris, rectangular, etc.). The question is how to deal with nonlinear chirps or signals utterly different from chirps. The following section tries to answer this issue.

Remark 2.3. *In the further part of this book, the Rényi entropy is used to measure the spectrogram concentration. Since this metric attains a minimum for the optimized time-frequency distribution, one can measure how well optimized the window used for the analysis is. Therefore, we use this parameter to estimate the quality of energy concentration in the spectrogram. The lower the entropy value, the better the properties of the time-frequency distribution we have achieved, understood as the concentration of energy around local maxima (putative signal components).*

2.5 ADAPTIVE SHORT-TIME FOURIER TRANSFORM

It may happen that the received signal is not a chirp. In practice, such circumstances are frequent, and an example of this was already shown in Figure 2.10b where the nonlinear pulse is shown. In radars or radio signal receivers, linear chirps occur mainly in electronic reconnaissance systems or passive radars when analyzing signals for other radars or using them as an illumination source.

Multicomponent signals, especially those with nonlinear FM, require window adaptation as their local time-bandwidth product varies. Thus, one window for the entire analysis is insufficient, and this statement seems to be clear when we look at Figure 2.13, where nonlinear FM is apparent. A homogeneous resolution does not allow focusing the energy around each pulse fragment.

As shown in (2.91), the Gaussian window with the optimal time spread σ_o equals the root of the inverse chirp rate. This statement fails when the signal is not a purely

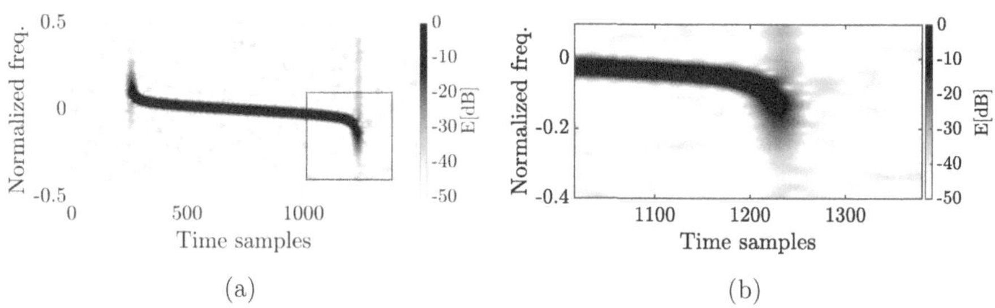

Figure 2.13 Nonlinearity in the frequency modulation in a radar pulse. (a) Spectrogram of the entire pulse with the selected area of interest; (b) close-up on the pulse end with a strong nonlinear frequency modulation.

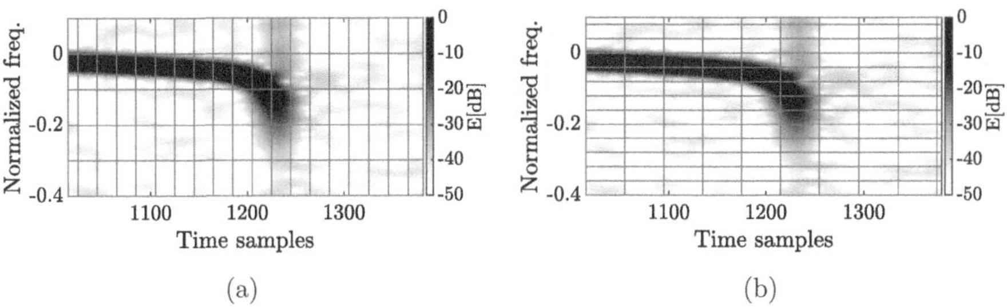

Figure 2.14 Graphical interpretation of why a single (a) narrow and (b) wide window is insufficient when analyzing nonlinear frequency modulation. The grid on the images simplistically represents the resolution defined by the Heisenberg-Gabor uncertainty.

linear frequency-modulated wave. Obviously, when multiple components of different natures exist in the received signal, their simultaneous optimization of energy distribution is challenging, albeit focusing on single-component signals with nonlinear FM makes the problem more tractable. Thus, let us make the STFT adaptive.

The reason why a single window cannot cope with nonlinear FM is illustrated in Figure 2.14, where simplified time-frequency grids using narrow and wide windows are shown. In such a scenario, (2.91) does not ensure the best energy distribution.

In the literature, we can distinguish plenty of works focusing on the optimization of energy distribution [52, 59–65] (these are only several examples to make the general overview, but the whole topic is wide). Although attempting to deal with this issue differently, scholars come down to the problem of adaptive STFT to the multi-window analysis. Instead of a fixed window, they make it time-varying and dependent on the signal chirp rate. Thus, we have $\alpha \to \alpha(t)$ and

$$\sigma_0(t) = \sqrt{\frac{1}{|\alpha(t)|}}, \quad \alpha(t) \neq 0. \tag{2.96}$$

Equation (2.96) is valid for the Gaussian window case. It can be interpreted as follows: the optimal window for each t is the squared root of the inverse of the FM rate at the same t. The way how α is estimated is different, and one of them is described in Section 2.10, but there is not only one correct technique to do so.

Although (2.78) admits any analysis window $h(t)$, let us stick to the Gaussian window whose optimal time spread can be easily regularized by the FM factor. The window adaptivity is feasible for any other bell-shaped function, but the literature does not provide a straightforward relation between the window and signal parameters. Consequently, the adaptive STFT using the adaptive Gaussian window governed by (2.96) reads as

$$F_x^{h_a}(t,\omega) = \int_{-\infty}^{\infty} x(\tau) \underbrace{\frac{1}{\sqrt{2\pi}\sigma(\tau-t)} e^{-\frac{(\tau-t)^2}{2(\sigma(\tau-t))^2}}}_{\text{adaptive Gaussian window}} e^{-j\omega(\tau-t)} \mathrm{d}\tau. \tag{2.97}$$

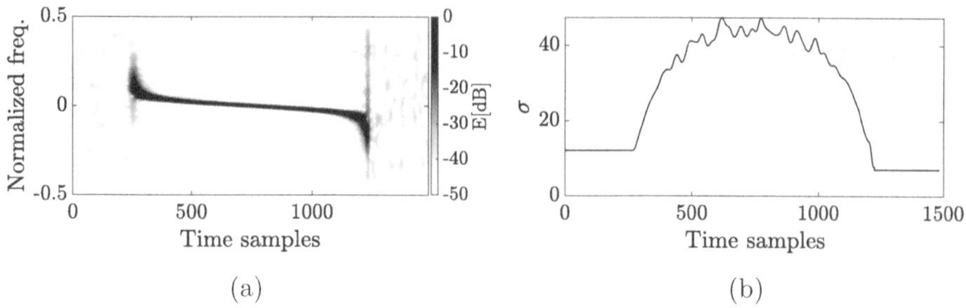

Figure 2.15 Results of the adaptive STFT. (a) Spectrogram of the adaptive STFT; (b) optimized Gaussian window-time spread in time.

The superscript in $F_x^{h_a}(t,\omega)$ (namely, h_a) differentiates between the adaptive STFT and the classical STFT. The energy distribution is defined in the same way as for the STFT

$$E_x^{h_a}(t,\omega) = \left|F_x^{h_a}(t,\omega)\right|^2. \tag{2.98}$$

The intuitive example of the adaptive STFT presenting the nonlinear frequency-modulated pulse already analyzed in this chapter is illustrated in Figure 2.15. In particular, Figure 2.15a delineates the spectrogram of the adaptive STFT with an improved and adaptive resolution. As a companion argument, Figure 2.15b shows the $\sigma_o(t)$ value. As can be seen, this curve resembles a parabola associated with the signal frequency rate. The frequency rate is high at the chirp's beginning and end, determining a narrow window (with small σ). The chirp rate is the smallest in the middle of the chirp, yielding a wider window (higher σ). The third-order Rényi entropy for the original spectrogram for Figure 2.10b is $H_R^{[3]}\left(E_x^h(t,\omega)\right) =$ 14.93 bits. After the adaptive window selection, this value dropped by almost one bit, giving $H_R^{[3]}\left(E_x^{h_a}(t,\omega)\right) = 14.08$ bits. The Rényi entropy reduction shows better concentration properties of the spectrogram (Figure 2.16).

To additionally illustrate the window impact on the time-frequency resolution, let us analyze noise with a strongly varying window-time spread $\sigma(t)$ (function character was arbitrarily chosen to visualize the problem). The spectrogram of noise and the accompanying $\sigma(t)$ are shown in Figure 2.17. The example clearly shows how

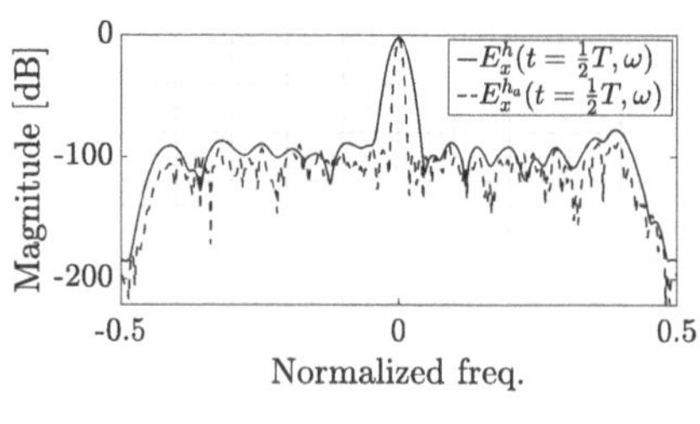

Figure 2.16 Spectrogram cross-sections for the signal center ($t = \frac{1}{2}T$).

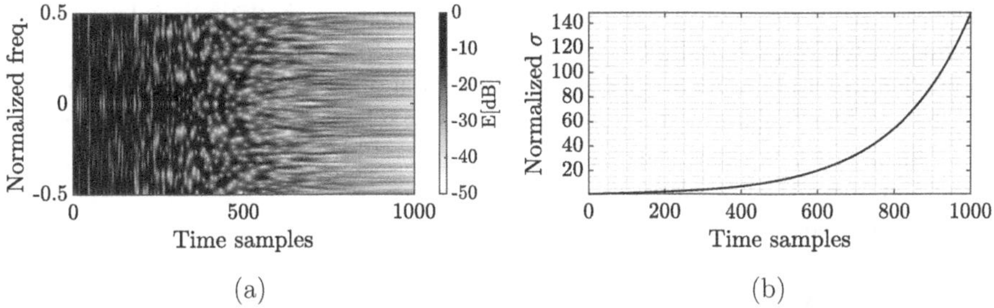

Figure 2.17 Depiction of how the window width leverages the noise time-frequency energy distribution. (a) Noise spectrogram; (b) assumed function for window width variability.

the window width impacts the spectrogram resolution. Narrow windows arrange the spectrogram cells longitudinally but narrowly (good time resolution at the expense of frequency resolution), which can be observed at the beginning of the spectrogram time axis. With increasing σ values, we observe a gradual flattening of these cells, leading to improved frequency resolution while worsening time resolution.

This section has shown that the adaptive STFT allows significant improvement in the spectrogram readability. However, the window width is not the only parameter to be fine-tuned when dealing with frequency-modulated signals. The Gaussian window (and generally, any other) is the function of time, thus making it possible to incorporate it with a chirp, leading to the so-called chirplet – a chirp with a bell-shaped envelope. The following section focuses on this property.

2.6 SHORT-TIME CHIRPLET TRANSFORM

The short-time chirplet transform (sometimes called a chirplet transform, shortly in this book) emerged as a natural evolution in time-frequency analysis, building upon earlier concepts and methods. Following Gabor's work on the STFT [66], the 1950s and 1960s saw radar and signal-processing advancements. Siebert [67] applied time-frequency methods to radar detection, utilizing Woodward's ambiguity function to analyze chirped signals for pulse compression. This approach highlighted the geometric shearing effects in the time-frequency plane caused by chirped signals.

In the 1980s, researchers like Grossman and Paul [68] formalized wavelet analysis using affine transformations, which became a precursor to chirplet theory. Around the same time, Papoulis introduced chirped signals as operators in the time-frequency domain, anticipating the broader applications of chirplets [69]. Jones and Parks [70] later linked these ideas to window selection and leakage minimization in time-frequency analysis, further incorporating sheared versions of standard window functions.

The chirplet transform was independently developed in the late 1980s and early 1990s. Mann and Haykin [71] extended these concepts by generalizing wavelets and STFTs into a multidimensional parameter space, allowing for affine transformations such as shearing, dilation, and translation in the time-frequency plane. This approach facilitated a unified representation encompassing both time-scale and time-frequency analyses. Subsequent work by Mihovilovic and Bracewell [72] introduced related chirplet concepts, emphasizing practical applications. The holistic overview of the chirplet transform was published in [73], where the authors analyze possible chirplets for nonstationary signal analysis. For instance, further developments making the transform adaptive can be found in [74, 75].

After the historical introduction to the topic, we can focus on the matter. So far, the STFT has been described with the accompanying bell-shaped window $h(t) \in \mathbb{R}$ (e.g., Gaussian). As summarized in the previous section devoted to the adaptive STFT, the window can be complex-valued and obtained by modulating the window by a complex waveform. In the chirplet transform, this waveform is a complex chirp.

Remark 2.4. *At this point, the reader should be informed that we made a significant step between the STFT and the chirplet transform, neglecting a piece of critical signal processing theory. There are no limitations to modulating the window using a harmonic term, and the example of such a complex-value window (windowed complex sinusoid waveform) is shown in Figure 2.18. The idea pioneered by Grossmann and Morlet [76] is an extension of the STFT dedicated to signal decomposition applications, where the analyzed signal is modeled as a sum of wavelets (window at carrier frequencies). Although the wavelet transform is widely used in plenty of signal processing-related applications, including radar signal processing [77–80], Figure 2.19 delineates why this tool is not a versatile approach. In some cases*

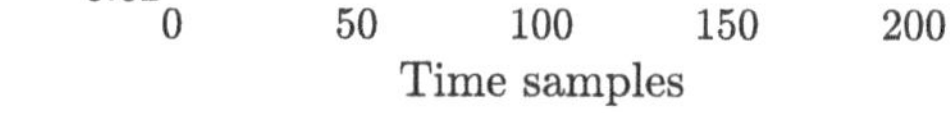

Figure 2.18 Wavelet example.

Figure 2.19 Energy distribution of the radar pulse processed using the wavelet transform.

where specific signal attributes need to be distinguished, the wavelet transform seems to be a perfect tool, e.g., when detecting signal components on a particular frequency. However, the radar pulse's energy distribution (spectrogram equivalent for the wavelet transform) from Figure 2.19 shows that this method is unsuitable for various radar signals. The pulse and its multipath copies cannot be intelligibly extracted as the resolution is ***heterogeneous****. Therefore, due to the limited scope of this book, the wavelet transform is not investigated, and the readers are encouraged to see the wavelet transform-related books, like [81–85].*

The short-time chirplet transform definition begins from the window ambiguity function definition, which, at first glance, can be seen as strange. The ambiguity function of a window in STFT analysis plays a crucial role in determining the resolution of the time-frequency distribution and is intuitive, especially for the chirplet transform. The ambiguity function has already been defined in (1.22). For window analysis, it takes on the following form:

$$\Psi_h(t,\omega) = \int_{-\infty}^{\infty} h(\tau)h^*(\tau - t)e^{-j\omega\tau}\mathrm{d}\tau, \tag{2.99}$$

where all symbols keep their meaning as stated before. A narrower ambiguity function in both t and ω results in better window localization in the time-frequency plane (precisely, the window-signal product), leading to higher resolution. Conversely, a broader ambiguity function indicates poorer resolution, as the window's energy spreads over a larger region in the time-frequency domain, as defined in the Heisenberg-Gabor trade-off (2.84). Such a conclusion is obvious and coincides with our previous divagations. However, the window ambiguity function geometry can alter interchangeably between t and ω and can also rotate. Although proposed almost three decades earlier, recently, the concept was aptly called a *directional kernel* [74].

Figure 2.20 illustrates the ambiguity function for different windows and the consequences in the chirplet analysis:

a. **Wide Real-Valued Window**: The ambiguity function is concentrated around zero in both time and frequency domains. This configuration provides good frequency resolution but limited time resolution.
b. **Narrow Real-Valued Window**: The function is stretched along the time axis, indicating improved time resolution at the expense of frequency resolution.
c. **Wide Complex-Valued Window with FM (Chirplet)**: Introducing linear FM tilts the ambiguity function. This effect enables the detection of signals with varying frequencies but introduces slanted artifacts.
d. **Narrow Complex-Valued Window with FM (Chirplet)**: Combining a narrow window with FM results in a stretched ambiguity function along the time axis with a slanted orientation. This configuration enhances the analysis of signals with rapidly changing frequency but limits frequency resolution.

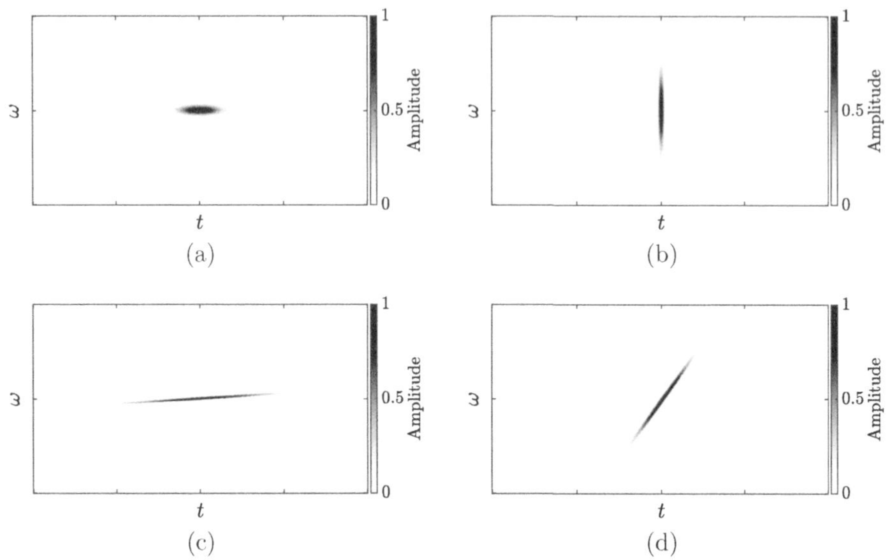

Figure 2.20 Simplistic visualization of the ambiguity function (2.99) for different windows. (a) Wide real-valued window; (b) narrow real-valued window; (c) wide complex-valued window with chirp (chirplet); (d) narrow complex-valued window with chirp.

FM chirp in the window introduces a slanted orientation in the ambiguity function, improving the ability to analyze signals with nonlinear frequency variations but also causing asymmetry in energy distribution.

The short-time chirplet transform with a fixed window (chirplet – window carrying the chirp) can be defined as

$$F_x^{h_c}(t,\omega) = \int_{-\infty}^{\infty} x(\tau) \underbrace{h_c^*(\tau - t) e^{-j\omega(\tau - t)}}_{\text{directional kernel}} \mathrm{d}\tau, \tag{2.100}$$

where

$$h_c(t) = h(t) e^{j\frac{\alpha t^2}{2}}, \tag{2.101}$$

is the chirplet. Assuming $h(t)$ being a Gaussian window, one has

$$h_c(t) = \frac{1}{\sqrt{2\pi}\sigma} e^{-\frac{t^2}{2\sigma^2}} e^{j\frac{\alpha t^2}{2}}, \tag{2.102}$$

with the symbol h_c denoting the chirplet.

Remark 2.5. *At this stage, we use the complex-valued window for the first time, explicitly using the notation with conjugation.*

Substituting (2.102) into (2.100) leads to the short-time chirplet transform with the Gaussian chirplet

$$F_x^{h_c}(t,\omega) = \int_{-\infty}^{\infty} x(\tau)\frac{1}{\sqrt{2\pi}\sigma}e^{-\frac{(\tau-t)^2}{2\sigma^2}}e^{-j\frac{\alpha(\tau-t)^2}{2}}e^{-j\omega(\tau-t)}\mathrm{d}\tau. \quad (2.103)$$

Since the window is conjugated, we change the sign in the window chirp rate.

Akin to the STFT and the adaptive STFT, the energy distribution equals the squared absolute value of the transform

$$E_x^{h_c}(t,\omega) = \left|F_x^{h_c}(t,\omega)\right|^2. \quad (2.104)$$

The method is perfect for both optimizing the spectrogram and finding the FM rate. In the following examples, we will analyze these two aspects.

The performance of the chirplet transform can be evaluated first with a synthetic linear chirp generated according to the model

$$y[n] = A_d w_t[n] \cdot \exp\left(j2\pi\left(\alpha\frac{n^2}{2} - f_n n\right)\right), \quad (2.105)$$

for $n = 0, \ldots, N-1$. The parameter A_d is the unitary amplitude, and $w_t[n]$ is a Tukey window with a cosine fraction equal to $\mathscr{T} = 0.1$, $\alpha = 500$ GHz/s denotes the chirp rate, and $f_s = 2.5$ MHz is the starting frequency of the pulse. The sampling rate equals $f_s = 10$ MSa/s and the pulse duration $T_{\mathrm{imp}} = 10\ \mu$s which gives 100 samples per pulse. This pulse was padded with 100 zeros on both sides, giving a reference waveform of a duration $T_w = 30\ \mu$s. Its multipath copy with delay $\tau = 25$ samples and amplitude $A_i = 0.5$ disturbed this synthetic direct signal and white circular Gaussian noise. The classical spectrogram using the STFT and the one using the chirplet transform are shown in Figure 2.21.

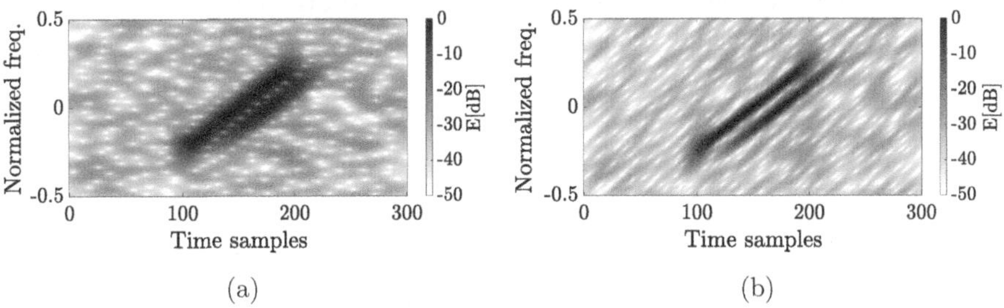

Figure 2.21 A comparison of the classical spectrogram and the chirplet-based spectrogram. In both cases, $\sigma = 8$ for the Gaussian window. (a) Spectrogram for the classical STFT with a real-valued Gaussian window, Rényi entropy equals $H_R^3\left(E_x^h(t,\omega)\right) = 13.99$ bits; (b) spectrogram for the chirplet transform with the chirplet optimized to the analyzed chirp, Rényi entropy equals $H_R^3\left(E_x^{h_c}(t,\omega)\right) = 12.95$ bits.

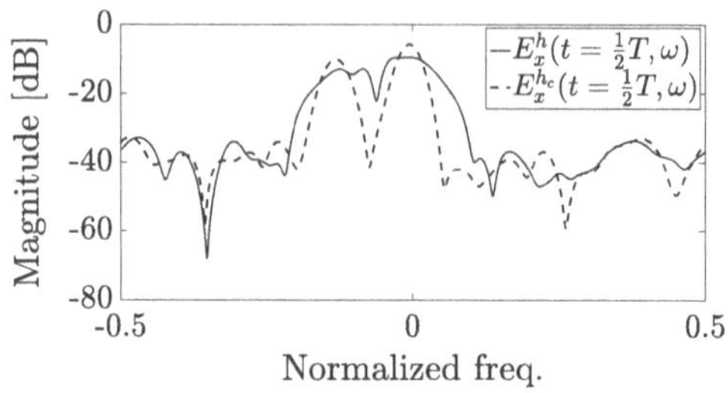

Figure 2.22 Cross-sections in the direct pulse center for the STFT-based and chirplet transform-based spectrograms.

The results exhibit a twofold improvement. Firstly, the resolution is significantly enhanced, leading to more concentrated energy on the instantaneous frequency ridges, both the direct pulse and the multipath copy (the one delayed with respect to the first chirp on the spectrograms). This effect is expressed using the Rényi entropy reduction of over 1 bit for the chirplet transform. Secondly, the multipath effect is more distinguishable in the chirplet transform spectrogram. The delayed pulse copy can be easily extracted from the spectrogram but is invalid for the classical STFT. The same effect can also be noticed when analyzing cross-sections in the center of the first (direct) chirp, as illustrated in Figure 2.22. The chirplet-based spectrogram has narrower peaks (both the direct pulse and its multipath copy), which are higher than the STFT-based approach.

The second radar application of the chirplet transform is to find the unknown chirp rate of the received pulse, especially in high noise. For this purpose, we can again use the Rényi entropy to measure the time-frequency energy distribution concentration. Computing this metric for each chirplet (precisely – window multiplied by chirps with different chirp rates) should be the minimum when window and signal chirp rates are equal. Indeed, this feature is clearly visible in two cases in Figure 2.23. The same simulated signal as in the previously analyzed case with a chirp rate of 500 GHz/s and two values of the SNR (10 and -10 dB) was processed, and its FM factor was estimated through the minimum of the entropy. As can be seen, the

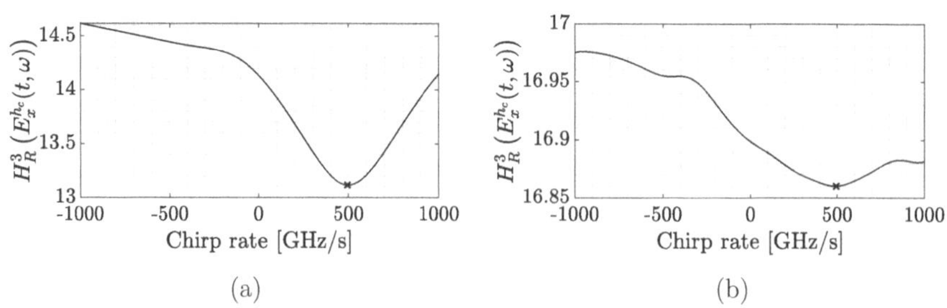

Figure 2.23 Rényi entropies as functions of the chirp rate – visualization of the unknown radar signal parameter estimation. The cross shows the minimum value in each plot. (a) Example for SNR $= 10$ dB; (b) example for SNR $= -10$ dB.

chirp parameters can be estimated despite a high noise level. Similar observations but using a different metric can be found in [74], where instead of the Rényi entropy, the authors used the chirp rate estimate (analyzed in Section 2.10). Nevertheless, the conclusion holds true for both cases, and readers are encouraged to see the cited work and references therein.

Although the chirplet transform efficacy is undoubtedly noticeable, a natural question is how to make this technique adaptive for nonlinear frequency-modulated signals, like nonlinear chirps. Two classes of adaptation can be distinguished. The first relies on changing the window chirp rate adaptively, assuming a fixed window size [74]. The adaptive chirplet transform can be written as

$$F_x^{h_{\mathrm{ac}}}(t,\omega) = \int_{-\infty}^{\infty} x(\tau)\frac{1}{\sqrt{2\pi}\sigma}e^{-\frac{(\tau-t)^2}{2\sigma^2}}e^{-j\frac{\alpha(\tau-t)(\tau-t)^2}{2}}e^{-j\omega(\tau-t)}\mathrm{d}\tau, \tag{2.106}$$

where $\alpha(t)$ is the window chirp rate function dependent on time and h_{ac} is the symbol of adaptive chirplet. In simple words, (2.106) acts as the STFT with a fixed window envelope (e.g., Gaussian window), but it adapts the window chirp rate to the temporal frequency rate of the signal. Naturally, the energy distribution reads as

$$E_x^{h_{\mathrm{ac}}}(t,\omega) = \left|F_x^{h_{\mathrm{ac}}}(t,\omega)\right|^2. \tag{2.107}$$

The second adaptive chirplet transform variant incorporates the preceding method with the adaptive STFT, making the chirplet transform doubly adaptive [75] (double adaptation refers to the adaptive window width and chirp rate). Merging (2.106) with (2.97) yields the double adaptive chirplet transform, which, with a Gaussian window, takes on the following form

$$F_x^{h_{\mathrm{dac}}}(t,\omega) = \int_{-\infty}^{\infty} x(\tau)\underbrace{\frac{1}{\sqrt{2\pi}\sigma(\tau-t)}e^{-\frac{(\tau-t)^2}{2\sigma(\tau-t)^2}}}_{\text{adaptive width}}\overbrace{e^{-j\frac{\alpha(\tau-t)(\tau-t)^2}{2}}}^{\text{adaptive rotation}}e^{-j\omega(\tau-t)}\mathrm{d}\tau, \tag{2.108}$$

where the optimization is governed twofold:

- using the adaptive window width controlled by $\sigma(t)$ – time spread parameter defined by (2.96),
- using the adaptive chirp modulation of the window $\alpha(t)$ – window chirp rate.

Its squared absolute value defines the energy representation in the time-frequency domain

$$E_x^{h_{\mathrm{dac}}}(t,\omega) = \left|F_x^{h_{\mathrm{dac}}}(t,\omega)\right|^2. \tag{2.109}$$

In (2.108) and (2.109), h_{dac} defines the double adaptive chirplet.

The already illustrated ambiguity functions for wide and narrow windows with and without a FM term presented in Figure 2.20 showed how the window can be

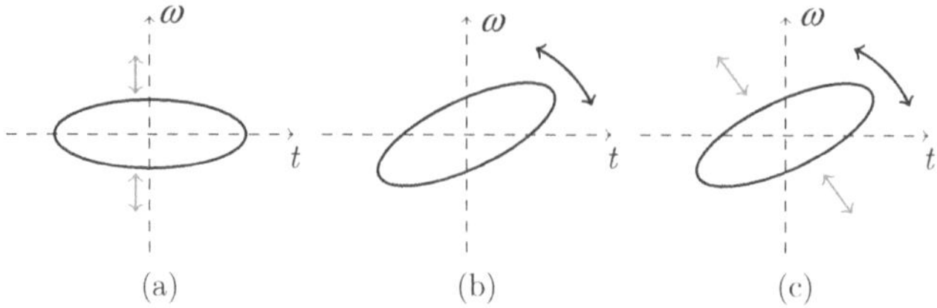

Figure 2.24 A simplified graphical interpretation of how the ambiguity function of the window behaves in each adaptive variant of the STFT. (a) Adaptive STFT; (b) adaptive chirplet transform; (c) double adaptive chirplet transform.

adapted to different signal characters. The difference between the adaptive STFT, adaptive chirplet transform, and double adaptive chirplet transform can be seen in Figure 2.24. From this illustration, we can deduce that the more precisely the window fits the signal, the better the resolution of the time-frequency representation. Indeed, this effect is visualized with two real-life examples.

The first example provides a comprehensive overview of all the transforms introduced thus far, including the STFT, adaptive STFT, adaptive chirplet transform, and double adaptive chirplet transform. The nonlinear pulse is an ideal waveform to demonstrate the main differences between these methods, with the results presented in Figure 2.25. A key assumption was that the FM rate is known at each moment of the chirp. In practice, this rate was estimated using an alternative method but is consistent with the value shown in Figure 2.15b.

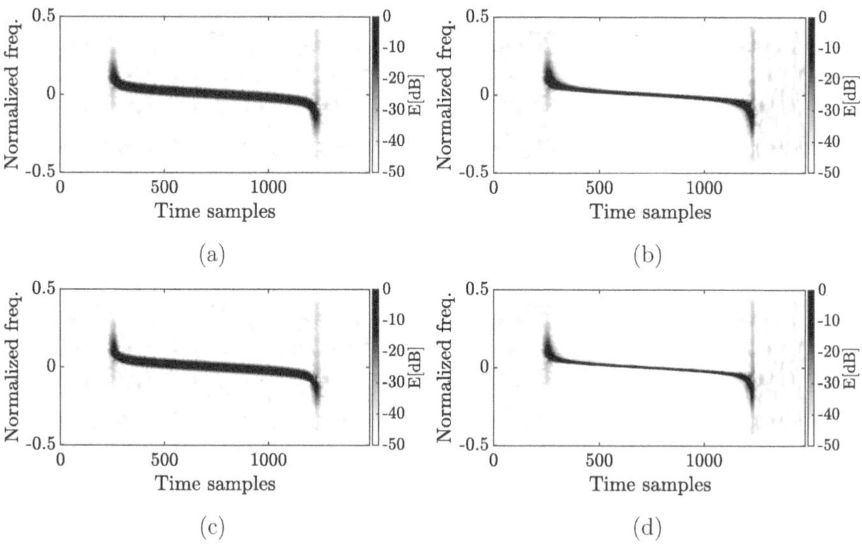

Figure 2.25 Nonlinear frequency-modulated radar pulse energy distributions for different transforms. (a) STFT; (b) adaptive STFT; (c) adaptive chirplet transform; (d) double adaptive chirplet transform.

Table 2.2
Third-Order Rényi Entropies for the Nonlinear Chirp in the Time-Frequency Domain

Method	Rényi Entropy
STFT	14.93
Adaptive STFT	14.09
Adaptive chirplet transform	14.07
Double adaptive chirplet transform	13.59

The nonlinear frequency-modulated radar pulse is represented by its energy distribution in the time-frequency domain, yet the spectrograms reveal significant differences. Firstly, there are noticeable variations in energy concentration around the main component. Secondly, the noise distribution surrounding the pulse, influenced by the instantaneous window parameters, differs among the visualizations. Notably, the double adaptive chirplet transform technique performs best by concentrating energy along the ensemble chirp. However, a numerical evaluation of the improvement in spectrogram clarity is necessary to quantify this enhancement.

The third-order Rényi entropies for the analyzed techniques are listed in Table 2.2. The conclusions support a general impression akin to the spectrogram cross-sections delineated in Figure 2.26. The double adaptation exhibits the strongest signal concentration along the instantaneous frequency ridge.

The final example addresses a more practical scenario in which the analyzed signal is a linear frequency-modulated radar pulse significantly affected by substantial multipath reflections, as illustrated in Figure 2.27. For the classical spectrogram, differentiation between the direct signals and the interference is challenging, especially for those copies appearing just after the direct pulse. Thanks to double adaptation, component separability is enhanced, allowing for more precise pulse extraction.

All the analyzed variations of the STFT are focused on modifications to the analysis window, including its width and modulation. These adaptations can affect certain aspects, improving the spectrogram's readability and higher separability of components. However, one might aim for an ideal time-frequency representation where the

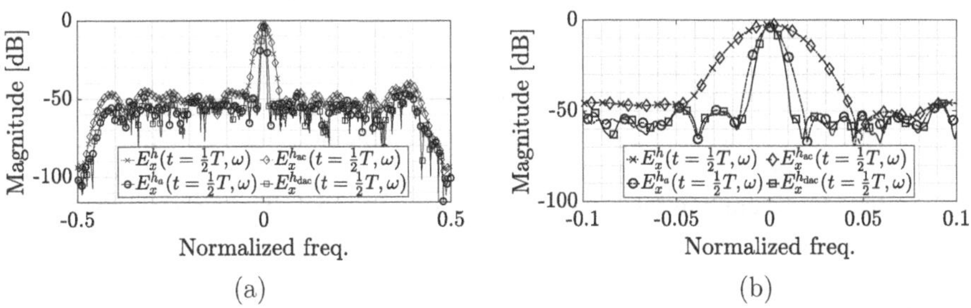

Figure 2.26 Cross-section through the nonlinear chirp center for all analyzed methods. (a) Original view; (b) close-up on the main peak.

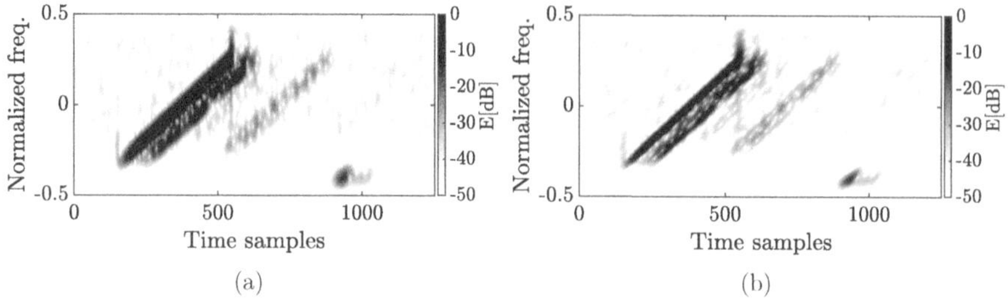

Figure 2.27 Practical utilization of the double adaptive chirplet transform – extracting the radar pulse from a highly interfered signal. (a) STFT-based spectrogram; (b) double adaptive chirplet transform-based spectrogram.

components are sharply localized on the spectrogram, allowing their parameters to be precisely estimated. Such a method would not only inherit the strengths of the presented approaches but also surpass them by energy concentration reinforcement. This technique is known as time-frequency reassignment, which we will explore in the next section.

2.7 TIME-FREQUENCY REASSIGNMENT

Time-frequency reassignment (referred to as a moving window method at its origin) has emerged as a transformative approach to improving the precision of time-frequency representations, addressing the inherent limitations of conventional methods such as the STFT. As observed in the STFT, adaptive STFT, and chirplet-based analysis, the ordinary STFT suffers from limited resolution, which is called the Heisenberg-Gabor uncertainty principle. Although we tried to address this barrier by modifying the window, the component separability was always hindered, and adhering signal modes could be entangled. Moreover, the presented adaptive transforms cannot always provide the optimal window, especially in cases where the multicomponent signal is analyzed, and each component has entirely different parameters (notably the FM rate). Therefore, the abovementioned limitation often results in blurred representations that fail to localize instantaneous signal characteristics accurately.

The concept of reassignment, introduced to overcome these challenges, fundamentally redefines how energy is distributed in the time-frequency plane. Instead of assigning energy to fixed grid points, reassignment methods relocate energy to coordinates that better reflect the actual dynamics of the analyzed signal. This relocation is achieved by leveraging the phase information of the STFT to estimate the center of gravity for energy distribution, thereby tremendously enhancing the localization of signal components.

The seminal work by Kodera, Gendrin, and de Villedary in the late 1970s laid the foundation for time-frequency representations by presenting a novel framework that reassigns energy based on instantaneous frequency and group delay estimates [50].

Their pioneering approach demonstrated how reassignment could sharpen spectrogram representations, enabling more accurate analysis of nonstationary signals. This breakthrough opened the door to numerous applications across diverse fields, including radar signal-processing, speech processing, biomedical signal analysis, and time-varying systems characterization.

This section seeks to elucidate the principles and methodologies underlying time-frequency reassignment, tracing its development from foundational theories to its modern-day radar applications. By examining the core ideas presented in the original work of Kodera et al., we aim to provide a comprehensive understanding of how reassignment transforms time-frequency analysis, offering unparalleled clarity and precision in representing complex signals.

It is worth noting that Kodera et al.'s works were dedicated to the STFT and spectrogram [50, 86]. Auger and Flandrin [87] expanded the reassignment technique to other time-frequency representations, including the Wigner-Ville or Rihaczek distributions, to name only a few. The considerations presented in the remainder part of this book are related solely to the STFT. However, it can be remembered that other time-frequency transforms can also be processed using the reassignment method, and the readers are encouraged to study them.

Remark 2.6. *Note that the reassignment technique allows for* ***readability*** *improvement, but not resolution. The STFT resolution is dictated by the window width and the signal character. The reassignment technique cannot improve the resolution, but can significantly improve the component separability.*

In [50], Kodera et al. proposed shifting the STFT values to their local centers of gravity (centroids). The STFT formula already presented in (2.78) can be generalized as the distribution of the magnitude $M_x^h(t,\omega)$ and phase $\Phi_x^h(t,\omega)$ as follows:

$$F_x^h(t,\omega) = \int_{-\infty}^{\infty} x(\tau)h(\tau-t)e^{-j\omega(\tau-t)}\mathrm{d}\tau = M_x^h(t,\omega)e^{j\Phi_x^h(t,\omega)}. \tag{2.110}$$

The centroids are associated with two fundamental signal parameters: instantaneous frequency and group delay. To begin with, let us define their time-domain definitions before delving into a more complex notation in the joint time-frequency representation. For a generic signal model with the amplitude $A(t)$ and the phase $\phi(t)$, we have

$$x(t) = A(t)e^{j\phi(t)}. \tag{2.111}$$

Its spectrum reads as

$$F_x(\omega) = M_x(\omega)e^{j\Phi_x(\omega)}, \tag{2.112}$$

with the spectrum magnitude $M_x(\omega)$ and phase $\Phi_x(\omega)$. The angular instantaneous frequency is given by the time-domain first-order phase derivative with respect to time

$$\hat{\omega}(t) = \frac{\mathrm{d}\phi(t)}{\mathrm{d}t}. \tag{2.113}$$

For the group delay, we can derive it as a signal spectrum derivative with respect to frequency

$$\hat{t}(\omega) = -\frac{\mathrm{d}\Phi_x(\omega)}{\mathrm{d}\omega}. \tag{2.114}$$

In the time-frequency domain, these quantities become slightly more complicated. Let us first derive the instantaneous frequency estimator in the time-frequency domain. Let $\phi(t) = \phi_x + \omega_x t$ define the signal phase and assume the constant amplitude so that $A(t) = \text{const}$. Therefore, the signal derivative with respect to time reads as

$$\frac{\mathrm{d}x(t)}{\mathrm{d}t} = \omega_x x(t). \tag{2.115}$$

From (2.78), we know two STFT equivalent definitions of the STFT. Their derivatives with respect to time yields

$$\frac{\partial F_x^h(t,\omega)}{\partial t} = \int_{-\infty}^{\infty} x(\tau)\frac{\mathrm{d}h(\tau-t)e^{-j\omega(\tau-t)}}{\mathrm{d}t}\mathrm{d}\tau = \int_{-\infty}^{\infty} \frac{\mathrm{d}x(\tau+t)}{\mathrm{d}t}h(\tau)e^{-j\omega\tau}\mathrm{d}\tau. \tag{2.116}$$

Next, we can group the terms

$$\begin{aligned}\int_{-\infty}^{\infty} x(\tau)\frac{\mathrm{d}h(\tau-t)}{\mathrm{d}t}e^{-j\omega(\tau-t)}\mathrm{d}\tau + \int_{-\infty}^{\infty} x(\tau)h(\tau-t)\frac{\mathrm{d}e^{-j\omega(\tau-t)}}{\mathrm{d}t}\mathrm{d}\tau = \\ \int_{-\infty}^{\infty} \frac{\mathrm{d}x(\tau+t)}{\mathrm{d}t}h(\tau)e^{-j\omega\tau}\mathrm{d}\tau,\end{aligned} \tag{2.117}$$

which leads to

$$-F_x^{\mathscr{D}h}(t,\omega) + j\omega F_x^h(t,\omega) = \int_{-\infty}^{\infty} \frac{\mathrm{d}x(\tau+t)}{\mathrm{d}t}h(\tau)e^{-j\omega\tau}\mathrm{d}\tau. \tag{2.118}$$

The first left-hand term in (2.118) is the STFT with the differentiated window, which is denoted as $F_x^{\mathscr{D}h}(t,\omega)$, where

$$\mathscr{D}h = \frac{\mathrm{d}h(t)}{\mathrm{d}t}. \tag{2.119}$$

After differentiation, the second left-hand factor in (2.118) gives the original STFT multiplied by the imaginary unit j and ω. Next, we can substitute (2.115) into (2.118), obtaining

$$-F_x^{\mathscr{D}h}(t,\omega) + j\omega F_x^h(t,\omega) = \omega_x \int_{-\infty}^{\infty} x(\tau+t)h(\tau)e^{-j\omega\tau}\mathrm{d}\tau, \tag{2.120}$$

where the last term is the ordinary STFT. Ultimately, we have

$$j\omega F_x^h(t,\omega) - F_x^{\mathscr{D}h}(t,\omega) = \omega_x F_x^h(t,\omega). \tag{2.121}$$

Dividing both sides by $F_x^h(t,\omega)$ and taking the imaginary part $\Im(\cdot)$ leads to the two-dimensional (time-frequency) estimator of the angular instantaneous frequency

$$\hat{\omega}_x(t,\omega) = \omega - \Im\left(\frac{F_x^{\mathscr{D}h}(t,\omega)}{F_x^h(t,\omega)}\right) = \omega - \Im\left(\tilde{\omega}_x(t,\omega)\right). \tag{2.122}$$

Simply put, (2.122) describes the angular instantaneous frequency distribution over the time-frequency plane and is given as the difference between ω and the imaginary part of the ratio of the STFT with the window derivative, and the original one with an unaltered window.

For the time-frequency group delay estimator, we define the following identity [88, 89]

$$\hat{t}_x(t,\omega) = t - \frac{\partial \Phi_x^h(t,\omega)}{\partial \omega}. \tag{2.123}$$

Taking the frequency derivative of the STFT gives

$$\frac{\partial}{\partial \omega} F_x^h(t,\omega) = \int_{-\infty}^{\infty} x(\tau+t)h(\tau)(-j\tau)e^{-j\omega\tau}d\tau, \tag{2.124}$$

which simplifies to

$$\frac{\partial}{\partial \omega} F_x^h(t,\omega) = -j\int_{-\infty}^{\infty} \tau x(\tau+t)h(\tau)e^{-j\omega\tau}d\tau = -jF_x^{\mathscr{T}h}(t,\omega). \tag{2.125}$$

where $\mathscr{T}h = t \cdot h(t)$ is the time-weighted window. Notice that $F_x^{\mathscr{T}h}(t,\omega)$ can be interpreted as the first-order moment of the STFT.

By the definition of the phase derivative

$$\frac{\partial}{\partial \omega}\phi(t,\omega) = \Im\left(\frac{\frac{\partial}{\partial \omega}F_x^h(t,\omega)}{F_x^h(t,\omega)}\right), \tag{2.126}$$

we get

$$\frac{\partial}{\partial \omega}\phi(t,\omega) = \Im\left(\frac{-jF_x^{\mathscr{T}h}(t,\omega)}{F_x^h(t,\omega)}\right). \tag{2.127}$$

Using the property $\Im(-jz) = -\Re(z)$, this becomes

$$\frac{\partial}{\partial \omega}\phi(t,\omega) = -\Re\left(\frac{F_x^{\mathscr{T}h}(t,\omega)}{F_x^h(t,\omega)}\right), \tag{2.128}$$

where $\Re(\cdot)$ is the real part of the complex number. Therefore, the group delay estimator is

$$\hat{t}_x(t,\omega) = t + \Re\left(\frac{F_x^{\mathscr{T}h}(t,\omega)}{F_x^h(t,\omega)}\right). \tag{2.129}$$

From our considerations, we see that the time-frequency group delay estimator is the sum of time t and the real part of the ratio of the two STFTs: the first computed using the window multiplied by the time ramp and the classical one.

Summing up, the following pair constitutes the reassignment operators

$$\begin{pmatrix} \hat{t}_x(t,\omega) \\ \hat{\omega}_x(t,\omega) \end{pmatrix} = \begin{pmatrix} t - \frac{\partial \Phi_x^h}{\partial \omega}(t,\omega) \\ \frac{\partial \Phi_x^h}{\partial t}(t,\omega) \end{pmatrix} = \begin{pmatrix} t \\ \omega \end{pmatrix} - \begin{pmatrix} -\Re\left(\frac{F_x^{\mathscr{T}h}(t,\omega)}{F_x^h(t,\omega)}\right) \\ \Im\left(\frac{F_x^{\mathscr{D}h}(t,\omega)}{F_x^h(t,\omega)}\right) \end{pmatrix} = \begin{pmatrix} t \\ \omega \end{pmatrix} - U_x^h(t,\omega). \tag{2.130}$$

Having (2.122) and (2.129), we can concentrate the spectrogram as follows [50, 87]:

$$R_x^h(t,\omega) = \int_{-\infty}^{\infty}\int_{-\infty}^{\infty} E_x^h(\tau,\Omega)\delta(t-\hat{t}_x(\tau,\Omega),\omega-\hat{\omega}_x(\tau,\Omega))\mathrm{d}\tau\mathrm{d}\Omega, \tag{2.131}$$

where $\delta(\cdot,\cdot)$ is a two-dimensional Dirac delta distribution. Equation (2.131) can be viewed as a post-processing operation that moves the time-frequency distribution energy ($E_x^h(t,\omega)$) to new coordinates according to $(t,\omega) \mapsto (\hat{t}_x(t,\omega),\hat{\omega}_x(t,\omega))$, where $\hat{t}_x(t,\omega)$ and $\hat{\omega}_x(t,\omega)$ are expected to be closer to the true support of the analyzed signal. Since the local energy is attracted to its local salients, the instantaneous frequency ridge of a particular component is usually called an *attractor*. Note that the reassignment method is non-invertible. As we relocate the energy (squared absolute value of the STFT), the phase content is lost, and the real-valued representation of the sharpened spectrogram is the final output. The graphical interpretation of the reassignment method is delineated, and the simulated example of the multicomponent signal is shown in Figure 2.28 and 2.29, respectively. The comparison between a classical spectrogram and its reassigned counterpart reveals tangible differences in time-frequency readability and energy localization. The classical spectrogram often suffers from smearing effects, where signal components are dispersed over a broader time-frequency area, leading to imprecise representation, especially for rapidly changing signals. In contrast, the reassignment technique sharpens the representation by relocating energy components to their proper instantaneous frequencies and times. This results in a more concentrated and accurate depiction of signal features, enabling better separation of overlapping components and improved tracking

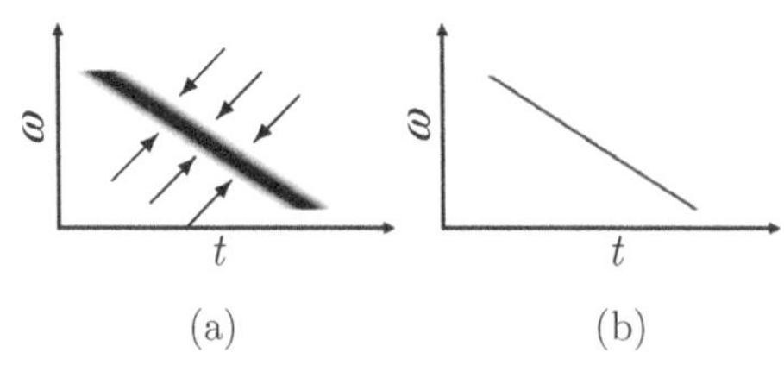

Figure 2.28 Time-frequency reassignment graphical interpretation. For the initial spectrogram (a) with limited readability, the energy is relocated to the instantaneous frequency ridge, making the spectrogram sharp (b).

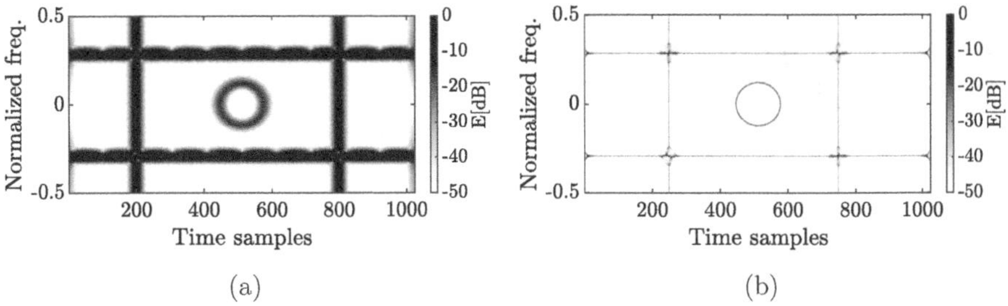

Figure 2.29 Time-frequency reassignment in action. (a) Spectrogram of the multicomponent simulated signal; (b) reassigned spectrogram.

of transient phenomena. The reassigned spectrogram thus proves particularly effective for signals with complex structures, where precise time-frequency localization is crucial.

Remark 2.7. *In the previous deduction, we introduced two variants of the analyzing window $h(t)$. The first one is the window derivative $\mathscr{D}h = \frac{dh(t)}{dt}$, and the second is the time-weighted window $\mathscr{T}h = t \cdot h(t)$. The Gaussian case of these windows in the form of its derivative and time ramp multiplication is shown in Figure 2.30. All the windows span symmetrically around the origin, which is pivotal for $\mathscr{T}h$, for which t is symmetrically centered around zero. Higher-order window derivatives and window-time-ramp products $\mathscr{T}^n h$ presented in the further part of this book are analogously obtained.*

The STFT-based reassignment result for the real-life pulse can be seen in Figure 2.31. The spectrogram of the already discussed pulse with strong interference (see Figure 2.31a) has overlapping multipath copies discarding unequivocal pulse extraction. The reassigned (concentrated) spectrogram is of significantly higher readability as the direct pulse is well-defined on the time-frequency plane. To do so, we first need to compute the reassignment operators. Figure 2.31c and d present the reassignment operators $V_x^h(t,\omega)$ (defined in (2.130)). They say how many bins the signal energy should be shifted (in frequency and time, respectively) to reach its local

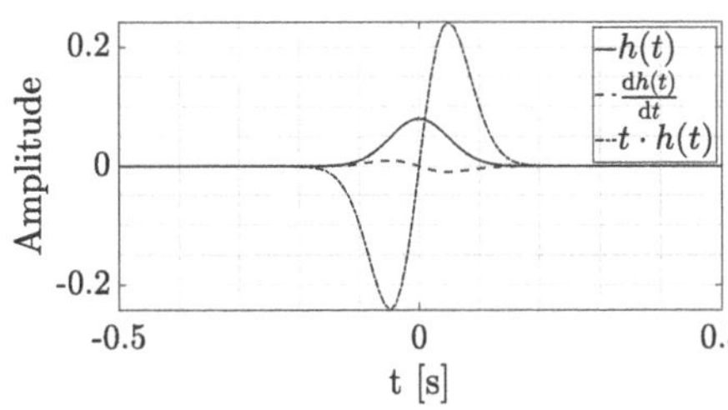

Figure 2.30 Gaussian window, its derivative (dashed), and the time-weighted window (dotted).

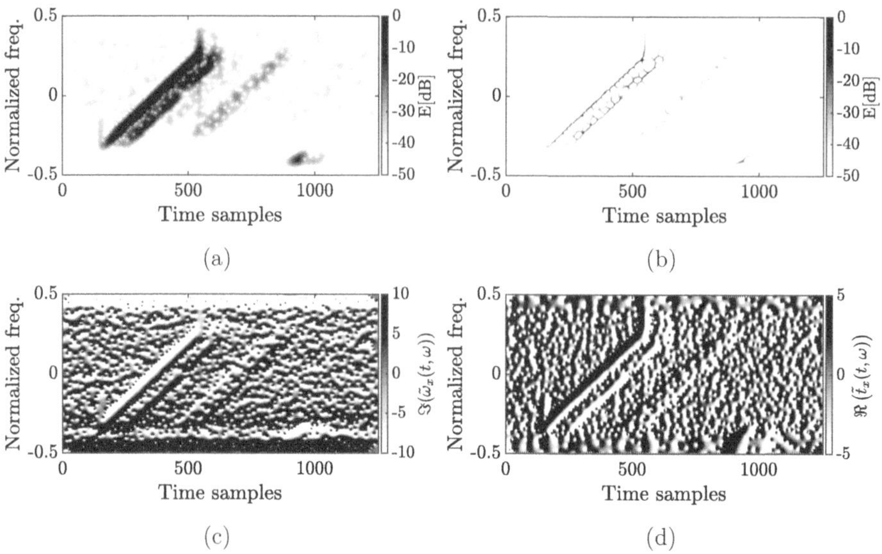

Figure 2.31 Time-frequency reassignment in action. (a) Original spectrogram; (b) reassigned spectrogram; (c) $-\Im\left(\tilde{\omega}_x(t,\omega)\right)$ – visualization of how to relocate the energy in the frequency domain; (d) $\Re\left(\tilde{t}_x(t,\omega)\right)$ – the same as (c), but for the time domain.

peak. Combined, they contribute to the reassignment operators (2.122) and (2.129) and allow for the signal concentration.

Radar pulses are not the only signals whose processing in the time-frequency domain allows a deep insight into their nature. Apart from multipath copies, radar pulses are rather single-component waveforms. In radars, we have another class of signals of a more complex character. They are called micro-Doppler signatures.

Remark 2.8. *More details on micro-Doppler signatures can be found in Section 3.2 in Chapter 3. Such a structure is dictated by the whole processing pipeline, which, in frequency-modulated continuous-wave radars, is oriented around two-dimensional signal-processing and requires a broader explanation. Here, we consider micro-Doppler data as a nonstationary signal resulting from the reflection of the transmitted signal from a target with multiple moving components.*

Micro-Doppler signatures represent a critical advancement in radar signal-processing, offering the ability to characterize dynamic target behaviors beyond conventional Doppler shifts. These signatures arise from micro-motions of target components, such as the rotation of helicopter blades, the swinging of limbs, or the vibrations of mechanical systems. By capturing the intricate modulations imparted to the radar signal, micro-Doppler analysis provides rich, high-dimensional features that enhance target identification, classification, and tracking capabilities [24].

The underlying principle of micro-Doppler phenomena is rooted in the decomposition of a target's motion into macro and micro-scale components. While traditional

Doppler shifts correspond to the translational velocity of the target, micro-Doppler effects encode information about periodic or oscillatory motions. This enables radar systems to infer valuable attributes such as gait patterns in human movement, machinery operating states, or even specific activities within cluttered environments.

Applications of micro-Doppler signatures span a wide range of domains, including security and surveillance [90], biomedical monitoring [91], and industrial diagnostics [92]. For instance, gait analysis using micro-Doppler has been employed in identifying individuals [93], while the detection of drones and their parameter estimation allow differentiation between them and birds [94]. In addition, the field has gained prominence in autonomous systems, where distinguishing between different objects or activities is paramount for decision-making processes [95].

Schematically, the micro-Doppler effect is illustrated in Figure 2.32. In the first example (Figure 2.32a), several movable elements, such as propellers, reflect the transmitted signal (solid arrow) across their available surfaces (dashed arrows). The rotation of the propellers introduces additional modulation, with each propeller individually affecting the signal. Beyond the primary echo from the drone's main body, distinct echoes from each propeller are also observed. A similar scenario is depicted in Figure 2.32b, where the torso of a walking human generates an echo modulated by its velocity. In contrast, the movement of the arms and legs, with their varying instantaneous velocities, contributes additional signals. These echoes exhibit distinct Doppler frequency shifts. Such a signal composition is an excellent example of the capabilities of time-frequency analysis.

Let us analyze the real-life micro-Doppler signature of a walking human. The signal was recorded using the XY-DemoRad microwave sensor system, developed by XY-Sensing Ltd. [96,97]. The radar recorded echoes from a walking human target using a bandwidth of $B = 1$ GHz, which corresponds to a range resolution of 15 cm (see (1.14)), a carrier frequency of $f_c = 24$ GHz, and a sweep repetition rate of $f_p = 1$ kHz. The experiment was conducted indoors, with the radar mounted on a tripod in a corridor approximately 2.5 m wide, 3 m high, and 15 m long. The

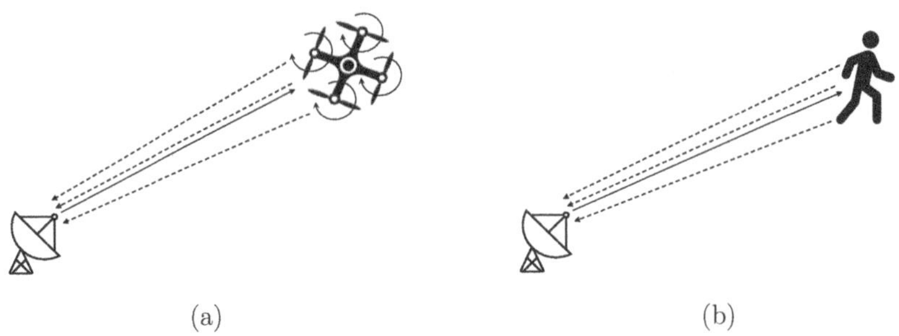

Figure 2.32 Micro-Doppler signal generation mechanism. (a) Flying drone, whose propellers individually modulate the reflected signal, introducing additional components; (b) walking person, whose torso introduces the Doppler shift proportionally to its velocity, but waving limbs modulate the signal according to their individual movements.

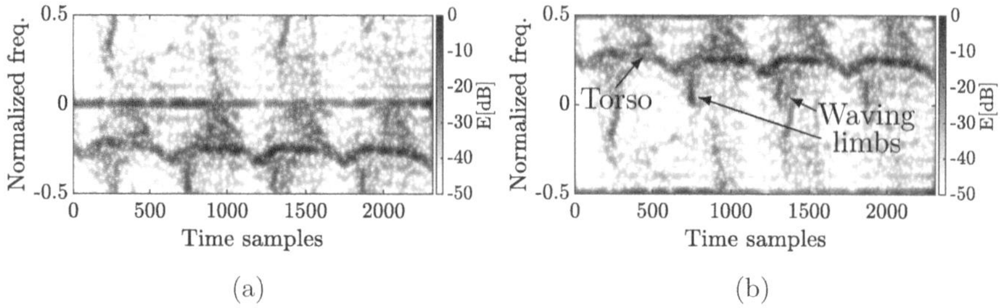

Figure 2.33 Real-life micro-Doppler signature of a walking person. (a) Original signal; (b) signal after modulation.

target walked in front of the radar at distances ranging from 1 m to 6 m. The echo of interest was extracted from the signal, giving the micro-Doppler signature of a walking human, which is shown in Figure 2.33a. As can be seen, the echo is aliased (the signal is sampled at a rate lower than twice its highest frequency, causing higher frequencies to fold back into lower ones and distort the signal) as the person was walking faster than the space was sampled. In pulse and FMCW radars, the pulse repetition rate determines the maximum unambiguous velocity of detected objects, as a higher pulse repetition rate allows the radar to correctly measure faster-moving targets without experiencing Doppler ambiguity. In the analyzed case, some parts of the signal jumped from negative to positive frequencies, making the spectrogram less clear. The signal was modulated for our analysis so that the micro-Doppler signature covers the entire frequency range, as shown in Figure 2.33b.

The spectrogram exhibits two main parts of the signature. The first one is the torso echo, which is apparent as an oscillating term at the normalized frequency of approximately 0.2. Additionally, we see echoes from waving arms and legs with a Doppler shift higher than that for the torso. They arise due to taking steps and waving arms during a walk. Such a micro-Doppler signature was post-processed to separate components from the spectrogram. The concentrated spectrogram using the reassignment method is shown in Figure 2.34. The component readability is intense, as each signal term strongly focuses on the local ridge. To numerically express the concentration,

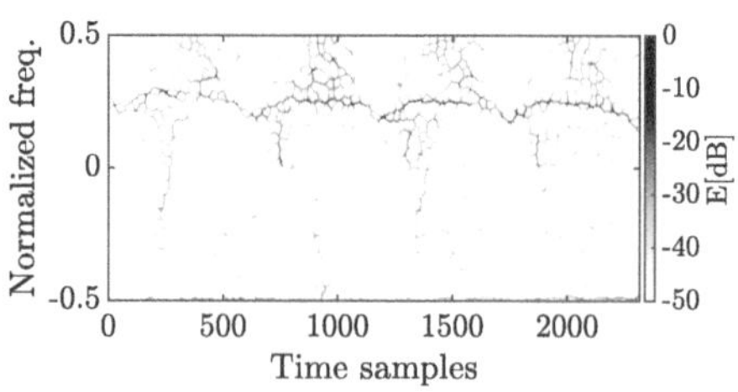

Figure 2.34 Reassigned micro-Doppler signature from Figure 2.33.

let us analyze the spectrograms using the Rényi entropy. For the original spectrogram, we obtain $H_R^{[3]}\left(E_x^h(t,\omega)\right) = 15.70$ bits, while for the reassigned spectrogram we obtain $H_R^{[3]}\left(R_x^h(t,\omega)\right) = 7.16$ bits, improving the metric of about 8 bits.

Although the reassignment method meaningfully streamlines time-frequency distribution interpretation as it concentrates the energy, one has to highlight its weaknesses. Firstly, the time-frequency reassignment does not differentiate between the noise and valuable components. In the presence of noise, time-frequency reassignment becomes sensitive to inaccuracies, as noise can disrupt the precise estimation of the instantaneous frequency and group delay estimates, leading to artifacts and misplacement of energy. At low SNRs, noise may dominate the time-frequency distribution, causing energy smearing and making it harder to distinguish useful signal components. As a companion argument, reassignment tends to reinforce spectral details, which can inadvertently amplify the visibility of noise-related artifacts, complicating signal interpretation. Figure 2.35 shows an example of such a phenomenon. A simulated chirp defined by (2.94) disturbed by a high noise giving SNR $= -10$ dB (see Figure 2.35a) has a barely distinguishable chirp. The concentration process did not improve the component readability but exacerbated it, making the chirp highly interfered with by surrounding concentrated noise.

At first glance, the time-frequency reassignment may be considered a super-resolution technique. In practice, it is not. Two (or more) closely located components, if they appear simultaneously at a given window position from which the Fourier transform is calculated, may become entangled with each other and interfere. Simulated and real-life examples of the resolution problem are shown in Figure 2.36. When signal modes adhere, we can observe artifacts in Figure 2.36b. This effect is because the reassignment operators (2.122) and (2.129) are impacted by two components at the same time. As a result, the concentrated distribution became distorted and impeded analysis in the area where these signals intersect on the plane. A similar observation comes from a real-life example when the multipath delay is shorter than the length of the analyzing window. In such a case, the direct pulse and its first multipath copy are entangled, manifesting as oblique and periodical connections between these components.

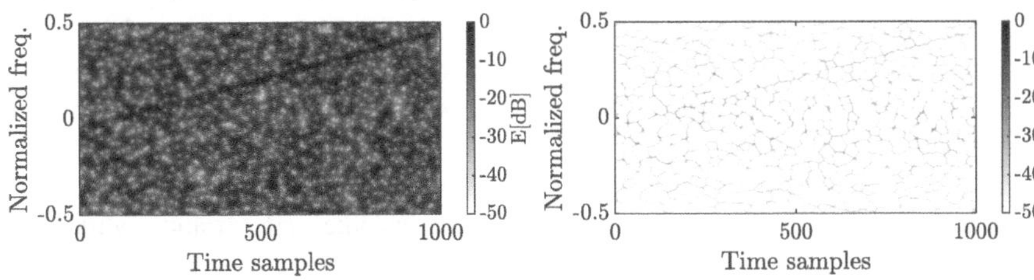

(a) (b)

Figure 2.35 Visualization of the reassignment performance in low SNR conditions. (a) Spectrogram of the noisy chirp; (b) reassigned spectrogram.

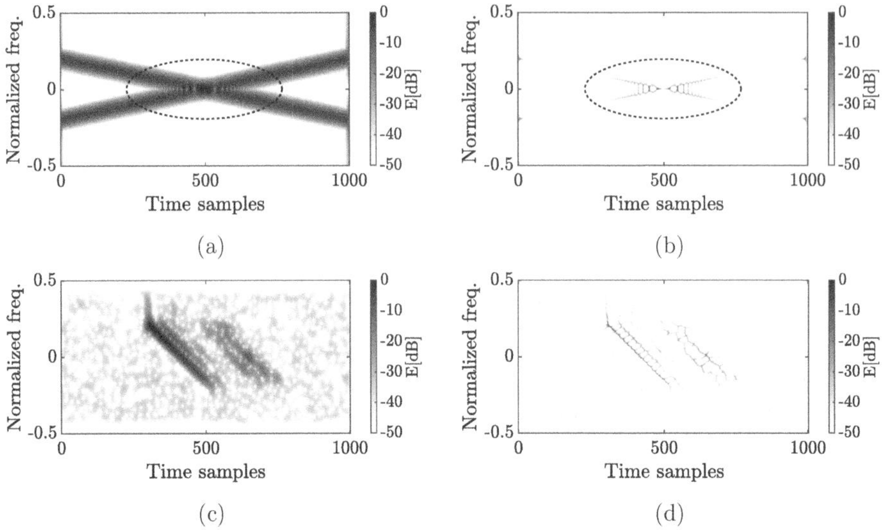

Figure 2.36 Signal component entanglement in time-frequency reassignment. (a) Two simulated intersecting chirps; (b) reassigned spectrogram of simulated chirps; (c) real-life radar pulse with a low-delayed multipath copy; (d) reassigned spectrogram of the radar pulse.

Although the reassignment technique renders a significant advancement in the field of time-frequency signal processing, hitherto described algorithms facilitate time-frequency energy concentration regardless of whether the signal is present or not. This feature is notably apparent in Figure 2.35 as the method does not differentiate between the useful components and noise. In the next section, we discuss the technique of suppressing the noise level in both the spectrogram and its sharpened (reassigned) version.

2.8 MULTITAPER STFT AND REASSIGNMENT

Hitherto, delineated methods for spectral analysis play a pivotal role in understanding signals' frequency and time-frequency characteristics. However, the problem not profoundly discussed so far is a trade-off between frequency resolution and variance, particularly when analyzing finite-length or noisy signals in the frequency domain. These limitations can obscure essential features of the spectrum, making reliable interpretation challenging.

Multitaper spectral analysis, introduced by Thomson in 1982 [98], was developed to address these challenges by reducing spectral leakage and variance while preserving resolution. Unlike single-taper methods, the multitaper approach employs a set of orthogonal tapering functions, known as Slepian sequences or discrete prolate spheroidal sequences, to minimize energy leakage outside a defined bandwidth. By averaging spectra obtained from multiple tapers, multitaper spectral analysis offers robust and high-resolution spectral estimates, even in the presence of noise or limited data. This approach has been extensively detailed and expanded upon in works such as those by Percival and Walden [99], offering a comprehensive theoretical and practical foundation for its use. The work of Riedel and Sidorenko [100] demonstrated the effectiveness of multiple tapers in achieving unbiased and reliable spectral estimates, even under severe conditions. Furthermore, the approach has been successfully applied in various fields.

In general, the multitaper spectrum proposed in [98] does not filter the input signal before Fourier analysis to reduce the noise level and improve spectral characteristics. Instead, it leverages multiple orthonormal windows multiplied by the signal before transforming it into the frequency domain. The results are averaged, yielding significant noise suppression and spectrum quality improvement [101].

The multitaper frequency-domain spectrum of the signal $x(t)$ can be described as follows:

$$\widehat{F}_x(\omega) = \frac{1}{K}\sum_{k=0}^{K-1}\left|\int_{-\infty}^{\infty} x(t)h_k(t)e^{-j\omega t}\mathrm{d}t\right|^2, \tag{2.132}$$

where K stands for the number of tapers (windows) $h_k(t)$. The multitaper spectrum can be summarized as

1. Generate K orthonormal windows $h_k(t)$, where $k = 0...K-1$.
2. Get the signal-window product for each k.
3. Compute the Fourier transform and its squared magnitude for each k.
4. Average the spectra.

The multitaper spectral analysis for the monochromatic signal with the constant normalized frequency of $f_n = 0.2$ for the SNR $= 0$ dB and $K = 4$ is shown in Figure 2.37.

When comparing the results of classical spectral analysis (Figure 2.37a), single-taper analysis (Figure 2.37b), and the multitaper method (Figure 2.37c), significant differences in spectral estimation quality can be observed. Classical spectral analysis, based on a single FFT computation, often suffers from high variance and spectral leakage, mainly when applied to short signals or in the presence of noise. Single-taper analysis, such as that using a Gaussian window, reduces spectral leakage to some extent but introduces additional distortions in the form of broadened side lobes, which can obscure closely spaced frequency components. In contrast, the multitaper method demonstrates superior performance by leveraging multiple orthogonal tapers to reduce variance and minimize spectral leakage. This approach results in sharper spectral peaks and improved robustness to noise. Additionally, multitaper analysis provides a more consistent estimation of the power spectral density, making it particularly effective for signals with complex or overlapping frequency structures. These advantages make the multitaper method a powerful tool for achieving accurate and reliable spectral analysis across various applications. Additionally, Figure 2.37d presents the orthonormal windows used in the analysis.

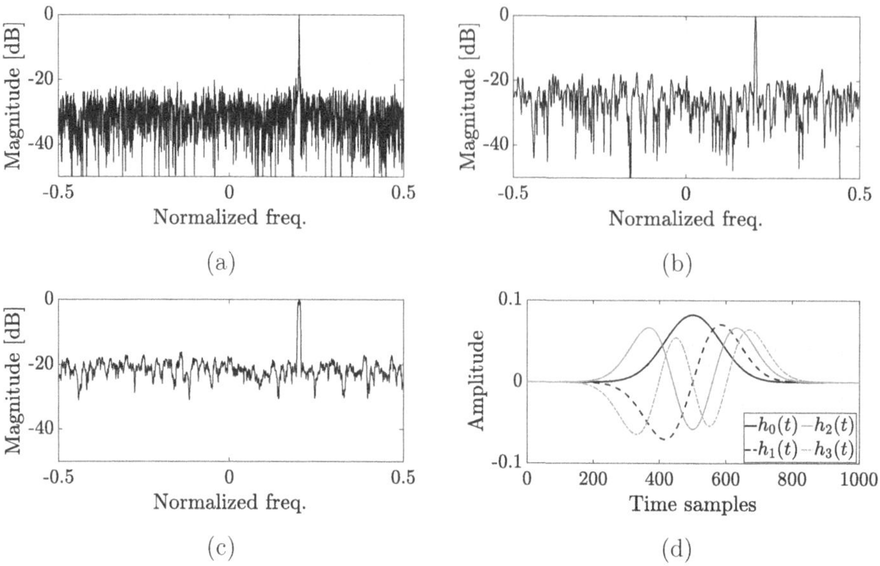

Figure 2.37 Multitaper spectral analysis; (a) Harmonic signal classical spectrum; (b) one taper spectrum; (c) four taper spectrum; (d) tapers.

Multitaper spectral analysis is a parametric technique, the most important being the number of tapers. This parameter directly affects the trade-off between spectral resolution, variance, and computational complexity. Increasing K improves the robustness of the spectral estimate by averaging across multiple independent realizations of the spectrum, thereby reducing variance. This operation makes the estimate more stable and reliable, particularly in the presence of noise or when analyzing short signals. However, more tapers deteriorate spectral resolution. Each taper contributes to a slightly broader spectral peak, as the energy from adjacent frequencies is included in the estimate. Therefore, selecting K involves balancing variance reduction and resolution preservation. Typically, this parameter is adjusted for a particular application, and it is reasonable to have it within the range of $K \in [2,6]$.

Additionally, increasing K generates a higher computational burden, as each taper requires separate computation of the Fourier transform. Thus, its choice should be tailored to the specific application, settling the trade-offs between variance, resolution, and computational efficiency.

The effect of how the number of tapers K affects the spectral estimate is delineated in Figure 2.38. In the example, the same signal and noise floor were used as in the previous case, whose results were shown in Figure 2.37. This comparison leads to straightforward observations as when the increasing K, we achieve a gradual variance decrease and a resolution inferior.

Let us stick to the window functions and their generation algorithm. At its origin, orthonormal windows were generated using the so-called Slepian sequence, also known as discrete prolate spheroidal sequences [98, 99]. As stated in this book

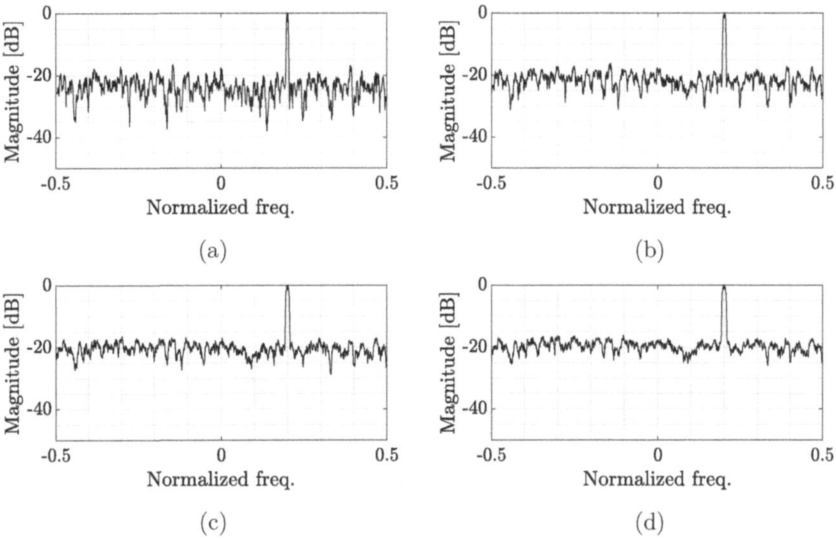

Figure 2.38 Influence of K on the spectral estimate. (a) $K=2$; (b) $K=4$; (c) $K=6$; (d) $K=8$.

and extensively in the literature, the Gaussian window ensures the optimal trade-off between the time and frequency resolution when applying to the STFT. The Hermite functions are a generalization of the Gaussian window. They fulfill the orthonormality requirement and can be used for multitaper analysis [89].

The Hermite function is defined as [102–105]

$$h_k(t) = \left((t-\mathscr{D})^k g\right)(t)/\sqrt{\pi^{1/2}2^k k!}, \tag{2.133}$$

with a Gaussian window $g(t) = \exp\left(-t^2/2\right)$ and $\mathscr{D}g(t)$ being its derivative. A practical implementation comes down to recursive computation

$$h_k(t) = g(t)H_k(t)/\sqrt{\pi^{1/2}2^k k!}, \tag{2.134}$$

where $H_k(t), k \in N$ is the Hermite polynomials obeying the following recursion

$$H_k(t) = 2tH_{k-1}(t) - 2(k-2)H_{k-2}(t), k \geq 2, \tag{2.135}$$

with initialization values $H_0(t) = 1$ and $H_1(t) = 2t$.

Remark 2.9. *The Hermite functions are particularly interesting in the Fourier analysis [106]. As a generalization of the Gaussian window, they preserve its key characteristics. Being eigenfunctions of the Fourier transform, they possess notable features when examined in the time-frequency domain and reveal distinctive behavior in this representation. They exhibit optimal energy concentration in specific time-frequency regions, such as circular or elliptical domains, yielding minimal*

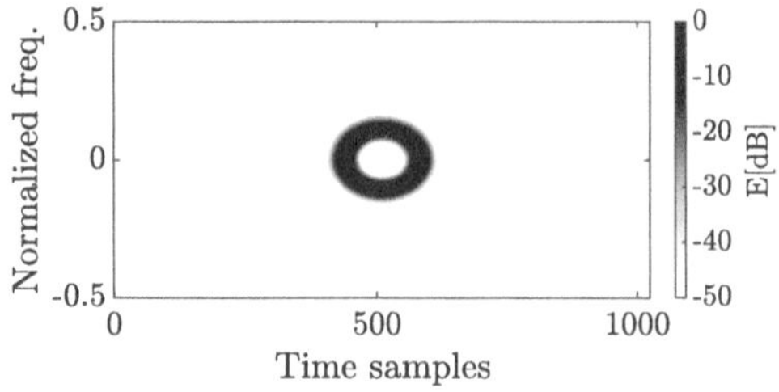

Figure 2.39 Hermite function spectrogram.

spectral leakage and high-resolution of spectrograms. Their orthogonality and ability to form a complete basis enable efficient decomposition and reconstruction of signals while preserving essential information. Hermite functions are beneficial in multitaper spectral analysis, where their time-frequency localization properties allow for precise representation of nonstationary signals while reducing variance and bias [105]. Furthermore, they are well-suited for reassignment techniques, enabling sharper localization of signal components and enhancing interpretability in complex time-frequency structures. A fun fact is that their spectrogram forms a ring on the time-frequency plane. The 25th order Hermite function's spectrogram is shown in Figure 2.39.

The multitaper spectral analysis for harmonic and stationary signals is an efficient approach to signal frequency content insight with a reasonable resolution-variance trade-off. As a result, the spectrum flattens, and simultaneously, we lose the distinctiveness of oscillations or local disturbances. The situation is more complicated for nonstationary signals, where simple spectral analysis is insufficient. Some readers can expect the transition from the multitaper spectrum to the multitaper spectrogram. Such works can be found in the literature, and some examples can be found in [105, 107–110]. Let us take a look at this technique's practical side.

For the already defined STFT (2.78) and the spectrogram (2.81), we calculate them with each orthonormal window $h_k(t)$ given by (2.134), which yields K spectrograms. Their average can be written as

$$\widehat{S}_x^h(t,\omega) = \frac{1}{K}\sum_{k=0}^{K-1} S_x^{h_k}(t,\omega), \tag{2.136}$$

where $S_x^{h_k}(t,\omega)$ is the spectrogram for the kth Hermite function as a window. Simply put, the multitaper spectrogram comes down to obtaining K ordinary spectrograms, each with an orthonormal window $h_k(t)$, and averaging the outcome. This simple yet efficient technique conserves properties of the multitaper spectrum analysis, where the noise flattening is feasible, leading to widening spectral peaks simultaneously. In the time-frequency domain, these properties diffuse to the two-dimensional space. From the orthonormality of Hermite functions, one has [89]

$$\int_{-\infty}^{\infty}\int_{-\infty}^{\infty} \frac{1}{2\pi} F_x^{h_m}(t,\omega) F_x^{h_{m'}}(t,\omega)^* \mathrm{d}t\mathrm{d}\omega = 0, \tag{2.137}$$

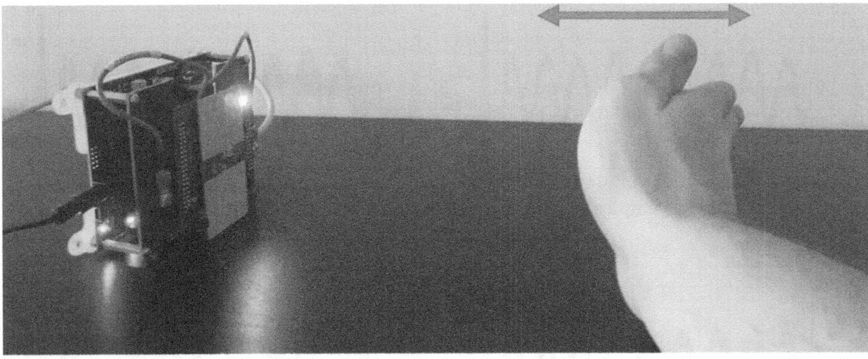

Figure 2.40 Measurement setup. (Reproduced from [93], licensed under CC BY 4.0.)

for any $m \neq m'$. Therefore, fluctuations resulting from noise existing in $F_x^{h_m}(t,\omega)$ and $F_x^{h_{m'}}(t,\omega)$ are mutually compensated, yielding noise suppression. Let us analyze the method's performance using a hand gesture micro-Doppler signature.

The XY-DemoRad system, already introduced in another micro-Doppler signal case context, was used for hand gesture signal recording as illustrated in Figure 2.40. Hand gesture recognition is a hot topic in Internet of Things applications. Recently, Google commercialized a radar microchip capable of recognizing human gestures to control a mobile phone [21]. Current trends show radars are great complementary remote control, diagnostics, and monitoring sensors. In this spirit, micro-Doppler signatures are of particular interest, and their precise analysis can pave the way for radars' implementations in future commercial equipment.

Let us analyze the micro-Doppler signature whose measurement setup is shown in Figure 2.40. The radar recorded the echo using a carrier frequency $f_c = 24$ GHz, signal bandwidth $B = 1$ GHz, and a sweep repetition rate $f_{\mathrm{SRR}} = 1$ kHz. The hand was waving approximately 50 cm from the antenna. The signal of a length $N = 2048$ samples was stored and processed offline.

Such a hand movement generated the echo of an oscillating instantaneous frequency. The signal was analyzed using the multitaper spectrogram technique, and the outcomes are presented in Figure 2.41. The subsequent pictures show the spectrograms for the Hermite functions in increasing order. As expected, spectrograms fluctuate more and more when k gets more extensive, and numerous local minima and maxima appear according to the window characteristic. They are notably apparent at the prominent component ridges, where the number of local minima equals k.

However, the final result is void of these fluctuations as they are averaged, leading to noise flattening. Although the main component is more expansive than for the original spectrogram (with one taper), as compared in Figure 2.42, the disturbances are smeared, giving a flat spectrogram.

At this point, one may wonder whether keeping the noise flattened but preserving the main components unaltered or improving their readability is possible. This outcome is obtainable when incorporating the reassignment technique with multitaper spectral estimation.

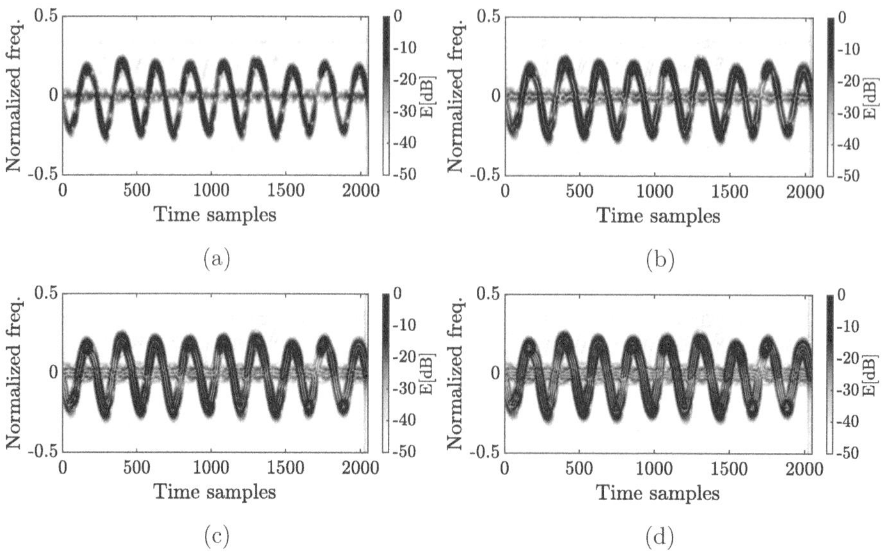

Figure 2.41 Spectrograms with orthonormal windows. (a) $k=0$; (b) $k=1$; (c) $k=2$; (d) $k=3$.

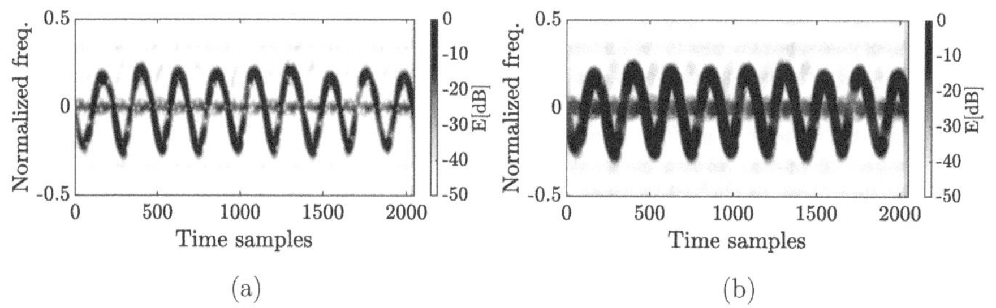

Figure 2.42 Comparison of the original spectrogram (a) and its multitaper version (b) for $K=4$.

In [111], Xiao and Flandrin proposed multitaper reassignment, allowing for both spectrogram concentration and multitaper noise suppression. The method was further developed, and its variants can be found in [112]. Multitaper reassignment serves a dual purpose. The first objective is to enhance the concentration of the signal spectrogram through time-frequency reassignment, improving chirp representation [50, 87]. The second goal involves utilizing multiple orthonormal tapers to average out noise, following the approach introduced by Thomson [98] and just described.

The multitaper time-frequency reassignment needs the original window and its two variants: its derivative and time-weighted form. The first and last requirements are easily fulfilled as the Hermite function is given as (2.134), and knowing the time support, one can easily get the window-time ramp product. For the window derivative, we have

$$\mathscr{D}h_k(t) = \mathscr{T}h_k(t) - \sqrt{2(k+1)}\,h_{k+1}(t), \tag{2.138}$$

where $\mathscr{T}h_k(t) = t \cdot h_k(t)$. Consequently, K reassignment operators are obtained

$$\hat{\omega}_x^{h_h}(t,\omega) = \omega - \Im\left(\frac{F_x^{\mathscr{D}h_k}(t,\omega)}{F_x^{h_k}(t,\omega)}\right). \tag{2.139}$$

$$\hat{t}_x^{h_k}(t,\omega) = t + \Re\left(\frac{F_x^{\mathscr{T}h_k}(t,\omega)}{F_x^{h_k}(t,\omega)}\right), \tag{2.140}$$

and for each $k \in [0, K-1]$ the reassigned spectrogram is defined as

$$R_x^{h_k}(t,\omega) = \int_{-\infty}^{\infty}\int_{-\infty}^{\infty} E_x^{h_k}(\tau,\Omega)\delta(t - \hat{t}_x^{h_k}(\tau,\Omega), \omega - \hat{\omega}_x^{h_k}(\tau,\Omega))\mathrm{d}\tau\mathrm{d}\Omega, \tag{2.141}$$

and averaged, giving the final concentrated multitaper spectrogram distribution with an averaged noise

$$\widehat{R}_x^{h}(t,\omega) = \frac{1}{K}\sum_{k=0}^{K-1} R_x^{h_k}(t,\omega). \tag{2.142}$$

This simple generalization of the time-frequency reassignment leads to meaningful spectrogram readability enhancement. Let us visualize the effect using the same micro-Doppler signature but employing the reassigned multitaper spectrogram. Consecutive reassigned spectrograms for $k = 0, 1, 2$, and 3 are shown in Figure 2.43. The window characteristic, apparent in Figure 2.41 as local extrema, is less visible for

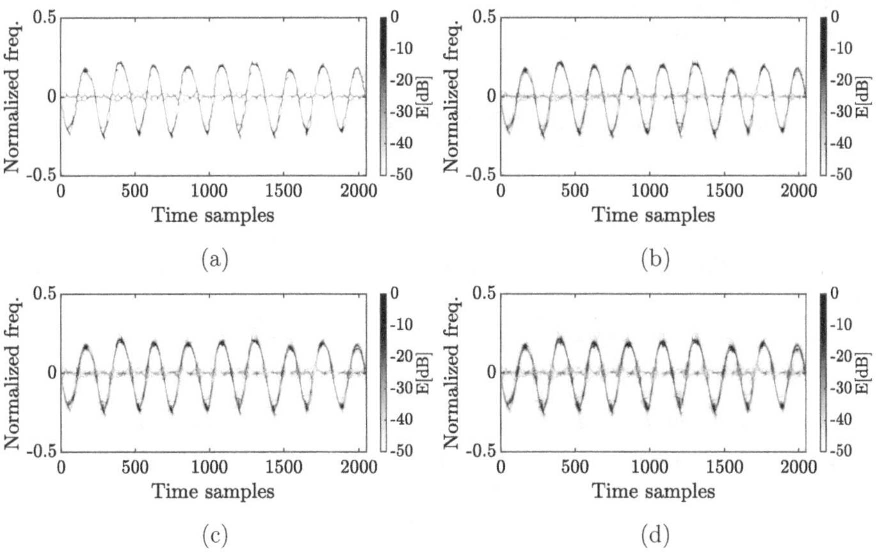

Figure 2.43 Reassigned spectrograms with orthonormal windows. (a) $k = 0$; (b) $k = 1$; (c) $k = 2$; (d) $k = 3$.

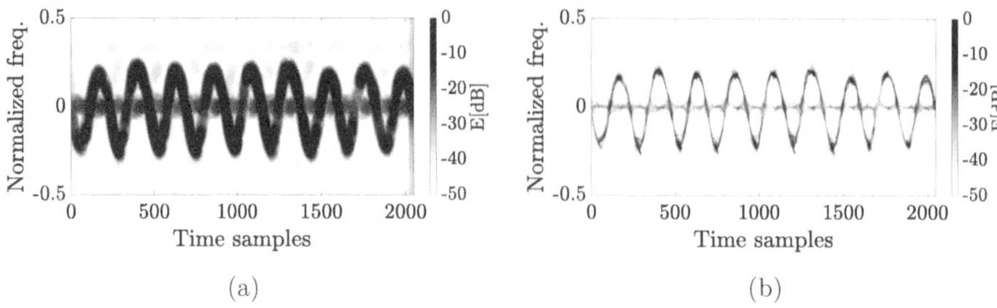

Figure 2.44 Comparison of the multitaper spectrogram (a) and the reassigned multitaper spectrogram (b) for $K = 4$.

the reassigned spectrograms. All spectrograms are generally well-concentrated, but finally, their mean shown in Figure 2.44 presents the benefits of averaging partial results. Comparing the multitaper spectrogram and its reassigned counterpart, we see a meaningful improvement in both signal sharpening and noise flattening.

The multitaper reassignment found several radar-related applications, including the walking human micro-Doppler signal analysis or forward scattering radar (the latter is investigated in detail in the next section) [113, 114]. Nonetheless, the reassignment (classical and multitaper) suffers from its non-invertibility as we accumulate the squared signal absolute value on the instantaneous frequency ridges. Such processing removes the phase content and limits further analysis to energy distribution without the possibility of retrieving the signal to its time-domain form. However, the STFT is the invertible transform (see (2.80)); thus, one can take advantage of this property and make the reassignment-like technique invertible as well. Such a technique is called synchrosqueezing and is described in the next section.

2.9 TIME-FREQUENCY SYNCHROSQUEEZING

So far, our considerations regarding the sharpened spectrograms have been limited to the reassignment technique, whose main objective is obtaining the concentrated energy distribution. Although readable with the signal energy focused on the signal ridges on the time-frequency plane, the reassigned spectrogram suffers from phase content loss. The outcome is, indeed, sharpened, but the final analysis comes down to the time-frequency distribution analysis without the possibility of reconstructing the signal into the time domain.

The synchrosqueezing technique was historically developed to overcome the limitations of the methods mentioned earlier (such as the spectrogram and reassigned spectrogram), which often suffer from resolution trade-offs and non-invertibility (in fact, synchrosqueezing was initially introduced as an extension of the continuous wavelet transform [115] and then redefined for the STFT [116]). The name synchrosqueezing comes from the way how this method works. The transform is *squeezed* (sharpened) by redistributing its values. It operates simultaneously for each time or frequency instants, and this is where the term *synchro* comes from.

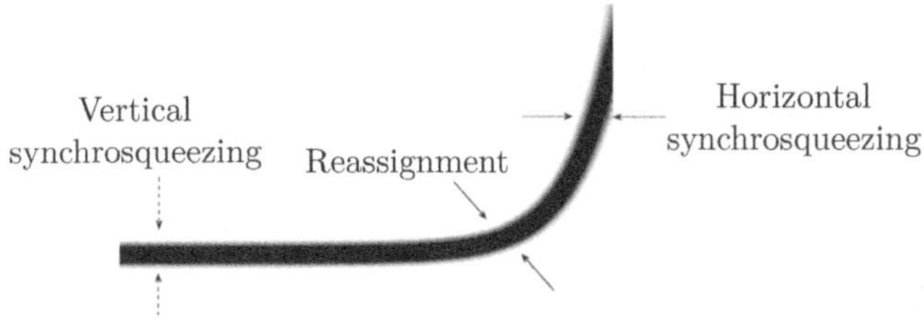

Figure 2.45 Visualization of how reassignment and synchrosqueezing relocate the distribution values. Vertical synchrosqueezing displaces the STFT vertically, i.e., along the frequency axis. Reassignment works in two domains, and horizontal synchrosqueezing allows transform values to be shifted alongside the time axis.

Synchrosqueezing, akin to reassignment, aims to displace STFT values into regions of high instantaneous frequency concentration, thus improving interpretability and enabling the extraction of intrinsic signal components. However, two distinctive features differentiate synchrosqueezing and reassignment. Firstly, reassignment relocates the energy while synchrosqueezing complex-valued transform points. Secondly, reassignment shifts the energy along both time and frequency axes simultaneously. Synchrosqueezing operates a displacement in solely one dimension (frequency or time). These two facets make synchrosqueezing invertible while maintaining concentration capabilities.

As mentioned, when applying the synchrosqueezing transform, one can displace the STFT values along the frequency or time axis. The resulting distributions are of inferior concentration compared to reassignment since only one dimension is concerned. Redistributing STFT points along the frequency axis is called *vertical synchrosqueezing* since typical time-frequency distributions are displayed with the frequency axis vertical. In contrast, *horizontal synchrosqueezing* adjusts STFT points along the time axis, offering better temporal localization and insight into transient components. Together, these approaches provide complementary tools for analyzing nonstationary signals. The main difference in terms of how each technique displaces the STFT values (neglecting that vertical and horizontal synchrosqueezing operate on complex values and reassignment on energy) is shown in Figure 2.45. Let us describe these techniques separately.

2.9.1 VERTICAL SYNCHROSQUEEZING

If we consider a practical problem of radar signal decomposition in the time or frequency domain (e.g., the one shown in Figure 2.4), we see the problem is nontrivial. In the example, multipath copies and noise interfered with the radar pulse we wish to extract. In the time domain, we do not know where precisely the signal envelope ends and where exactly the direct pulse can be cut off to prune long-lasting multipath copies. Notwithstanding, although we know the precise pulse duration, we cannot filter out noise from the entire bandwidth through time-domain pulse pruning. On the other hand, frequency-domain filtering cannot remove the multipath propagation effect, as each pulse copy occupies the same bandwidth. Finally, the STFT, as the joint time-frequency representation, allowed for differentiation between the useful signal modes and disturbances and noise. The problem, notwithstanding, occurs when two signal components are very close to each other on the time-frequency plane. Due to the limited resolution stemming from the Heisenberg-Gabor uncertainty principle, we cannot simultaneously define the time and frequency signal parameters, and this is where the reassignment technique assists. Nevertheless, the reassignment method allows solely **analysis** of time-frequency representation due to the phase content loss. Signal synthesis, decomposition, and mode retrieval are not the case of reassignment; thus, an invertible transform is needed.

Especially, signal decomposition can be obtained under a set of assumptions imposed on the signal $x(t)$, with the analysis carried out using a Gaussian window $h(t)$ [88]. First, it is assumed that each component $x_k(t)$ is weakly frequency-modulated. More precisely, there exists a sufficiently small parameter ε such that, for every t, the following bounds hold

$$\sigma^2 \left|\phi_k''(t)\right| \leq \varepsilon, \quad \left|a_k'(t)\right| \leq \varepsilon \phi_k'(t), \tag{2.143}$$

where σ denotes the time spread of the Gaussian window. Second, the signal components are assumed to be well separated in the frequency domain. This condition is expressed in terms of the effective bandwidth of the Gaussian window, which is given by the interval $[-\Delta, \Delta]$, where $\Delta = \frac{\sqrt{2\log(2)}}{\sigma}$. Under this assumption, the instantaneous frequencies of distinct components satisfy

$$\left|\phi_k'(t) - \phi_l'(t)\right| \geq 2\Delta, \tag{2.144}$$

for all t and for $k \neq l$. Since the instantaneous frequencies are ordered, i.e., $\phi_k'(t) < \phi_{k+1}'(t)$ for all t, the above condition can always be satisfied by selecting an appropriate value of σ.

Under these hypotheses, the following properties hold. First, for any (t, ω) such that $\left|\omega - \phi_k'(t)\right| \leq \Delta$, there exists an index k for which the estimated instantaneous frequency satisfies

$$\left|\hat{\omega}_x(t,\omega) - \phi_k'(t)\right| \leq C\varepsilon, \tag{2.145}$$

where C is a constant. Second, the reconstruction error associated with the recovery of $x_k(t)$ by integrating the transform coefficients in a neighborhood of the instantaneous frequency curve $F_x^h(t, \phi_k'(t))$ tends to zero as $\varepsilon \to 0$. These results demonstrate that the parameter ε controls the accuracy of the estimation, while the window width σ determines the degree of separation between components in the time-frequency domain.

Vertical synchrosqueezing is a "poorer" reassignment defined as

$$V_x^h(t,\omega) = \int_{-\infty}^{\infty} F_x^h(t,\Omega)\delta(\omega - \hat{\omega}_x(t,\Omega))\,\mathrm{d}\Omega, \tag{2.146}$$

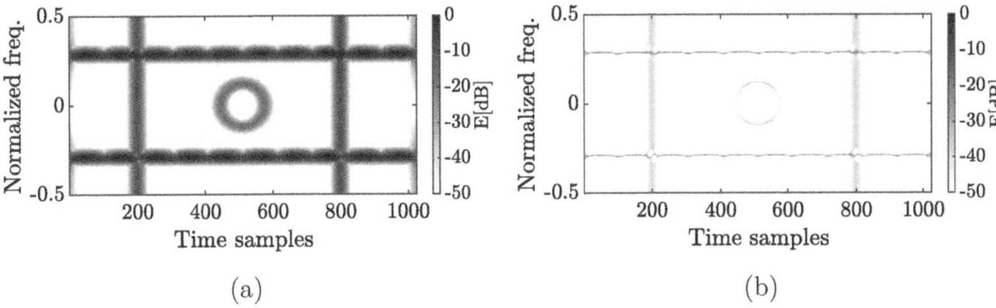

Figure 2.46 Spectrogram of the simulated signal (a) and its vertical synchrosqueezing counterpart (b).

where the STFT values are displaced only in the frequency domain. The instantaneous (angular) frequency estimator $\hat{\omega}_x(t,\omega)$ was already defined by (2.122). Form (2.146), the signal can be reconstructed as [117]:

$$\hat{x}(t) = \frac{1}{2\pi h^*(0)} \int_{-\infty}^{\infty} V_x^h(t,\omega)\mathrm{d}\omega, \tag{2.147}$$

where $h(0) \neq 0$ and, as shown in the further part of this section, refers to the entire distribution and particular regions, making signal components tractable to be extracted. Equation (2.147) comes down to the coherent distribution integration over the frequency. In the further part of this book, the result of the synchrosqueezing transform is presented as the spectrogram; namely, its squared absolute value is depicted (see (2.81)). A comparison of the multicomponent simulated signal spectrogram and the one vertically synchrosqueezed is shown in Figure 2.46. The observation from this picture is clear. Vertical synchrosqueezing restricts concentration only for the instantaneous frequency characterized by a low variance (upper and lower parts of the Hermite function and two harmonic terms below and above it). In other cases, the concentration fails.

The rationale for this effect follows from (2.122). We derived $\hat{\omega}_x(t,\omega)$ under the premise of a simple signal model. Therefore, any frequency-modulated signal with a meaningful modulation rate cannot be sufficiently displaced to occupy the signal ridge. It is tractable to show that the FM makes (2.122) biased with the bias proportional to higher-order phase terms. Last but not least, the amplitude also affects the result, and its leverage must be respected [117].

If we deem the signal to be described by the second-order amplitude and phase modulation, we get

$$x(t) = \mathscr{A} e^{-\frac{(t-t_x)^2}{2T_x^2}} e^{j\left(\phi_0 + \omega_x t + \frac{\alpha_x t^2}{2}\right)}, \tag{2.148}$$

where $\mathscr{A} e^{-\frac{(t-t_x)^2}{2T_x^2}}$ is the Gaussian envelope with $\mathscr{A}$ = conts., T_x is the Gaussian function spread, t_x is the time shift, and ϕ_0, ω_x, and α are previously defined terms: constant phase, angular frequency, and FM rate, respectively. The derivative of (2.148) leads to

$$\frac{\mathrm{d}x(t)}{\mathrm{d}t} = x(t)(q_x t + p_x), \tag{2.149}$$

where $q_x = -1/T_x^2 + j\alpha_x$ and $p_x = t_x/T_x^2 + j\omega_x$. If we put (2.149) into (2.118), we have

$$-F_x^{\mathcal{D}h}(t,\omega) + j\omega F_x^h(t,\omega) = \int_{-\infty}^{\infty} (x(\tau+t)q_x(\tau+t) + p_x)h(\tau)e^{-j\omega\tau}\mathrm{d}\tau, \tag{2.150}$$

and consequently

$$j\omega F_x^h(t,\omega) - F_x^{\mathcal{D}h}(t,\omega) = F_x^h(t,\omega)\,(q_x t + p_x) + q_x F_x^{\mathcal{T}h}(t,\omega). \tag{2.151}$$

Dividing both sides by $F_x^h(t,\omega)$ exhibits how (2.122) is biased when $\alpha \neq 0$

$$j\omega - \frac{F_x^{\mathcal{D}h}(t,\omega)}{F_x^h(t,\omega)} = q_x t + p_x + q_x \frac{F_x^{\mathcal{T}h}(t,\omega)}{F_x^h(t,\omega)}. \tag{2.152}$$

Fourer et al. in [117] presented how this issue can be addressed. Another similar work was done by Oberlin et al. [118]. Both articles are extensive, and curious readers are encouraged to see these in the original. In [117], the authors derived two explicit formulas allowing for precise instantaneous frequency estimation under FM conditions leading to the second-order vertical synchrosqueezing. It was achieved using higher-order STFT derivatives and the signal model as in (2.148). In turn, [118] structured the instantaneous frequency estimator comprising the chirp rate estimator in the time-frequency domain. Both approaches allow FM bias mitigation, and STFT values shift to the signal ridge and work similarly. In short, Fourer et al. [117] proposed to differentiate $F_x^{\mathcal{D}h}(t,\omega)$ with respect to time or angular frequency leading to linear systems giving two estimators $\hat{\omega}_x^{[\omega]}(t,\omega)$ and $\hat{\omega}_x^{[t]}(t,\omega)$ defined respectively

$$\hat{\omega}_x^{[\omega]}(t,\omega) = \omega + \Im\left(\frac{F_x^{\mathcal{D}\mathcal{T}h}(t,\omega)F_x^{\mathcal{T}h}(t,\omega) - F_x^{\mathcal{D}h}(t,\omega)F_x^{\mathcal{T}^2h}(t,\omega) + F_x^{\mathcal{T}h}(t,\omega)F_x^h(t,\omega)}{\left(F_x^{\mathcal{T}h}(t,\omega)\right)^2 - F_x^{\mathcal{T}^2h}(t,\omega)F_x^h(t,\omega)}\right), \tag{2.153}$$

$$\hat{\omega}_x^{[t]}(t,\omega) = \omega - \Im\left(\frac{F_x^{\mathcal{D}^2h}(t,\omega)F_x^{\mathcal{T}h}(t,\omega) - F_x^{\mathcal{D}\mathcal{T}h}(t,\omega)F_x^{\mathcal{D}h}(t,\omega)}{F_x^{\mathcal{D}h}(t,\omega)F_x^{\mathcal{T}h}(t,\omega) - F_x^{\mathcal{D}\mathcal{T}h}(t,\omega)F_x^h(t,\omega)}\right), \tag{2.154}$$

where the STFTs $F_x^h(t,\omega)$, $F_x^{\mathcal{D}h}(t,\omega)$, $F_x^{\mathcal{D}^2h}(t,\omega)$, $F_x^{\mathcal{T}h}(t,\omega)$, $F_x^{\mathcal{T}^2h}(t,\omega)$, and $F_x^{\mathcal{D}\mathcal{T}h}(t,\omega)$ are computed with the windows $t \mapsto h(t), \frac{\mathrm{d}h(t)}{\mathrm{d}t}, \frac{\mathrm{d}^2h(t)}{\mathrm{d}t^2}, t\cdot h(t), t^2\cdot h(t)$, $t\cdot\frac{\mathrm{d}h(t)}{\mathrm{d}t}$, respectively.

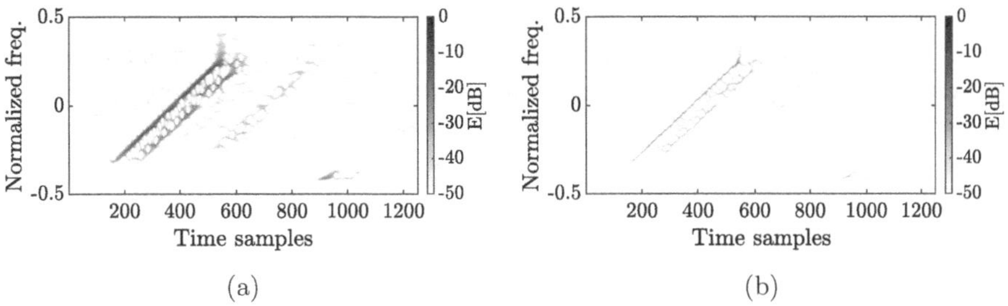

Figure 2.47 Spectrogram of the vertically synchrosqueezed STFT. (a) Using (2.122); (b) using (2.156).

In turn, Oberlin et al. [118] proposed a compact solution taking into account the modulation operator $\hat{\alpha}_x(t,\omega)$ (Section 2.10 seeks to elucidate this topic in more detail), building up the following estimator:

$$\hat{\omega}_x^{[\alpha]}(t,\omega) = \hat{\omega}_x(t,\omega) - \hat{\alpha}_x(t,\omega)(t - \hat{t}_x(t,\omega)), \tag{2.155}$$

where

$$\hat{\alpha}_x(t,\omega) = \Im\left(\frac{F_x^{\mathscr{D}^2h}(t,\omega)F_x^h(t,\omega) - \left(F_x^{\mathscr{D}h}(t,\omega)\right)^2}{F_x^{\mathscr{D}h}(t,\omega)F_x^{\mathscr{T}h}(t,\omega) - F_x^{\mathscr{D}\mathscr{T}h}(t,\omega)F_x^h(t,\omega)}\right), \tag{2.156}$$

is the chirp rate estimator in the time-frequency domain.

A real-life example of the vertical synchrosqueezing technique performance and highlighting how the signal model influences the concentration capabilities is shown in Figure 2.47. The energy distributions show an enormous difference in component concentration. The analyzed pulse in Figure 2.47a has a non-negligible FM, which affects the simple estimator as derived in (2.152). The effect of non-concentrated chirp is meaningful. The spectrogram becomes sharp for the entire chirp duration when considering the FM and including it in the instantaneous frequency estimate using (2.156). Besides visual aspects, the difference can be compared using the Rényi entropy, which equals 11.68 and 8.43 bits for the images in Figure 2.47, respectively.

Remark 2.10. *As highlighted, a valuable feature of synchrosqueezing is its invertibility. A gory detail, however, between the ordinary STFT and its synchrosqueezed form is the component concentration, which has a meaningful impact on the reconstruction quality when specific modes are retrieved. In particular, the distribution with a reinforced component separation settles the spectrogram area occupied by the single term. Concentrated spectrogram leverages a more straightforward definition of a time-frequency mask and the inverse transform. A time-frequency mask is a binary distribution having a value of* 1 *whenever the signal is useful and desired to be extracted and* 0 *otherwise. There are plenty of methods for ridge detection for mode retrieval [119–126], but each amounts to the separation of the ridge and consequently binary mask design.*

A well-established example of a mode retrieval algorithm for the synchrosqueezed STFT, whose foundational concept was introduced in [127], is used in the examples presented in the sequel of this chapter. The approach involves identifying the optimal instantaneous frequency curve $\Omega(t)$ within the time-frequency distribution by maximizing the signal energy, subject to a smoothness constraint imposed through total variation regularization [128]

$$\hat{\Omega} = \underset{\Omega}{\operatorname{argmax}} \int_{-\infty}^{\infty} \left| V_x^h(t, \Omega(t)) \right|^2 \mathrm{d}t - \kappa \int_{-\infty}^{\infty} \left| \frac{\mathrm{d}\Omega(t)}{\mathrm{d}t} \right|^2 \mathrm{d}t, \tag{2.157}$$

where κ controls the trade-off between energy maximization and the smoothness of $\Omega(t)$. For each extracted component $\Omega_i(t)$, the corresponding time-domain signal can be reconstructed using the expression:

$$\hat{x}(t) = \frac{1}{2\pi h(0)} \int_{\Omega_i(t)-\frac{\Delta\omega}{2}}^{\Omega_i(t)+\frac{\Delta\omega}{2}} V_x^h(t,\omega)\mathrm{d}\omega, \tag{2.158}$$

where Δ_ω specifies the integration band around the ridge, effectively defining the binary mask.

Before moving further to higher-order vertical synchrosqueezing, we can examine signal reconstruction performance for the second-order vertical synchrosqueezing. Let our goal be to filter out multipath interference of the radar pulse. For the already presented chirp, whose spectrogram (concentrated using the second-order synchrosqueezing technique) is shown in Figure 2.47b, the dominant component was extracted using (2.157). The detected signal component is superimposed on the spectrogram in Figure 2.48a. Next, the binary mask was designed around the signal ridge, as delineated in Figure 2.48b. The mask covers the entire pulse duration and is spread to the whole distribution length (also embracing some noise).

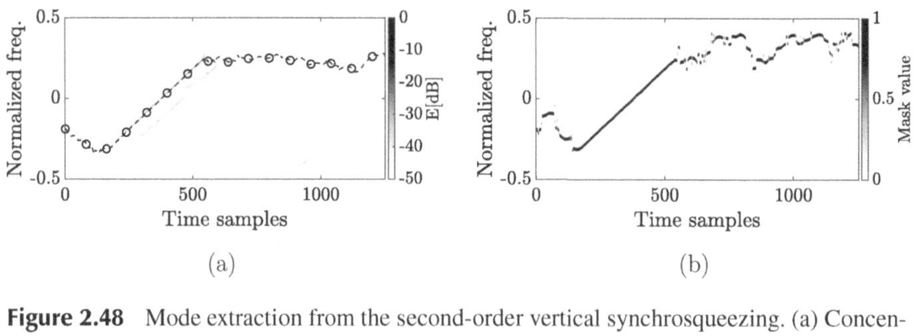

Figure 2.48 Mode extraction from the second-order vertical synchrosqueezing. (a) Concentrated spectrogram with the ridge extracted using (2.157) superimposed as a dashed line with circles; (b) time-frequency binary mask.

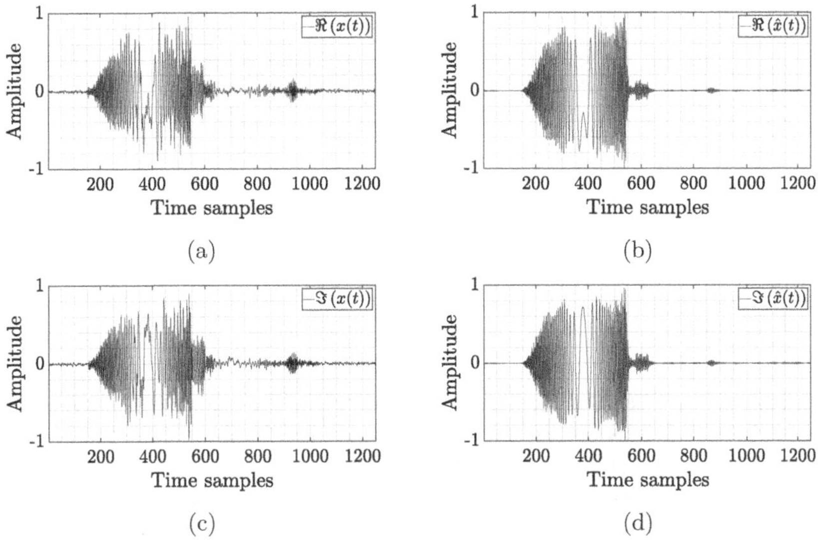

Figure 2.49 Second-order vertical synchrosqueezing signal filtering results. (a) Real part of the original signal; (b) real part of the filtered signal; (c) imaginary part of the original signal; (d) imaginary part of the filtered signal.

Signal retrieval comes down to integrating the sharpened STFT whenever the mask equals 1 according to (2.158). As a result, we achieved the time-domain signal and compared its real and imaginary evolutions in Figure 2.49. Also, the filtered signal spectrogram is presented in Figure 2.50.

The filtering result shows the ability to retrieve signal components and cut off noise and interference at the same frequency. This feature is not the case for typical filters, which remove everything beyond the passband. In Figure 2.50, we observe a piece of disturbance resulting from noise in the mask area. It can be fine-tuned by choosing a mask spread in the implemented algorithm.

Another practical application of the second-order synchrosqueezing in radars is target echo signal extraction in passive radars working in the forward scatter mode.[3]

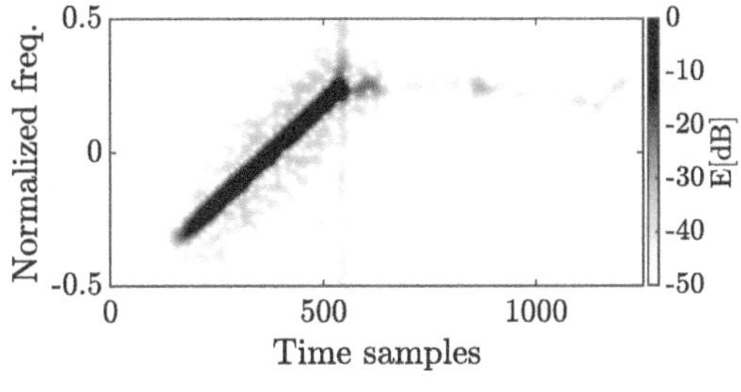

Figure 2.50 Filtered pulse spectrogram.

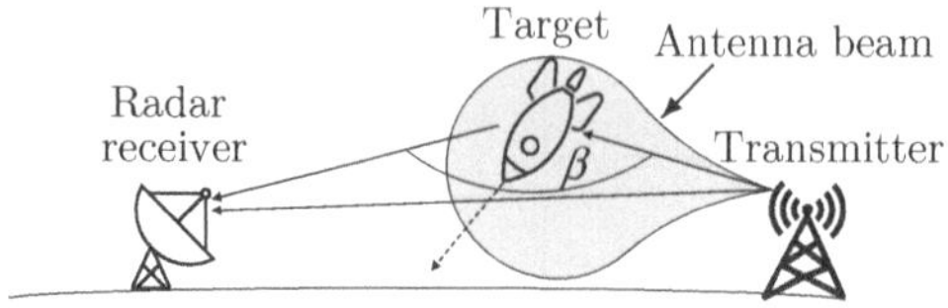

Figure 2.51 Forward scatter radar geometry. The forward scatter effect occurs when the target crosses the baseline – path between the transmitter and the receiver, i.e., when the $\beta \approx 180^\circ$.

As mentioned in Chapter 1, passive radar systems, unlike conventional radar, do not emit their own signals but exploit existing transmissions from third-party sources, such as broadcast, cellular, or satellite signals. One particularly interesting operational mode of passive radar is the forward scatter mode. In this configuration, the target crosses the baseline between the transmitter and the receiver, creating a geometry where the target heavily scatters the signal from the transmitter forward. The forward scatter geometry shown in Figure 2.51 offers unique advantages in target detection. As the target passes through the forward scatter region (mainly when the bistatic angle between the impinging and reflected signal $\beta \approx 180^\circ$), a pronounced signal enhancement is produced due to constructive interference between direct and scattered waves. This phenomenon, often referred to as the *forward scatter effect*, makes forward scatter radar especially suitable for detecting stealthy or low radar cross-section targets, which are otherwise difficult to observe using traditional radar techniques.

Such a signal was acquired during one of the North Atlantic Treaty Organization (NATO) trials under the acronym APART-GAS 2019 (Active PAssive Radar Trials Ground-based, Airborne, Sea-borne) (see [129] and references herein). The passive radar recorded the television signal (Digital Video Broadcasting – Terrestrial (DVB-T)) from the occasional transmitter located 27.8 km from the receiver. It emitted a DVB-T signal with the power of 10.4 kW on a frequency equal to 184.5 MHz and with 7 MHz bandwidth. The small aircraft passed through the baseline, allowing forward scatter phenomenon observation, which was processed and delineated in Figure 2.52.

The original spectrogram shown in Figure 2.52a presents a single component whose instantaneous frequency is almost constant in the first half and then rapidly falls to stabilize again. Such a frequency trajectory is a typical shape of the target echo in the forward scatter geometry, and when the frequency passes 0, the target crosses the baseline. The vertically synchrosqueezed spectrogram and the same with the detected signal component are shown in Figure 2.52b and c, respectively. After signal filtering, the spectrogram is presented in Figure 2.52d.

The filtered signal retains a clean echo from the target, free from disruptive noise, while maintaining both phase and amplitude characteristics. The time-domain signals (both real and imaginary parts) are illustrated in Figure 2.53. This refined signal

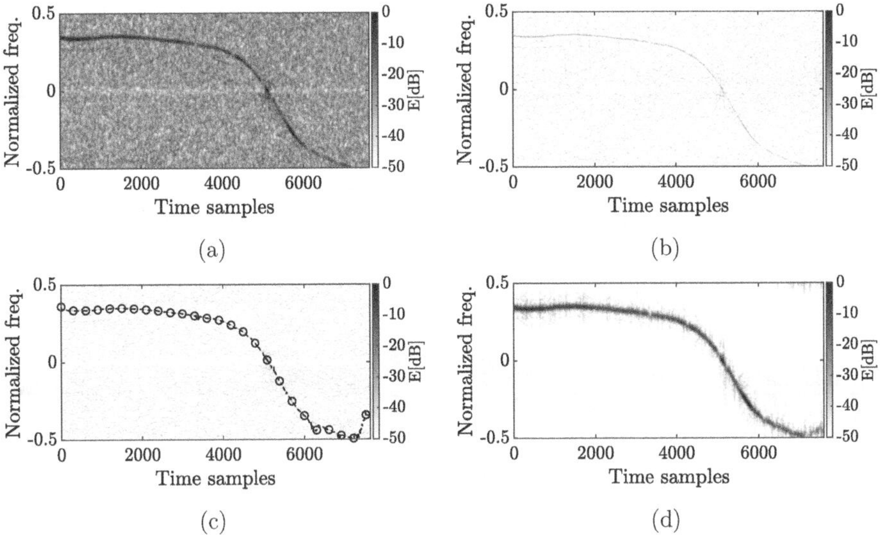

Figure 2.52 Passive forward scattering signal extraction results using second-order vertical synchrosqueezing. (a) Original spectrogram; (b) spectrogram after concentration; (c) the same as (b) but with superimposed detected signal component; (d) filtered signal spectrogram.

is suitable for further analyses, such as estimating parameters or kinematic properties of non-cooperative targets. Also, the approach proves valuable in differentiating between multiple targets crossing the baseline simultaneously or in scenarios where multipath propagation causes significant interference.

So far, the considerations we have limited to the second-order synchrosqueezing stemming from the signal model defined by the quadratic phase (see (2.148)). However, this approach has some limitations, especially noticeable when analyzing strongly nonlinear FM. For example, Figure 2.54 shows the problem of STFT value displacement when a strong nonlinear FM occurs. The method's ability to accurately reallocate time-frequency energy is compromised, leading to distorted representations and reduced reliability in such scenarios.

To figure out the problem's origin, we can be inspired by (2.152), where the chirp rate biases a simple, instantaneous frequency estimator. A similar case but for higher-order terms can be seen in Figure 2.54, where we can observe the third- and higher-order phase terms apart from linear FM. If we contemplate this problem, a straightforward solution turns out to be an extension of the signal model to a higher-order polynomial in the phase and amplitude. Indeed, such a methodology was explored in [130–132]. Let us quickly pass through the derivations; further details can be found in the referenced articles.

The Nth-order polynomials generally define signal amplitude and the phase. Defining the generic signal model as

$$x(t) = A_x(t)e^{j\phi_x(t)}, \tag{2.159}$$

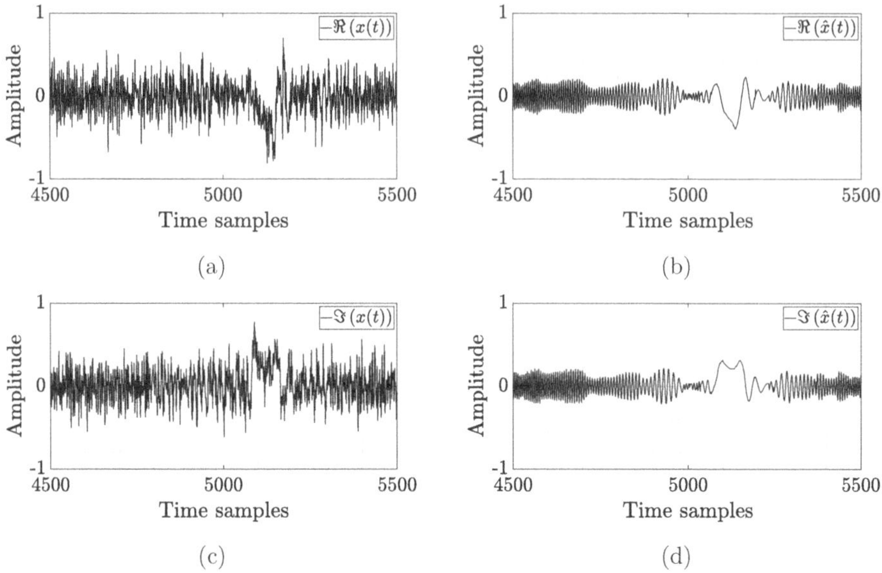

Figure 2.53 Time-domain results of passive forward scattering signal extraction using second-order vertical synchrosqueezing. (a) Real part of the original signal; (b) real part of the filtered signal; (c) imaginary part of the original signal; (d) imaginary part of the filtered signal.

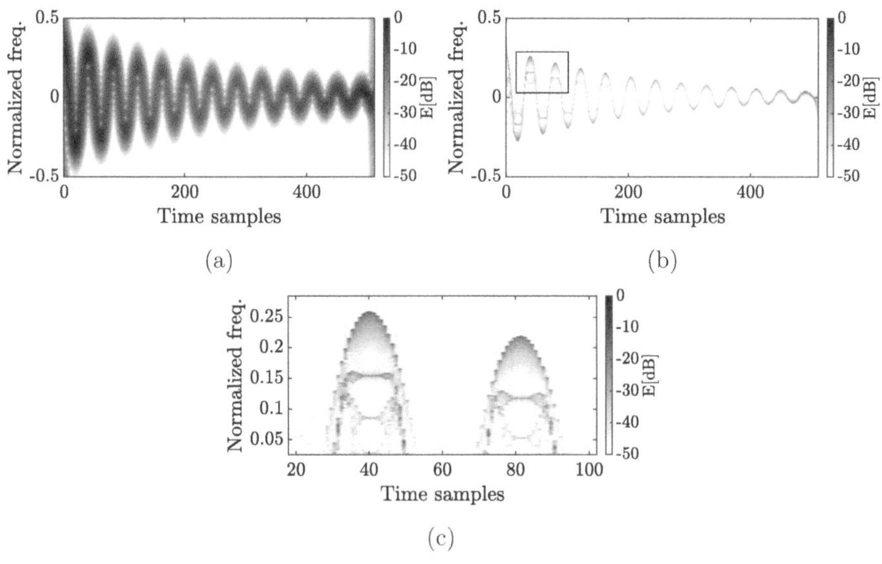

Figure 2.54 Limitations of the second-order vertical synchrosqueezing. (a) Spectrogram of the signal with a highly oscillating frequency; (b) second-order vertically synchrosqueezed spectrogram; (c) close-up on the rectangle.

one can define its amplitude and phase, respectively

$$\log(A(t)) = \sum_{n=0}^{4} \frac{\alpha_n t^n}{n!}, \tag{2.160}$$

$$\phi(t) = \sum_{n=0}^{4} \frac{\beta_n t^n}{n!}. \tag{2.161}$$

To circumvent this problem on the limited number of pages (derivations can be tremendously extensive) and make it doable in practical applications, let us assume $N = 4$, which is a reasonable value. By doing so, we can define the Gaussian-based amplitude as

$$\mathscr{A}_x \cdot \exp\left(-\frac{(t-t_x)^2}{2T_x^2} - \Delta_x \frac{(t-t_x)^3}{3T_x^2} - \Sigma_x \frac{(t-t_x)^4}{4T_x^2}\right), \tag{2.162}$$

$\mathscr{A}_x$ is the constant envelope, t_x is the time shift, T_x is the duration, and components $-\frac{(t-t_x)^2}{2T_x^2}$, $-\Delta_x \frac{(t-t_x)^3}{3T_x^2}$, and $-\Sigma_x \frac{(t-t_x)^4}{4T_x^2}$ are responsible for the second, third, and fourth-order amplitude modulation component, respectively. The phase can then be described as

$$\phi_x(t) = \varphi_x + \omega_x t + \alpha_x t^2/2 + \beta_x t^3/3 + \gamma_x t^4/4, \tag{2.163}$$

where φ_x, $\omega_x, \alpha_x, \beta_x$, and γ_x are the initial phase, angular instantaneous frequency, second-, third-, and fourth-order modulation terms, respectively. Including (2.162) and (2.163) in the signal model, its derivative reads as

$$\frac{\mathrm{d}x(t)}{\mathrm{d}t} = \left(p_x t^3 + q_x t^2 + r_x t + s_x\right) x(t), \tag{2.164}$$

where

$$p_x = \frac{-\Sigma_x}{T_x^2} + j\gamma_x, \tag{2.165}$$

$$q_x = \frac{-\Delta_x + 3\Sigma_x t_x}{T_x^2} + j\beta_x, \tag{2.166}$$

$$r_x = -\frac{1 - 2\Delta_x t_x + 3\Sigma_x t_x^2}{T_x^2} + j\alpha_x, \tag{2.167}$$

$$s_x = \frac{t_x - \Delta_x t_x^2 + \Sigma_x t_x^3}{T_x^2} + j\omega_x. \tag{2.168}$$

Substituting (2.164) into (2.118) and grouping the formula, one has

$$\begin{aligned} F_x^{\mathscr{D}h}(t,\omega) = &- p_x F_x^{\mathscr{T}^3 h}(t,\omega) \\ &- (3p_x t + q_x) F_x^{\mathscr{T}^2 h}(t,\omega) \\ &- (3p_x t^2 + 2q_x t + r_x) F_x^{\mathscr{T} h}(t,\omega) \\ &- (p_x t^3 + q_x t^2 + r_x t + s_x - j\omega) F_x^{h}(t,\omega), \end{aligned} \tag{2.169}$$

To estimate parameters p_x, q_x, r_x, and s_x, let us build a linear system allowing for derivations of signal parameters. Differentiating (2.169) once with respect to ω yields

$$\begin{aligned} F_x^{\mathscr{D}\mathscr{T}h}(t,\omega) + F_x^h(t,\omega) = &-p_x F_x^{\mathscr{T}^4 h}(t,\omega) \\ &-(3p_x t + q_x) F_x^{\mathscr{T}^3 h}(t,\omega) \\ &-(3p_x t^2 + 2q_x t + r_x) F_x^{\mathscr{T}^2 h}(t,\omega) \\ &-(p_x t^3 + q_x t^2 + r_x t + s_x - j\omega) F_x^{\mathscr{T}h}(t,\omega). \end{aligned} \tag{2.170}$$

In general, the $(n-3)$th-order derivative (for $n \geq 4$) of (2.169) with respect to ω can be obtained as

$$\begin{aligned} F_x^{\mathscr{T}^{n-3}\mathscr{D}}(t,\omega) + (n-4) F_x^{\mathscr{T}^{n-3}}(t,\omega) = &-p_x F_x^{\mathscr{T}^n}(t,\omega) \\ &-(3p_x t + q_x) F_x^{\mathscr{T}^{n-1}}(t,\omega) \\ &-(3p_x t^2 + 2q_x t + r_x) F_x^{\mathscr{T}^{n-2}}(t,\omega) \\ &-(p_x t^3 + q_x t^2 + r_x t + s_x - j\omega) F_x^{\mathscr{T}^{n-3}}(t,\omega), \end{aligned} \tag{2.171}$$

where $F_x^{\mathscr{T}^n\mathscr{D}}(t,\omega)$ is a combined window derivative multiplied by the nth order time ramp. Using (2.169), (2.170), and the two next derivatives of (2.170) with respect to ω according to (2.171), the following linear system can be written:

$$\begin{bmatrix} F_x^h(t,\omega) & F_x^{\mathscr{T}h}(t,\omega) & F_x^{\mathscr{T}^2 h}(t,\omega) & F_x^{\mathscr{T}^3 h}(t,\omega) \\ F_x^{\mathscr{T}h}(t,\omega) & F_x^{\mathscr{T}^2 h}(t,\omega) & F_x^{\mathscr{T}^3 h}(t,\omega) & F_x^{\mathscr{T}^4 h}(t,\omega) \\ F_x^{\mathscr{T}^2 h}(t,\omega) & F_x^{\mathscr{T}^3 h}(t,\omega) & F_x^{\mathscr{T}^4 h}(t,\omega) & F_x^{\mathscr{T}^5 h}(t,\omega) \\ F_x^{\mathscr{T}^3 h}(t,\omega) & F_x^{\mathscr{T}^4 h}(t,\omega) & F_x^{\mathscr{T}^5 h}(t,\omega) & F_x^{\mathscr{T}^6 h}(t,\omega) \end{bmatrix}$$

$$\times \begin{bmatrix} -p_x t^3 - q_x t^2 - r_x t - s_x + j\omega \\ -3p_x t^2 - 2q_x t - r_x \\ -3p_x t - q_x \\ -p_x \end{bmatrix} = \begin{bmatrix} F_x^{\mathscr{D}h}(t,\omega) \\ F_x^{\mathscr{D}\mathscr{T}h}(t,\omega) + F^h(t,\omega) \\ F_x^{\mathscr{D}\mathscr{T}^2 h}(t,\omega) + 2F^{\mathscr{T}h}(t,\omega) \\ F_x^{\mathscr{D}\mathscr{T}^3 h}(t,\omega) + 3F^{\mathscr{T}^2 h}(t,\omega) \end{bmatrix}, \tag{2.172}$$

which leads to (if the inverse matrix can be obtained)

$$\begin{bmatrix} v_1 \\ v_2 \\ v_3 \\ v_4 \end{bmatrix} = \begin{bmatrix} -p_x t^3 - q_x t^2 - r_x t - s_x + j\omega \\ -3p_x t^2 - 2q_x t - r_x \\ -3p_x t - q_x \\ -p_x \end{bmatrix}$$

$$= \begin{bmatrix} F_x^h(t,\omega) & F_x^{\mathcal{T}h}(t,\omega) & F_x^{\mathcal{T}^2h}(t,\omega) & F_x^{\mathcal{T}^3h}(t,\omega) \\ F_x^{\mathcal{T}h}(t,\omega) & F_x^{\mathcal{T}^2h}(t,\omega) & F_x^{\mathcal{T}^3h}(t,\omega) & F_x^{\mathcal{T}^4h}(t,\omega) \\ F_x^{\mathcal{T}^2h}(t,\omega) & F_x^{\mathcal{T}^3h}(t,\omega) & F_x^{\mathcal{T}^4h}(t,\omega) & F_x^{\mathcal{T}^5h}(t,\omega) \\ F_x^{\mathcal{T}^3h}(t,\omega) & F_x^{\mathcal{T}^4h}(t,\omega) & F_x^{\mathcal{T}^5h}(t,\omega) & F_x^{\mathcal{T}^6h}(t,\omega) \end{bmatrix}^{-1} \times \begin{bmatrix} F_x^{\mathcal{D}h}(t,\omega) \\ F_x^{\mathcal{D}\mathcal{T}h}(t,\omega) + F_x^h(t,\omega) \\ F_x^{\mathcal{D}\mathcal{T}^2h}(t,\omega) + 2F_x^{\mathcal{T}h}(t,\omega) \\ F_x^{\mathcal{D}\mathcal{T}^3h}(t,\omega) + 3F_x^{\mathcal{T}^2h}(t,\omega) \end{bmatrix}. \tag{2.173}$$

Hence, the following estimators can be obtained recursively:

$$\hat{p}_x(t,\omega) = -v_4, \tag{2.174}$$

$$\hat{q}_x(t,\omega) = -3\hat{p}_x\tilde{t}(t,\omega) - v_3, \tag{2.175}$$

$$\hat{r}_x(t,\omega) = -3\hat{p}_x[\tilde{t}(t,\omega)]^2 - 2\hat{q}_x\tilde{t}(t,\omega) - v_2, \tag{2.176}$$

$$\hat{s}_x(t,\omega) = -\hat{p}_x[\tilde{t}(t,\omega)]^3 - \hat{q}_x[\tilde{t}(t,\omega)]^2 - \hat{r}_x\tilde{t}(t,\omega) - v_1 + j\omega, \tag{2.177}$$

where

$$\tilde{t}(t,\omega) = \frac{F_x^{\mathcal{T}h}(t,\omega)}{F_x^h(t,\omega)}. \tag{2.178}$$

Their imaginary parts give $\hat{\gamma}_x(t,\omega), \hat{\beta}_x(t,\omega), \hat{\alpha}_x(t,\omega)$, and $\hat{\omega}_x(t,\omega)$, respectively

$$\hat{\gamma}_x(t,\omega) = \Im\left(\hat{p}_x(t,\omega)\right), \tag{2.179}$$

$$\hat{\beta}_x(t,\omega) = \Im\left(\hat{q}_x(t,\omega)\right), \tag{2.180}$$

$$\hat{\alpha}_x(t,\omega) = \Im\left(\hat{r}_x(t,\omega)\right), \tag{2.181}$$

$$\hat{\omega}_x(t,\omega) = \Im\left(\hat{s}_x(t,\omega)\right). \tag{2.182}$$

Remark 2.11. *Another but adequate derivations for the high-order synchrosqueezing are described in [130]. The readers are encouraged to get familiar with the corresponding approach in the referenced work.*

Estimators (2.179)–(2.181) are of our interest in Section 2.10. Now, we can consider the influence of a higher-order instantaneous frequency estimator in the vertical synchrosqueezing performance. If we deem higher-order phase terms, the instantaneous frequency estimator becomes robust on local FMs, especially those nonlinear. The vertically synchrosqueezed spectrogram using the higher-order instantaneous frequency estimator is shown in Figure 2.55. The spectrogram strongly focuses the frequency ridge for the entire signal duration, which is not the case for simple estimators.

Of course, the concentration capabilities, although their visual advantages streamline spectrogram perception, enhance signal decomposition. The Rényi entropy for the spectrogram and the vertically synchrosqueezed spectrograms obtained using the

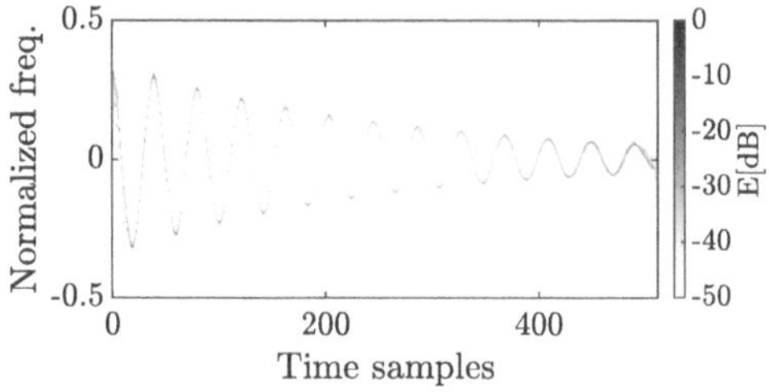

Figure 2.55 Concentrated spectrogram of signal with a highly oscillating frequency. The analysis used a high-order instantaneous frequency estimator in the time-frequency domain, leading to precise STFT value displacement to the signal ridge.

first, second, and higher-order instantaneous frequency estimators is 15.25 bits, 9.07 bits, 9.02 bits, and 7.66 bits, respectively. No wonder the better the spectrogram is concentrated, the easier its decomposition. The detected signal component for each analyzed case is superimposed on the spectrogram in Figure 2.56. The first 100 samples are the most interesting, as the oscillations decrease with time, but the signal beginning exhibits a high-frequency variance hampering the component extraction capabilities. The outcomes definitely show the performance of the high-order instantaneous frequency estimator relative to concentration capabilities. For

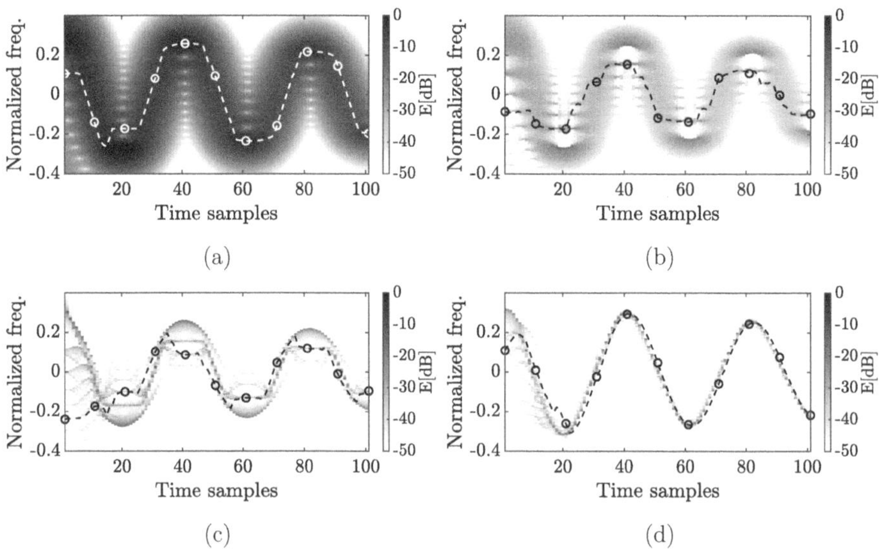

Figure 2.56 Signal component extraction outcomes depending on the instantaneous frequency estimator order. The detected signal mode is superimposed as the dashed line with circles. (a) Original spectrogram; (b) first-order estimator; (c) second-order estimator; (d) higher-order estimator.

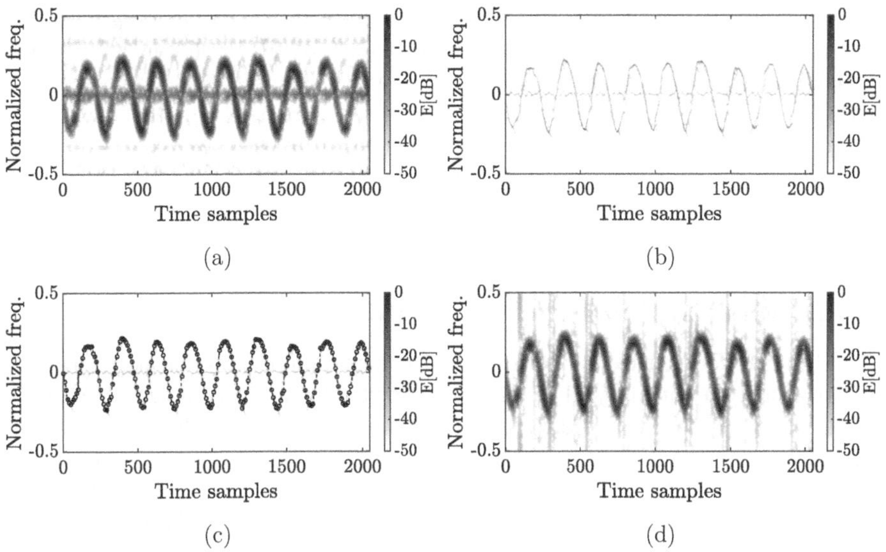

Figure 2.57 Real-life micro-Doppler gesture decomposition using high-order vertical synchrosqueezing. (a) Spectrogram; (b) vertically synchrosqueezed spectrogram; (c) the same as (b) but with a superimposed detected mode; (d) spectrogram of the reconstructed signal component. (Reproduced from [93], licensed under CC BY 4.0.)

lower-order instantaneous frequency estimators, a noticeable bias becomes evident. This bias manifests as a lack of alignment between the STFT values and the actual instantaneous frequency curve of the signal. Instead of being tightly concentrated along the putative frequency trajectory, the spectral components are shifted from their correct positions. Additionally, interference patterns emerge, further degrading the accuracy and quality of signal extraction, making precise interpretation of time-frequency characteristics more challenging. However, one has to remember the increasing computational complexity of higher-order estimators. The practical applications must consider the trade-off between the concentration capabilities and the processing complexity.

Utilization of the high-order synchrosqueezing can be visualized using the real-life micro-Doppler signal we have already analyzed in Section 2.8. As delineated in Figure 2.57, signal decomposition using vertical synchrosqueezing allowed removing clutter (stationary echoes without the Doppler shift) [93]. As a result, a pure echo of a waving hand was obtained.

The vertical synchrosqueezing discussion here offers theoretical insights and practical examples, illustrating its utility in radar applications. While this section outlines the fundamental principles and key methods, it does not claim to cover the topic exhaustively. Readers are encouraged to explore the rich body of literature for a deeper and more comprehensive understanding of vertical synchrosqueezing and its diverse applications.

Nevertheless, the material herein should equip readers with a firm foundational grasp of the technique and its practical implications. By demonstrating how vertical synchrosqueezing can be effectively used for signal analysis in radar systems, we aim to bridge the gap between theoretical advancements and real-world implementations.

In the following subsection, we shift our focus to horizontal synchrosqueezing. This complementary approach further enhances time-frequency analysis by addressing spikes and transients analysis challenges, offering additional tools for advanced signal decomposition and feature extraction.

2.9.2 HORIZONTAL SYNCHROSQUEEZING

The last STFT-based technique for signal concentration analyzed in this book is a horizontal variant of synchrosqueezing. Some scholars sometimes refer to this technique as time-reassigned synchrosqueezing since *synchro* has explicit time-related meaning and expresses simultaneous relocation (*squeezing*) of transform values for the whole time axis. For the sake of simplicity, let us refer to this technique simply as horizontal synchrosqueezing. In horizontal synchrosqueezing, the STFT distribution is refined so that its values are simultaneously shifted (reassigned) along the time axis for the entire frequency axis. Tracking of signal dynamics is viable, particularly when analyzing rapidly evolving temporal structures. By effectively reducing the smearing of signal components in time, the method allows for precise identification and reconstruction of time-domain features. This method was introduced to address the limitations of traditional synchrosqueezing techniques by enhancing temporal precision rather than focusing solely on frequency readability.

As one can expect, in horizontal synchrosqueezing, we make use of (2.129) and combine it with (2.131), getting rid of STFT values' relocation along the frequency axis. The general definition reads as

$$H_x^h(t,\omega) = \int_{-\infty}^{\infty} F_x^h(\tau,\omega)\delta\left(t - \hat{t}_x(\tau,\omega)\right)\mathrm{d}\tau. \tag{2.183}$$

We define the spectrogram of the horizontally concentrated representation as the squared modulus, as in all previous cases.

Remark 2.12. *To derive the inverse horizontal synchrosqueezing transform, let us slightly alter the STFT definition. So far, we have used (2.78), which involves integrating the signal $x(t)$ modulated by a window function $h(\tau - t)$ centered at time t. In this formulation, the Fourier transform is computed with respect to the local window shift t, capturing the frequency components around a specific time segment of the signal. Conversely, the second STFT formulation can be defined as (the difference in the exponent)*

$$F_x^h(t,\omega) = \int_{-\infty}^{\infty} x(\tau)h(\tau - t)e^{-j\omega\tau}\mathrm{d}\tau. \tag{2.184}$$

Equation (2.184) shifts the perspective by interpreting the transform as integrating over the entire signal for the window position at each time t. Here, the computation captures how the signal's content aligns with a frequency component ω over the entire domain relative to a particular window position.

In essence, the difference between the two STFT definitions (2.78) and (2.184) lies in whether the Fourier transform is computed with respect to the shift of the window or the entire signal, offering complementary ways of analyzing localized or global frequency content.

We can obtain the ISTFT from (2.184) through the marginalization of $F_x^h(t,\omega)$ over time [133]

$$\int_{-\infty}^{\infty} F_x^h(t,\omega)\mathrm{d}t = \int_{-\infty}^{\infty}\int_{-\infty}^{\infty} x(\tau)h(\tau-t)e^{-j\omega\tau}\mathrm{d}t\mathrm{d}\tau = \int_{-\infty}^{\infty}\int_{-\infty}^{\infty} x(\tau)h(u)e^{-j\omega\tau}\mathrm{d}u\mathrm{d}\tau =$$
$$\int_{-\infty}^{\infty} x(\tau)e^{-j\omega\tau}\mathrm{d}\tau \int_{-\infty}^{\infty} h(u)\mathrm{d}u = F_x(\omega)F_h(0), \tag{2.185}$$

where

$$F_x(\omega) = \int_{-\infty}^{\infty} x(t)e^{-j\omega t}\mathrm{d}t, \tag{2.186}$$

denotes the Fourier transform of $x(t)$. Next, from (2.185), one can compute the Fourier transform of $x(t)$ as

$$F_x(\omega) = \frac{1}{F_h(0)^*}\int_{-\infty}^{\infty} F_x^h(t,\omega)\mathrm{d}t, \tag{2.187}$$

and the signal reconstruction formula can be derived by applying the inverse Fourier transform as

$$x(t) = \frac{1}{2\pi F_h(0)^*}\int_{-\infty}^{\infty}\int_{-\infty}^{\infty} F_x^h(\tau,\omega)e^{j\omega t}\mathrm{d}\tau\mathrm{d}\omega. \tag{2.188}$$

The presented deduction can also be applied to (2.183), yielding

$$\int_{-\infty}^{\infty} H_x^h(t,\omega)\mathrm{d}t = \int_{-\infty}^{\infty}\int_{-\infty}^{\infty} F_x^h(\tau,\omega)\delta\left(t-\hat{\tau}_x(\tau,\omega)\right)\mathrm{d}\tau\mathrm{d}t$$
$$= \int_{-\infty}^{\infty} F_x^h(t,\omega)\mathrm{d}t = F_x(\omega)F_h(0). \tag{2.189}$$

Horizontal synchrosqueezing in action is shown in Figure 2.58. This is an already analyzed multicomponent signal, allowing for a straightforward comparison of different STFT concentration techniques. The conventional spectrogram exhibits a smeared and less localized energy distribution across time-frequency components.

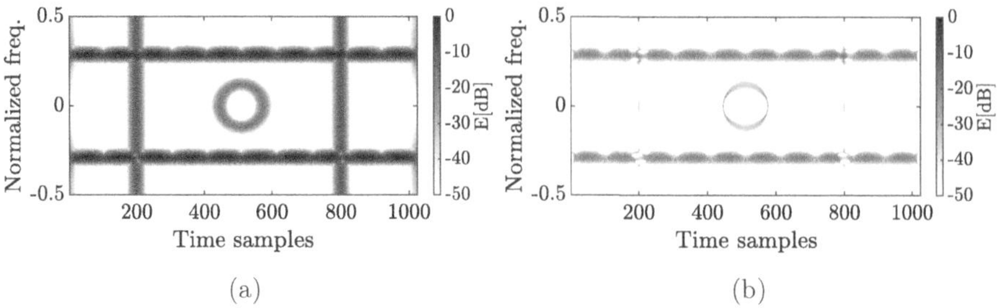

Figure 2.58 Horizontal synchrosqueezing spectrogram. (a) Original energy distribution; (b) horizontally synchrosqueezed spectrogram.

As discussed in this book, the spectrogram tends to blur the structure of signal components, making precise identification challenging. However, the horizontally synchrosqueezed spectrogram demonstrates the effectiveness of this method in concentrating the STFT distribution on transients and bursts. The synchrosqueezed representation provides a sharper, more compact depiction, where better readability of spikes and improved signal interpretation are viable. This enhanced localization is particularly useful for analyzing signals with rapid pulses and enormous variance of the instantaneous frequency.

The real-life example illustrating the horizontal synchrosqueezing performance is presented in Figure 2.59. The analyzed signal was recorded using the XY-DemoRad

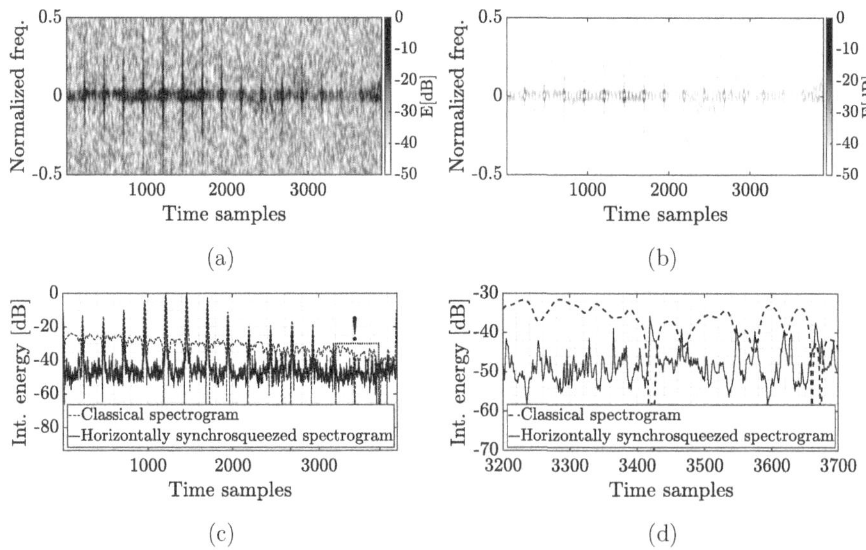

Figure 2.59 Horizontal synchrosqueezing performance in real-life signal analysis. (a) Spectrogram of the helicopter echo; (b) horizontally synchrosqueezed spectrogram; (c) plots of the integrated energy from the spectrogram and its concentrated counterpart; (d) close-up on the selected part of (c).

(its picture is shown in Figure 2.40) and contains the echo of the helicopter preparing for the flight. The helicopter was stationary on the ground, approximately 50 meters away from the radar. The measurement was done during start-up before the flight. The results of horizontal synchrosqueezing-based processing are illustrated in Figure 2.59.

The rotor blades of a helicopter generate a distinctive crest of vertical stripes on a spectrogram when analyzing their radar reflections due to the periodic nature of their rotational motion. As the blades sweep through space, they create a series of periodic Doppler shifts in the reflected radar signal. Each blade acts as a moving reflector, and its speed relative to the radar changes cyclically during each rotation. This effect can be observed in Figure 2.59a. These periodic speed variations cause modulations in the frequency of the reflected signal, leading to spectral components that align as vertical patterns on the spectrogram. The rotor rotation frequency and the number of blades determine the separation between these stripes.

The horizontally synchrosqueezed spectrogram in Figure 2.59b provides a more explicit depiction of the timing of each spike, corresponding to the half-rotations of the helicopter blades. The distribution shown in Figure 2.59c is obtained by integrating the energy along the frequency axis, where the integrated energy reaches the local salients at each half-turn. The close-up of the selected plot part (dotted rectangle in Figure 2.59c) is illustrated in Figure 2.59d. The result reveals the superiority of the concentrated spectrogram over the classical one, as the weak echo was correctly focused, leading to the estimation of the rotational speed of the helicopter rotor tractable [134].

In its primary definition, horizontal synchrosqueezing is efficient for spikes and rapid transients. When the variance of signal instantaneous frequency decreases, we can observe a drop in efficacy in the STFT concentration. In other words, horizontal synchrosqueezing does not ensure sufficient concentration when the signal frequency varies. Similar observations (but with the opposite influence of the signal FM rate) were done for the already discussed vertical synchrosqueezing, where chirps required leveraging high-order instantaneous frequency estimators. For horizontal synchrosqueezing, the deduction preserves conserved, and the derivations paving the way for higher-order synchrosqueezing were described in [118, 133].

As such, the second-order horizontal synchrosqueezing aims to relocate the STFT values from the coordinates (t, ω) to $(t_x^{[2]}, \omega)$ expected to be closer to the instantaneous frequency curve. Assuming the second-order polynomial defining the signal phase and amplitude as in (2.148)

$$x(t) = \mathscr{A} e^{-\frac{(t-t_x)^2}{2T_x^2}} e^{j\left(\phi_0+\omega_x t+\frac{\alpha_x t^2}{2}\right)}, \tag{2.190}$$

the instantaneous frequency reads as

$$\omega = \omega_x + \alpha_x t_x^{[2]}, \tag{2.191}$$

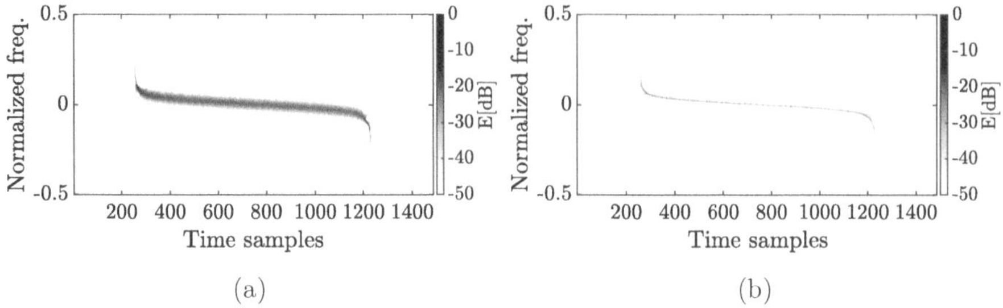

Figure 2.60 Comparison of the first- (a) and the second-order (b) horizontal synchrosqueezing with the example of the real-life radar pulse with nonlinear frequency modulation.

which leads to the second-order group delay estimator in the time-frequency domain [133]

$$\hat{t}_x^{[2]}(t,\omega) = \begin{cases} \frac{\omega - \hat{\omega}_x(t,\omega) + \Im(\hat{q}_x(t,\omega)\tilde{t}_x(t,\omega))}{\hat{\alpha}_x(t,\omega)} & \text{if } \hat{\alpha}_x(t,\omega) \neq 0 \\ \hat{t}_x(t,\omega) & \text{otherwise} \end{cases}, \tag{2.192}$$

where $\hat{t}_x(t,\omega)$ is defined by $\hat{t}_x(t,\omega) = \Re\left(\tilde{t}_x(t,\omega)\right)$ where

$$\tilde{t}_x(t,\omega) = t + \frac{F_x^{\mathscr{T}h}(t,\omega)}{F_x^h(t,\omega)}, \tag{2.193}$$

(as defined in (2.129)), and $\hat{\alpha}_x(t,\omega) = \Im\left(\hat{q}_x(t,\omega)\right)$ is the chirp rate estimator.

A comparative example of the first- and second-order horizontal synchrosqueezing can be seen in Figure 2.60. The radar pulse with a nonlinear FM exhibits sharp ends resulting from the spectrum leakage and the nonlinearity in the instantaneous frequency curve. For the first-order horizontal synchrosqueezing from Figure 2.60a, signal transients are well-sharpened, but the chirp is of a moderate concentration. However, if we consider the FM and use (2.192), the concentrated distribution strongly focuses on the whole chirp including the transients at the signal ends and the part with a linear FM. In Figure 2.60, the Rényi entropy is 10.19 bits for Figure 2.60a, whereas for Figure 2.60b it is reduced to 7.33 bits.

Analogously, as for vertical synchrosqueezing, one can derive higher-order group delay estimators in the time-frequency domain [135]. Extending (2.191) to higher-order polynomials, thus putting there (2.164), we can obtain the third- and fourth-order estimators. The derivation of higher-order group delay estimators has been omitted here, as it closely parallels the earlier derivations for higher-order vertical synchrosqueezing presented in this chapter. These methods have also been extensively documented in the existing literature. Readers interested in a deeper exploration are encouraged to derive these estimators independently, leveraging the conceptual framework outlined in the previous sections. This exercise not only reinforces an understanding of the underlying theory but also highlights the remarkable structural similarities between the vertical and horizontal approaches.

So far, we have based our considerations on two basic parameters related to signals: time and frequency. However, we have also dealt with sometimes enigmatically described *frequency modulation parameters*. They define the signal phase changeability at each (t, ω) coordinate and are a valuable tool for nonstationary and multicomponent signal analysis in the time-frequency domain.

2.10 SIGNAL PARAMETER ESTIMATION

Indeed, the radar signal parameter estimation is far-reaching and encompasses plenty of algorithms, e.g. [79, 136–148], to list only a few. These algorithms could theoretically be categorized as more or less accurate, more or less efficient, or generally superior or inferior. However, such classifications would remain relative, as no universal standard for evaluating these methods exists. While computational complexity can be assessed, it does not always directly correlate with computation time, which heavily depends on factors like the computing platform and implementation technique. Moreover, a fast algorithm may lack precision or exhibit high variance. Algorithms can also operate in various domains, including time, frequency, and time-frequency. Effective comparison typically requires sticking to a particular domain, although certain transforms, such as the wavelet transform, fall outside these categories by analyzing signal distribution in the time-scale domain. Additionally, some algorithms come with inherent assumptions, such as constant amplitude or the presence of a single dominant component, which could serve as further criteria for classification and characterization.

Besides readable signal representation and decomposition, practical systems like radars and radio signal receivers seek to estimate signal (or target) phase parameters. The principal approach comes down to the computation of the phase derivatives with respect to time. The first-order phase derivative gives the frequency; the second-order gives the chirp rate, and so on. Although this approach seems straightforward and efficient, it fails when analyzing multicomponent and noisy waveforms. To visualize this constraint, we can first deem a noiseless case for the simple chirp defined previously by (2.93). Its definition reads as

$$x(t) = \exp\left(j2\pi\left(\frac{1000t^2}{2}\right)\right). \tag{2.194}$$

From this equation, we can deduce that the phase is quadratic, instantaneous frequency changes linearly, and thus, the FM rate remains constant. The spectrogram of this chirp with some additive noise was illustrated in Figure 2.12d. The time-domain parameters of this signal are delineated in Figure 2.61. This trivial example illustrates the essential signal phase parameters using its derivatives. However, this simple approach is noise-sensitive. When we disturb this waveform with relatively low noise, giving the SNR = 20 dB, we obtain the outcomes as in Figure 2.62.

From Figures 2.61 and 2.62, it is clear that even a tiny portion of noise highly disturbs signal parameter estimation, especially those related to higher-order phase derivatives. One faces another limitation when analyzing multicomponent signals, as

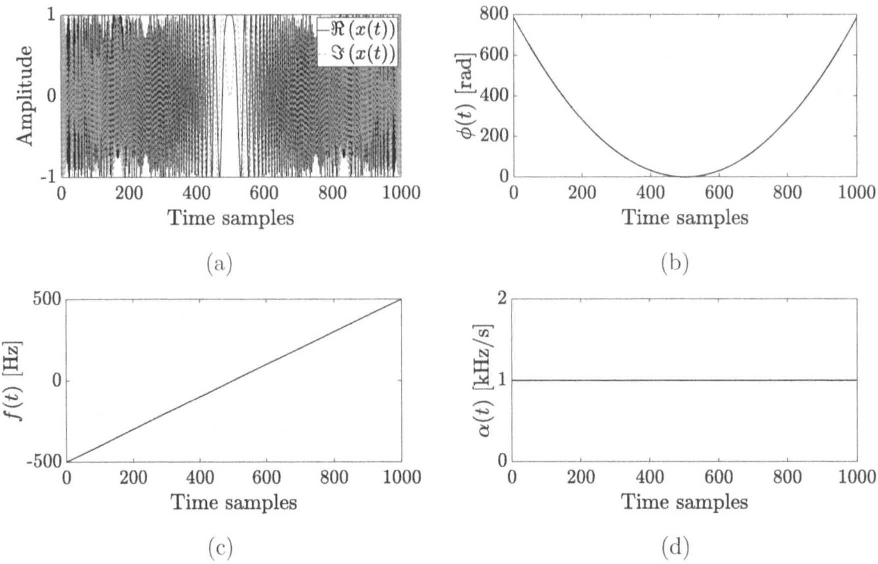

Figure 2.61 Noiseless linear chirp and its time-domain parameters. (a) Time-domain signal (real and imaginary parts); (b) phase; (c) frequency; (d) frequency modulation rate.

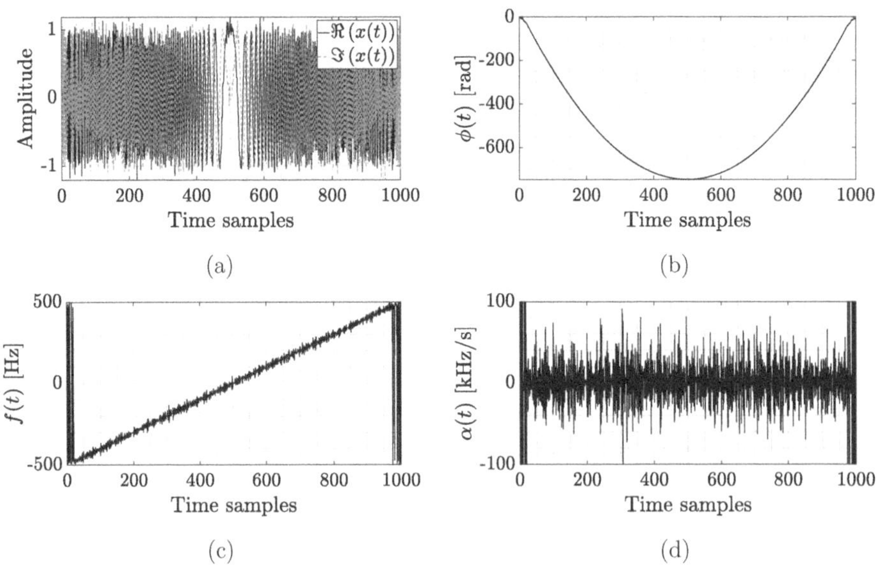

Figure 2.62 Linear chirp embedded in noise and its time-domain parameters. (a) Time-domain signal (real and imaginary parts); (b) phase; (c) frequency; (d) frequency modulation rate.

their phase derivatives do not allow for differentiation between signal components. Thus, estimating real-world radar signals where noise and multipath propagation occur almost always requires more sophisticated approaches.

Remark 2.13. *This is a quick digression of one of the most precise signal parameter estimation methods called the extended generalized chirp transform (EGCT) [149]. The method was used in passive radars using other pulse radars as illuminators [150]. The EGCT method surpasses the most commonly used techniques in terms of accuracy due to the matched filtering principle applied in this technique (see Section 1.3.7).*

Assuming that $x(t)$ is the investigated signal, the matched filtering process can be performed using (1.21) where $g(t) = \exp(-j\Phi_{\boldsymbol{P}}(t))$ is the impulse response of the matched filter. The phase coefficients vector is composed of the terms $\boldsymbol{P} = [a_0, a_1, ..., a_N]$, and the phase is expressed as polynomial functions as

$$\Phi_{\boldsymbol{P}}(t) = a_0 + \sum_{n=1}^{N} \frac{a_n}{n} t^n. \tag{2.195}$$

To estimate the phase model, a maximum likelihood estimator can be applied for the vector given by (2.195), which can be defined as

$$\hat{\boldsymbol{P}} = \underset{a_n}{\operatorname{argmax}} |\xi(\tau)|, \tag{2.196}$$

where $\hat{\boldsymbol{P}}$ gives the estimated values of a_n in (2.195). In practical considerations, this problem comes down to formulating a K-points set of values in which the component a_n will be searched. The main limitation of this method stems from its iterative nature. If a_n extent is not roughly assessed, K increases, thus making the processing obstructively long.

The essential remedy for the high computational complexity is to use the Fourier transform for signal correlations. The matched filtering can be performed in both time (as a convolution of the signal) and frequency domains (as a multiplication of their spectra and then a calculation of the inverse Fourier transform); therefore, the processing time can be reduced using an FFT algorithm. Examples of this technique can be found in [150, 151], where nonlinear frequency-modulated radar pulses were analyzed, and their parameters were successfully estimated.

The example of the EGCT map presenting the magnitude of matched filtering for different matched filter parameters is depicted in Figure 2.63a. The signal (2.194) was disturbed by the noise giving the $SNR = -10$ dB. The black peak in the middle of the illustrated EGCT map exhibits the estimated chirp parameters. They were correctly estimated, although the noise was stronger than the chirp.

Czarnecki, who continued the work of Barnes [153], Kodera [50], and Nelson [154], pioneered signal parameter estimation in the time-frequency domain [152]. All of these scholars examined STFT properties in an attempt to find estimators of signal parameters on the time-frequency plane. Let us follow their train of thought.

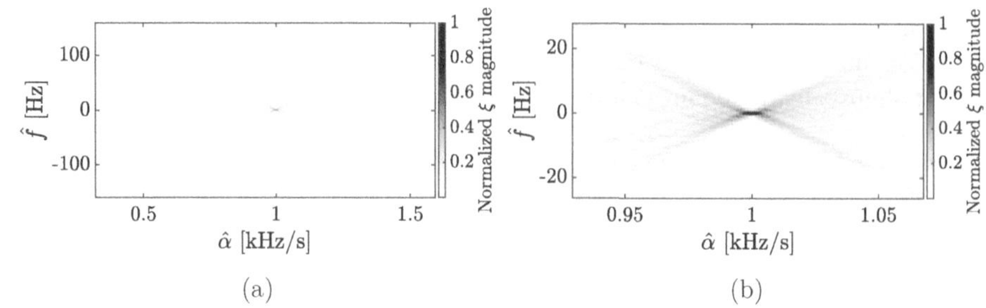

Figure 2.63 Example of the EGCT map as a precise tool for radar pulse parameter estimation. (a) Full distribution; (b) close-up on the peak.

In order to facilitate the discussion of key dependencies relevant to the subsequent problem description, let us recall the STFT definition already defined in (2.110)

$$F_x^h(t,\omega) = \int_{-\infty}^{\infty} x(\tau)h(\tau - t)e^{-j\omega(\tau - t)}\mathrm{d}\tau = M_x^h(t,\omega)e^{j\Phi_x^h(t,\omega)}. \tag{2.197}$$

By expressing the STFT result in its complex form, the following expression is obtained:

$$\vartheta_x^h(t,\omega) = \log\left(F_x^h(t,\omega)\right) = \Lambda_x^h(t,\omega) + j\Phi_x^h(t,\omega), \tag{2.198}$$

where $\Lambda_x^h(t,\omega) = \log\left(M_x^h(t,\omega)\right)$. Although the gradient (i.e., the set of partial and mixed derivatives) of the complex phase of the STFT has been widely explored in the literature, a comprehensive review of this work lies beyond the scope of this book. Instead, this book focuses on key parameters that will aid in analyzing radar signals. Additional details can be found in [50, 153–156].

For the sake of simplicity, the signal under consideration is regarded to have a constant envelope, which will be generalized later. Additionally, it is postulated that $M_x^h(t,\omega) > 0$, ensuring a real-valued $\Lambda_x^h(t,\omega)$. Under these premises, Barnes [153] defined the instantaneous bandwidth of a component, which was later extended in [152] as the *signed channelized instantaneous bandwidth*, expressed as

$$\Sigma_x^h(t,\omega) = \frac{\partial \Lambda_x^h(t,\omega)}{\partial t} = \Re\left(\frac{\partial \log\left(F_x^h(t,\omega)\right)}{\partial t}\right). \tag{2.199}$$

In (2.199), $\Sigma_x^h(t,\omega)$ is measured in $\frac{\mathrm{rad}}{\mathrm{s}}$. Furthermore, a related distribution called the *signed local group duration* was proposed in [152]. This distribution involves the derivative with respect to angular frequency and is defined as

$$\mho_x^h(t,\omega) = -\frac{\partial \Lambda_x^h(t,\omega)}{\partial \omega} = -\Re\left(\frac{\partial \log\left(F_x^h(t,\omega)\right)}{\partial \omega}\right). \tag{2.200}$$

The signed local group duration is measured in rad $\cdot$ s. Both $\Sigma_x^h(t,\omega)$ and $\mho_x^h(t,\omega)$ represent local signal characteristics on the time-frequency plane, offering information

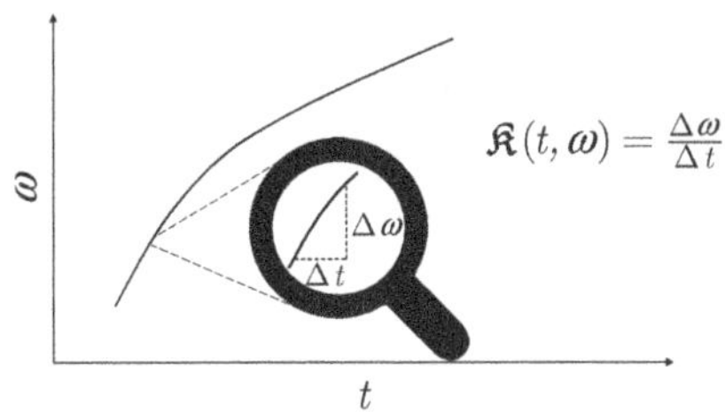

Figure 2.64 Simplified interpretation of the chirp rate estimation in the time-frequency domain.

at each (t,ω) coordinate. Precisely, $\Sigma_x^h(t,\omega)$ reflects the instantaneous bandwidth component dominating a particular channel at time t and angular frequency ω. On the other hand, $\mho_x^h(t,\omega)$ can be interpreted as the duration of a signal component within a given frequency channel delayed by t. For infinitesimal values of $\Delta t \to 0$ and $\Delta\omega \to 0$, assuming a constant signal amplitude, the local chirp rate can be approximated by the familiar ratio $\Delta\omega/\Delta t$. In the time-frequency domain, this estimate is described by

$$\mathfrak{K}(t,\omega) = \frac{\Sigma_x^h(t,\omega)}{\mho_x^h(t,\omega)} = -\frac{\partial \Lambda_x^h(t,\omega)}{\partial t} \Big/ \frac{\partial \Lambda_x^h(t,\omega)}{\partial \omega}. \tag{2.201}$$

This equation represents the chirp rate distribution, expressed in units of $\frac{1}{s^2}$, Hz/s or Hz^2. Graphical representations of this estimate are referred to as *accelerograms*, and its graphical interpretation is depicted in Figure 2.64. As such, $\mathfrak{K}(t,\omega)$ allows for the estimation of the ratio between the frequency increase and the time increase for each distribution point.

Remark 2.14. *We can notice an interesting feature from the presented derivations: the phase signal parameters can also be derived from the STFT amplitude (but under some limitations discussed later in this section). The phase-magnitude relationship in the STFT was investigated in [157].*

The spectrogram and the corresponding accelerogram are presented in Figure 2.65. The latter distribution shows the phase acceleration for each (t,ω) point. A grayscale intensity (associated with the chirp rate estimate) corresponds to the phase acceleration, so one can notice, e.g., black color when the signal frequency increases and white for the opposite frequency curve direction.

Estimator (2.201) has two drawbacks. Firstly, it can be biased when the amplitude varies. Secondly, there is an uncertainty in the vicinity of the close attractor (instantaneous frequency ridge), i.e., when the estimate should be precise. The effect can be noticed when analyzing signal ridges in Figure 2.65b and stems from the uncertainty following 0/0 division at ridges. Generally, any function derivative changes its sign and passes through 0 for the function ridge. In (2.201), we have two derivatives, both of which equal 0 at the STFT ridges. Their ratio gives the uncertainty, and the

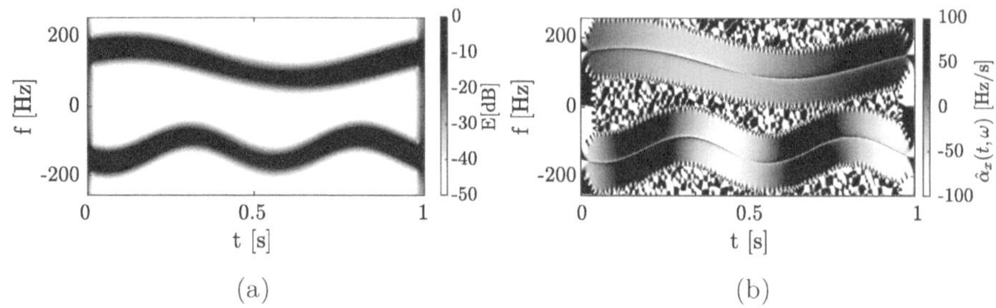

Figure 2.65 Spectrogram (a) and the phase accelerogram (b) of the simulated signal.

accelerogram is affected at each salient point. This also has another interpretation related to the window and its modifications, but its explanation needs the following considerations described in [117].

As already defined in (2.148), the signal can be written as the second-order phase-amplitude model as

$$x(t) = \mathscr{A} e^{-\frac{(t-t_x)^2}{2T_x^2}} e^{j\left(\phi_0 + \omega_x t + \frac{\alpha_x t^2}{2}\right)}, \tag{2.202}$$

where all symbols keep their meaning, as in (2.148). If we regard the amplitude as constant, i.e., $T_x \to \infty$, then

$$\frac{\mathrm{d}x(t)}{\mathrm{d}t} = j(\alpha_x t + \omega_x)x(t), \tag{2.203}$$

and

$$j\omega F_x^h(t,\omega) - F_x^{\mathscr{D}h}(t,\omega) = F_x^h(t,\omega)\left(\alpha_x t + \omega_x\right) + q_x F_x^{\mathscr{T}h}(t,\omega). \tag{2.204}$$

Multiplying by $F_x^h(t,\omega)^*$ and taking the real part gives (using basic complex number operations)

$$\hat{\alpha}_x(t,\omega) = \frac{\Re\left(F_x^{\mathscr{D}h}(t,\omega)F_x^h(t,\omega)^*\right)}{\Im\left(F_x^{\mathscr{T}h}(t,\omega)F_x^h(t,\omega)^*\right)}, \tag{2.205}$$

which is the equivalent of (2.201). Here, we can also explain the uncertainty around the instantaneous frequency ridge from another perspective. As we already know, $F_x^{\mathscr{D}h}(t,\omega)$ and $F_x^{\mathscr{T}h}(t,\omega)$ denote the STFTs with the differentiated and the time-weighted windows, respectively. If we compare these altered windows (see Figure 2.30), they both equal 0 in their centers, which is the second interpretation of the uncertainty. In other words, an indeterminate form arises at each instantaneous frequency ridge, resulting in numerical instabilities in its vicinity. In order to address this issue and extend the method to cases where the constant amplitude constraint is not met, higher-order phase derivatives were utilized [117]. High-order STFT derivatives were already defined when deriving robust instantaneous frequency estimators for vertical synchrosqueezing (see Section 2.9.1). As such, the chirp rate estimator can be obtained either using the partial derivatives with respect to time or frequency.

Each approach leads to the linear system, whose solution gives the instantaneous frequency and chirp rate estimators.

- Estimators using the STFT partial derivatives with respect to time.
  From (2.151), we have

$$F_x^{\mathscr{D}h}(t,\omega) = -q_x F_x^{\mathscr{T}h}(t,\omega) - (q_x t + p_x - j\omega) F_x^h(t,\omega). \tag{2.206}$$

  Differentiating (2.206) with respect to t once leads to

$$F_x^{\mathscr{D}^2h}(t,\omega) = -q_x F_x^{\mathscr{D}\mathscr{T}h}(t,\omega) - (q_x t + p_x - j\omega) F_x^{\mathscr{D}h}(t,\omega). \tag{2.207}$$

  Equations (2.206) and (2.207) build a linear system the same as (2.172), whose solution gives the already defined instantaneous frequency estimator (2.156), and the chirp rate estimator

$$\hat{\alpha}_x^{[t]}(t,\omega) = \Im\left(\frac{F_x^{\mathscr{D}^2h}(t,\omega)F_x^h(t,\omega) - \left(F_x^{\mathscr{D}h}(t,\omega)\right)^2}{F_x^{\mathscr{D}h}(t,\omega)F_x^{\mathscr{T}h}(t,\omega) - F_x^{\mathscr{D}\mathscr{T}h}(t,\omega)F_x^h(t,\omega)}\right). \tag{2.208}$$

- Estimators using the STFT partial derivatives with respect to frequency.
  Differentiating (2.206) with respect to ω yields

$$F_x^{\mathscr{D}\mathscr{T}h}(t,\omega) + F_x^h(t,\omega) = (q_x t + p_x - j\omega) F_x^{\mathscr{T}h}(t,\omega). \tag{2.209}$$

  Again, combining (2.206) and (2.209) into the linear system gives (2.153) and the second chirp rate estimator

$$\hat{\alpha}_x^{[\omega]}(t,\omega) = \Im\left(\frac{F_x^{\mathscr{D}\mathscr{T}h}(t,\omega)F_x^h(t,\omega) + \left(F_x^h(t,\omega)\right)^2 - F_x^{\mathscr{T}h}(t,\omega)F_x^{\mathscr{D}h}(t,\omega)}{\left(F_x^{\mathscr{T}h}(t,\omega)\right)^2 - F_x^{\mathscr{T}^2h}(t,\omega)F_x^h(t,\omega)}\right). \tag{2.210}$$

Both (2.208) and (2.210) solve the vulnerability of the time-varying amplitude and numerical instabilities near the instantaneous frequency ridge. Their efficacy, computational complexity, and precision are very similar; thus, they can be used interchangeably. Their comparison is shown in Figure 2.66.

It follows from the presented derivations that the accelerograms are void of instabilities near the ridges, allowing more precise signal parameter estimation. Also, they are almost identical, equally efficient, and of the same processing complexity. They can be used in the same way and with the very same outcomes.

Remark 2.15. *Estimators (2.208) and (2.210) can be derived from (2.201) using the L'Hôpital's rule as a known approach to uncertainty elimination. Derivations and details, however, leading to the same formulas as described, can be found in [74].*

For multicomponent signals with unknown parameters, one can estimate them using the so-called chirp rate profile [74, 158], which integrates the spectrogram's energy for each chirp rate value $r \in \mathbb{R}$ [158]

$$\mathscr{R}(r) = \int_{-\infty}^{\infty}\int_{-\infty}^{\infty} E_x^h(t,\omega)\delta(r - \hat{\alpha}_x(t,\omega))\mathrm{d}t\mathrm{d}\omega. \tag{2.211}$$

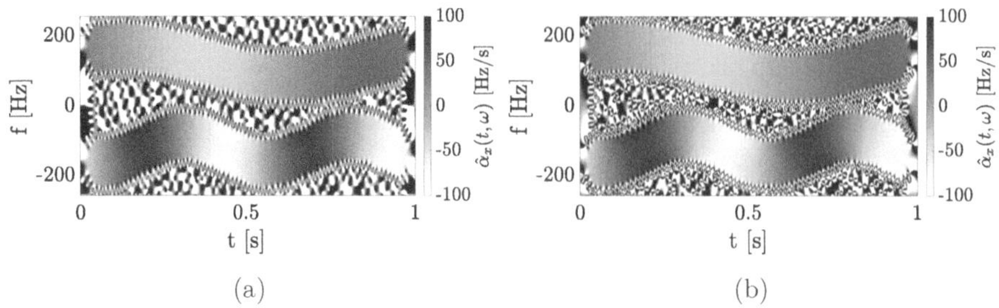

Figure 2.66 Comparison of two accelerograms. (a) Using (2.208); (b) using (2.210).

This is an efficient way to extract dominant chirp rates when several exist in the analyzed signals. For example, we can consider a two-chirp case defined by (2.95) whose spectrogram was shown in Figure 2.12f. Assuming the SNR = 20 dB, the accelerogram (using (2.210)) and the profile (2.211) are shown in Figure 2.67. Two prominent peaks reveal the chirp rate values and allow signal parameter estimation when the modulation rate is unknown. The salient points appear at the chirp rate of 500 and 1000 Hz/s, respectively, which coincides with the signal model.

Remark 2.16. *The accuracy of chirp rate estimation is also sensitive to the analysis of window parameters. Its efficacy can be enhanced (or inferred) when selecting the window width appropriately (or inappropriately). A combination of the chirp rate estimation with adaptive STFT and its application can be found, e.g., in [74, 75].*

In real-life radar applications, chirp rate estimation can be successfully used for non-cooperative radar signal parameter assessment. The signal whose parameters can be revealed publicly (both due to the civilian nature of the radar and the fact that these parameters are no longer in use due to the radar renovation after signal acquisition) is the nonlinear frequency-modulated radar pulse analyzed several times in this chapter. The accelerogram (using (2.208)) and the corresponding profile (2.211) are shown in Figure 2.68.

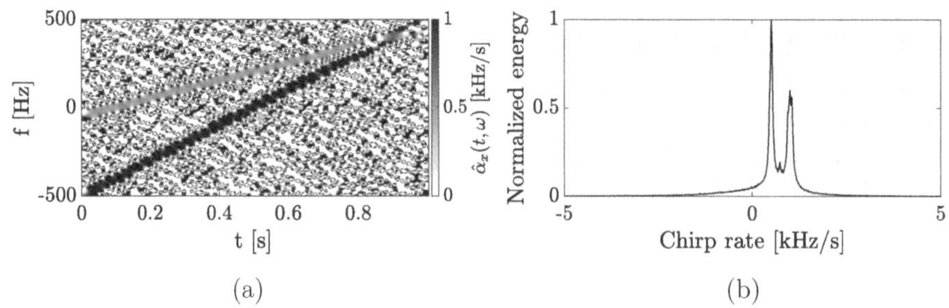

Figure 2.67 Chirp rate estimation results for the simulated signal. (a) Accelerogram of the signal composed of two different chirps; (b) profile (2.211) of the same signal.

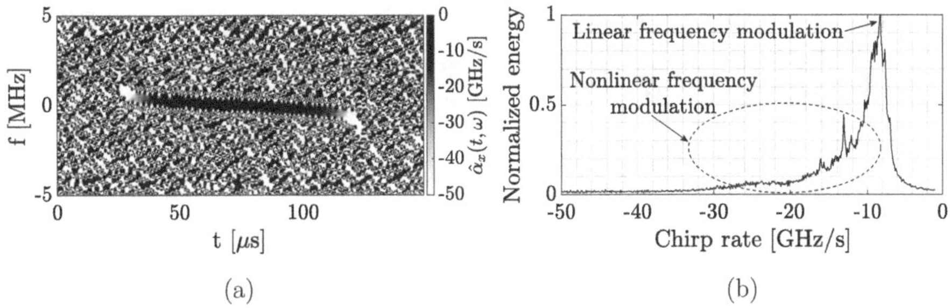

Figure 2.68 Chirp rate estimation of the real-life radar pulse with nonlinear frequency modulation. (a) Accelerogram of the radar pulse with nonlinear frequency modulation; (b) profile (2.211).

The results lead to the following observations. Firstly, the dominant chirp rate equals $\hat{\alpha}_x = -8.26$ GHz/s. This value coincides with the EGCT analysis result in Figure 2.69. Secondly, the profile exhibits a tail from the peak toward approximately -30 GHz/s, which follows from nonlinear FM at the pulse ends, thus making the estimator biased.

The bias can be easily proven using the STFT properties. For example, for the fourth-order signal model with the amplitude and phase given by (2.162) and (2.163), respectively, one has (2.169), which apparently differs from (2.206) by the third- and fourth-order terms. The observation shows why high-order signal parameter estimators were derived. In our previous considerations, we already obtained estimators $\hat{\beta}_x(t,\omega)$ and $\hat{\gamma}_x(t,\omega)$ using the derivatives with respect to frequency. In turn, similar considerations, but using derivatives with respect to time, were presented in [131, 132]. The reader is encouraged to delve into mixed derivatives for analytical practice and to examine these novel estimators' properties. This exercise not only fosters a deeper understanding of the theoretical framework but also allows for profound insights into the estimator's behavior under different conditions.

In practice, high-order estimators are noise-sensitive. Also, numerical instabilities resulting from the complicated nature of the estimators and the need to, e.g., invert matrices, make them less useful than chirp rate estimators. A reasonable

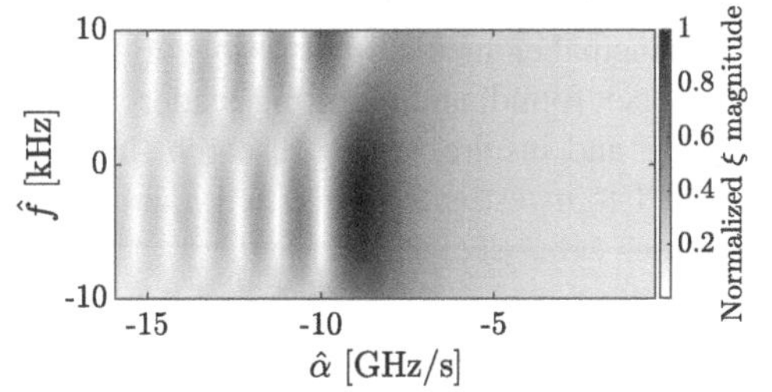

Figure 2.69 EGCT analysis results for the real-life nonlinear frequency-modulated radar pulse.

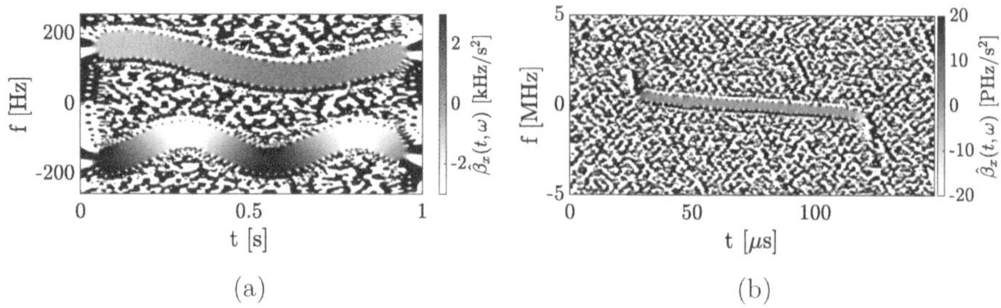

Figure 2.70 Examples of the third-order phase parameter estimation in the time-frequency domain. (a) Simulated two-cosine signal; (b) radar pulse.

maximum phase order suitable for processing in the time-frequency domain reaches 3, in specific very high SNR examples, 4.

Let us close this section with two examples of $\hat{\beta}_x(t,\omega)$ as the illustrative presentation of the high-order signal parameter estimation in the time-frequency domain. The simulated and real-life examples are shown in Figure 2.70. For the simulated case (Figure 2.70a), the result can be confirmed by calculating the third-order phase derivative of the time-domain signal. The precise model is unknown for the real-life case, but one can notice high-value points at the pulse ends, suggesting that higher-order estimators can be used, as $\hat{\beta}_x(t,\omega)$ can still be biased.

In summary, the introduced estimators of the STFT gradient are a compelling framework for nonstationary and multicomponent signal parameter estimation. STFT phase gradient-based methods allow for precise capture of nonstationary signal variability, enabling better representation of their structure. In radar applications, this translates into more accurate target tracking, movement assessment, and signal-to-noise discrimination, which is crucial in dynamic and noisy environments. As presented using both simulated and real-life examples, these techniques can successfully be applied in practical considerations like pulse chirp rate estimation. Also, as shown in previous sections, chirp rate estimate can be used for adaptive chirplet transforms and is part of the instantaneous frequency or group delay estimators (see (2.155) and (2.192)).

So far, our considerations have primarily focused on improving the spectrogram readability or estimating signal parameters in the time-frequency domain. Although these aspects are worth dozens or hundreds of pages of additional descriptions, we exhausted the topic in principal elements. Many publications related to the analyzed problems have recently been found, and inquiring readers will undoubtedly find adequate articles to explore and inspire further research. In the next section, we go back to the classical STFT to investigate its geometry and the fascinating properties behind this transformation.

2.11 STFT AND SPECTROGRAM MORPHOLOGY

So far, we have mainly discussed transforms and methods that allow spectrogram sharpening and component separability improvement. The presented methods were

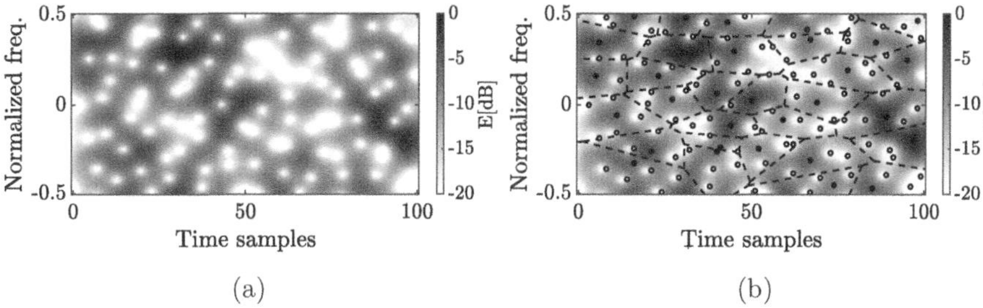

Figure 2.71 Depiction of the complex white Gaussian noise signal spectrogram and its local maxima and zeros. (a) Spectrogram; (b) spectrogram with superimposed local maxima (black dots), zeros (black circles), and Voronoi cells (dashed polygons).

primarily used for signal parameter estimation and sub-signal separation. In this section, we discuss the geometrical features of the STFT and their utilization for radar signal decomposition.

Before moving to radar signals, we can focus on noise to stress out the main spectrogram geometric attributes. The STFT (and the spectrogram as its squared modulus) of noise has local salient points and zeros (see Figure 2.71a). They naturally form local peaks with the magnitude surpassing their vicinity. Zeros, in turn, are points where the spectrogram locally equals 0. In a general view, a local maximum can be interpreted as a single logon – a Gaussian-modulated function used to achieve efficient signal representation with balanced time and frequency resolution [89]. Common impacts create interference and zeros in the middle of each neighboring salient feature. One can notice that zeros surrounding each maxima form polygons of regular shapes, which, by definition, can be defined by a Voronoi tessellation [89, 159, 160]. So-called Voronoi cells with local maxima (black dots) and zeros (black circles) for the corresponding noise signal are depicted in Figure 2.71. It is apparent that the Voronoi cells have edges covering (or in the vicinity of) the structure of zeros, and they form a homogeneous distribution of patches (mostly hexagonal). The maxima distribution can be described using the Poisson process [89] and statistically described [160].

Another, but equally interesting observation stems from the analysis of spectrogram zeros. Voronoi tessellations, based on points attached to spectrogram zeros, are shown in Figure 2.72a. Displayed polygons create similar shapes, mostly hexagonal with edges, however, overlapping local salient. Their cohesion is notably apparent when superimposing the Voronoi cells with the reassigned spectrogram, as presented in Figure 2.72b. Therefore, we see regularity in local maxima and zeros for white Gaussian noise, and the latter are more numerous than the former.

In particular, zeros carry more information about the signal that one can see at first glance [122, 161–163]. Flandrin expressed the STFT as a representation of a complex function $z = \omega + jt$. For the Gaussian window with unit energy defined as

$$h(t) = \pi^{-1/4} e^{-t^2/2}, \tag{2.212}$$

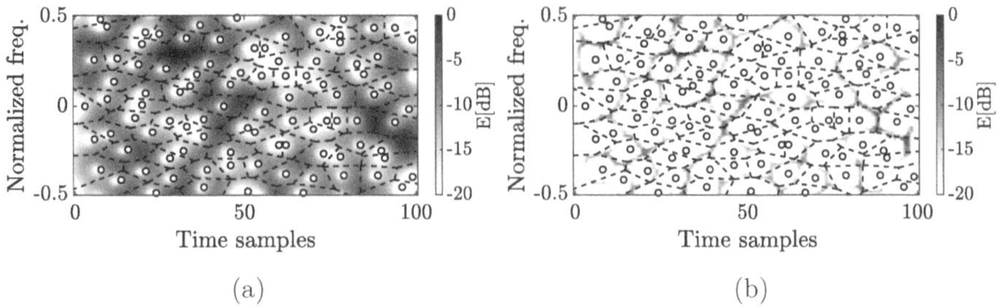

Figure 2.72 Voronoi tessellation based on spectrogram zeros. (a) Spectrogram with superimposed zeros and voronoi cells attached to them; (b) the same as (a) but for the reassigned spectrogram.

the STFT reads as

$$F_x^h(t,\omega) = e^{-\frac{|z|^2}{4}} \mathscr{F}_x(z), \tag{2.213}$$

where $\mathscr{F}_x(z)$ is the Bargmann transform of order 2 [164], permitting the following Weierstrass-Hadamard factorization

$$\mathscr{F}_x(z) = z^m \exp\{P_2(z)\} \prod_{n=1}^{\infty} \left(1 - \frac{z}{z_n}\right) \exp\left\{\frac{z}{z_n} + \frac{1}{2}\left(\frac{z}{z_n}\right)^2\right\}, \tag{2.214}$$

with zeros $z_n = \omega_n + jt_n$ of the Bargmann transform, $m \in \mathbb{N}$ corresponds to a possible m-fold zero at the plane origin, and $P_2(z) = C_0 + C_1 z + C_2^2$ is the second-order complex-valued polynomial with the Bargmann transform zeros defined as $z_n = \omega_n + jt_n$. Additionally, $P_2(z) = C_0 + C_1 z + C_2 z^2$ represents a second-degree complex-valued polynomial.

From the presented contemplations, it follows that the zeros of the STFT (and also the spectrogram) and the Bargmann transform coincide, and the spectrogram characterizes distinct properties of the time-frequency representation and can be determined by their distribution.

Considering the same noise example as previously, we can plot out other geometric attributes of spectrograms. To do so, we can use the Delaunay triangulation [159], which stands out because it maximizes the minimum angles of the triangles, creating a mesh of triangles that are as equilateral as possible. Another defining feature is that no point from the set lies inside the circumcircle of any triangle in the triangulation. If we connect the spectrogram zeros using the Delaunay triangulation, we get the result as in Figure 2.73a, with a triangle stellar composition as in Figure 2.73b. The graph formed by the triangles exhibits a lattice whose edges fall into a given range. The stationarity of the white Gaussian noise results in the homogeneity of the location of the spectrogram zeros, whereas the appearance of any coherent structure (in this case - any signal) violates this assumption. This effect is apparent in Figures 2.73c and d, where the linear chirp “disturbs” randomness of noise. The evident elongation of triangles’ edges enables leveraging this feature for signal decomposition.

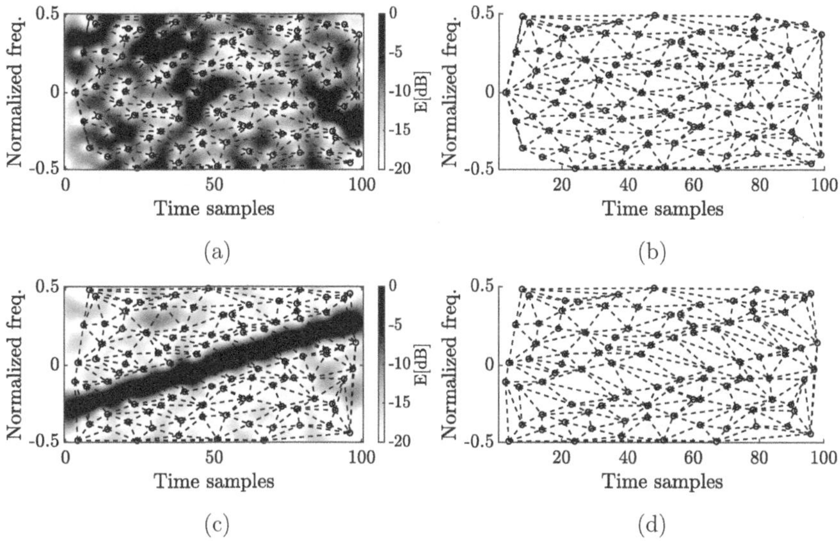

Figure 2.73 Distributions of zeros and their Delaunay triangulation. (a) Spectrogram with superimposed zeros and their triangulation; (b) a mesh composed of triangles for noise; (c) noise plus chirp – zeros and their Delaunay triangulation; (d) a stellar for the noise plus chirp case composed of zeros triangulation.

Flandrin observed that for white noise, the Euclidean distance between adjacent zeros is bounded by $L_{\max} \sim 2.16$, with a probability of exceeding 2 being 10^{-3}. Any triangle edge exceeding this threshold covers the spectrogram fraction putatively containing a signal other than Gaussian noise. Since spectrogram zeros determine the signal distribution on the time-frequency plane, triangles with at least one edge longer than $L_{\max}$ can be used for signal retrieval. As such, in the algorithm proposed in [161], adjacent spectrogram zeros are connected using Delaunay triangulation, forming triangles with edge lengths in the range $L_{\min} < |e_{m,n}| < L_{\max}$ for white noise, and outside this extent in the presence of a signal.[4] Based on triangle edge length constraints, only those triangles with at least one edge length outside the noise range are selected. Neighboring triangles with edges beyond the noise length range are connected, defining a binary mask (1 for detected components, 0 for noise). Applying this binary mask to the original STFT distribution selects the components, which are then reconstructed using the inverse STFT defined by (2.80).

The practical utilization of this method in radar applications can be illustrated using the chirp disturbed by multipath propagation. Pulse retrieval to its pure form, void of noise and fluctuations resulting from reflections from stationary obstacles, is shown in Figure 2.74.

Firstly, we see a mesh superimposed on the spectrogram in Figure 2.74a. As can be observed, the presence of a coherent structure in the form of a strong linear chirp results in the elongation of the edges of the triangles above a given threshold. The

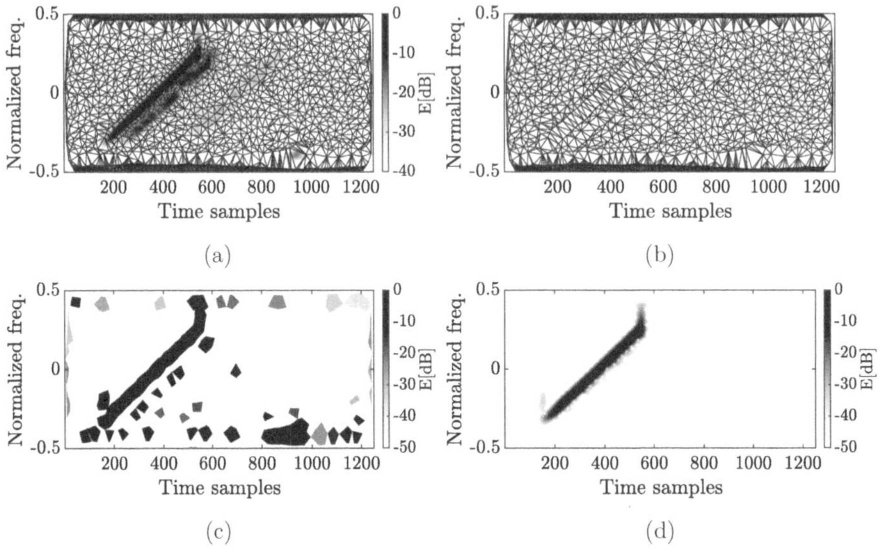

Figure 2.74 Depiction of the Delaunay triangulation-based real-life radar signal retrieval. (a) Spectrogram of the radar pulse disturbed by multipath interference with a triangulation stellar superimposed; (b) mesh of triangles; (c) selected triangles fulfilling $L_{\min} > |e_{m,n}| > L_{\max}$ condition on the triangle edge length; (d) spectrogram after filtering.

effect can be more distinguishable in Figure 2.74b, where only the mesh is depicted. Entangled triangles satisfying the imposed length of the edges allowed the signal domains to be extracted. Each detected domain is displayed in Figure 2.74c. The collection of entangled triangles with the highest energy (black color) was used for a binary mask design. The mask, whose structure was already discussed with vertical synchrosqueezing, is a binary distribution with the value of 1 for any (t, ω) instant containing the signal we want to extract and 0 otherwise. In our case, the mask precisely covers the strongest domain in Figure 2.74c, which corresponds to the radar pulse. The signal was filtered, and its spectrogram is shown in Figure 2.74d.

The filtered signal lacks noise and multipath copies. The extracted pulse is pure, as shown in the time-domain plots in Figure 2.75. The outcomes are pretty much similar to those obtained for synchrosqueezing (for comparison, see Figure 2.49). In both cases, disturbances vanish, giving a pulse with all its phase-magnitude relations.

The displayed example reveals how the inherent geometry of the spectrogram can be used for signal decompositions. For lower SNRs, the efficacy drops, but some remedies have been proposed for robust signal retrieval [122, 165]. These works draw a perspective for autonomous utilization of spectrogram zeros for signal extraction. The spectrogram triangulation-based approach is highly informative and leverages extractions of various signal types as a completely different signal decomposition method compared to those discussed so far. This situation is contrastive to vertical synchrosqueezing, where the less FM, the better the reconstruction quality.

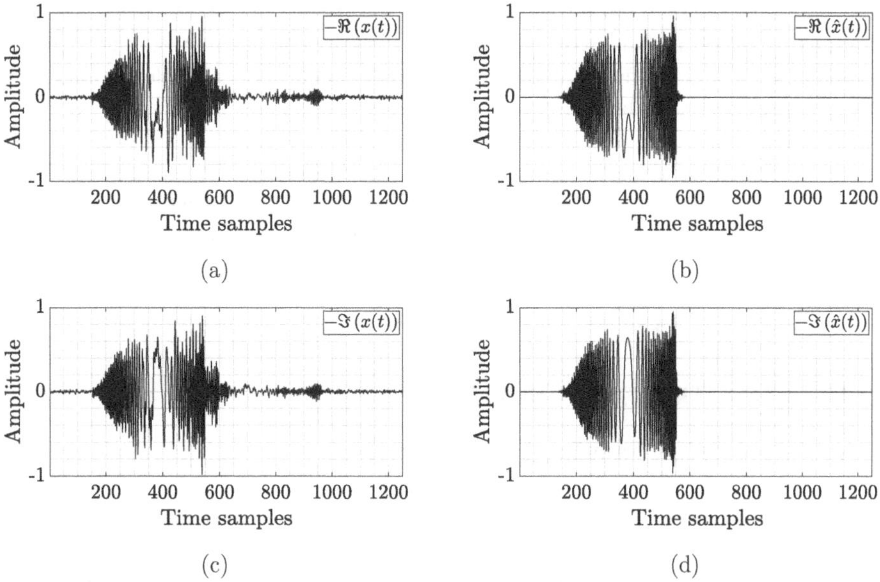

Figure 2.75 Time-domain signal filtering results. (a) Real part of the original signal; (b) real part of the filtered signal; (c) imaginary part of the original signal; (d) imaginary part of the filtered signal.

Since spectrogram zero triangulation does not favor any signal type *a priori*, one can quickly recover transients, chirps, and oscillations.

2.12 SUMMARY

This chapter has introduced several basic signal analysis methods. Starting from time-domain signals, we passed through the frequency analysis to conclude that it is a powerful tool for spectral characteristic determination. Further, we introduced the time-frequency concept, which is superior for nonstationary and multicomponent signal analysis. Next, we discussed different STFT variants whose adaptivity improved component separability. Further considerations led to reassignment and synchrosqueezing. While the former is non-invertible but the most efficient in energy concentration on the time-frequency plane, the latter admits signal decomposition and filtering. Signal parameter estimation was the next topic described, and it revealed a meaningful opportunity to assess the phase gradients on the time-frequency plane. Finally, we evaluated the spectrogram as a geometric identity and formulated an efficient signal decomposition tool. All these methods were practically used in radar applications and abundantly examined using real-life examples, and most of them are available under an open-source license.

The presented methods certainly do not cover all possible aspects of signal-processing in the time-frequency domain. However, as indicated by the book's title, they are an excellent introduction, outlining the horizons, possibilities, and limitations of STFT and related techniques. The ability to decompose and estimate parameters of multicomponent and nonstationary signals is a tangible advantage of these approaches, and the author hopes that readers will make abundant use of these concepts.

In radar systems and other applications, multidimensional signals are often encountered. For instance, in multi-antenna systems, each antenna receives a one-dimensional signal (time series), which can be organized in matrix or spatial structures. Consequently, signal analysis becomes multidimensional, a topic explored in Chapter 3.

2.13 CHAPTER IN A NUTSHELL

The most important information from this chapter:

- The Fourier transform is a powerful tool for spectral analysis. It allows for a deep insight into the frequency content of any signal.
- Nonstationary and multicomponent signals may require more sophisticated approaches like time-frequency signal analysis; the most prominent example is the abundantly advocated STFT.
- The STFT (and spectrogram – its squared magnitude) suffers from the Heisenberg-Gabor uncertainty principle, which constrains component separability.
- The STFT readability can be reinforced using optimized windows that fit local signal parameters.
- To further improve the spectrogram readability, one can use time-frequency reassignment and synchrosqueezing. The former works better when the goal is the energy concentration, while the latter is invertible and allows signal decomposition.
- From the STFT invertibility, signal decomposition algorithms can be deduced. Signal retrieval from its multicomponent mixture is a beneficial STFT property.
- When struggling with real-life signals, noise is one of the crucial limitations coming from the environment and the local oscillator. Multitaper processing can be employed to alleviate its influence, but it needs to be remembered that it is not a noise removal method, but rather the technique that allows for its smearing.
- The presence of multiple signal components can severely hinder the accurate estimation of individual parameters. Time-frequency analysis offers a solution by enabling the extraction of each phase component, even when the signal is a complex mixture of distinct sub-signals.
- STFT and the accompanying spectrogram have interesting geometric properties. Zeros of the spectrogram determine the signal being analyzed, paving the way for efficient signal denoising and decomposition.

2.14 EXERCISES

1. Generate a real sinusoidal signal with unit amplitude and frequency 100 Hz and plot the magnitude and phase of its spectrum. Then generate a complex sinusoidal signal with the same amplitude and frequency and compare the spectra.
2. Implement your own version of the FFT using the radix-2 algorithm.
3. Generate the magnitude spectrum of an amplitude-modulated signal. Use a harmonic signal as the carrier and another harmonic signal with a different frequency as the modulating signal. Observe how the spectrum differs from that of a simple harmonic signal.
4. Generate the magnitude spectrum of a frequency-modulated signal (e.g., the radar pulse generated in the exercises at the end of Chapter 1). Compare the results with the harmonic signal and the amplitude-modulated signal.
5. Write your own implementation of the STFT using any analysis window.
6. Write a procedure that computes the STFT using a filter bank. Prepare a filter bank for all analyzed frequencies and filter the signal through each filter. Compare the results with the Fourier transform-based implementation. Analyze the influence of filter length on the results.
7. Compute the spectrogram of the signals generated in Exercises 3 and 4. Compare the results with the spectrogram obtained using the toolbox at `https://github.com/kabratkiewicz`.
8. Using your implementation from Exercise 5 or the toolbox implementation, analyze how the spectrogram changes for different window widths when analyzing a signal with linear FM. For each window width, compute the third-order Rényi entropy and plot the results. Change the modulation rate and repeat the analysis. Add Gaussian noise with different levels and compare the results.
9. For a signal with linear FM, select the window chirp rate and width so that the spectrogram energy is maximally concentrated. Repeat the task for a sinusoidally frequency-modulated signal. Verify whether the observations are consistent with the Rényi entropy values.
10. Given an analytical expression for any window (e.g., the Gaussian window or the Hamming window), derive formulas for the first-order and second-order derivatives. Implement a procedure that generates the window, its derivatives, the window multiplied by first-order and second-order time ramps, and mixed cases where the derivative is multiplied by a time ramp.
11. Compute and plot the spectrograms for the windows generated in Exercise 10.
12. Using the spectrograms for the modified windows, compute the reassignment operators and obtain the concentrated spectrogram.
13. Using the Rényi entropy of the concentrated spectrogram, evaluate how the window width affects concentration quality. Compare these results with the entropy computed for the classical spectrogram.
14. Generate two identical signals with linear FM. Compute the spectrogram. Vary the center frequency of one signal so that the two components

gradually separate in frequency. Observe at what point they can be separated. Repeat the experiment by shifting one signal in time. Evaluate the effect of different noise levels and window widths. Test an optimal window minimizing the Rényi entropy.

15. Repeat the analysis from Exercise 14 using the concentrated spectrogram.
16. Using the script `run_multitaper.m` from the author's repository, plot the averaged spectrogram of noise for different numbers of tapers. Observe what happens when the number of averaged spectrograms tends to infinity.
17. For any signal with sinusoidal FM, compute the spectrogram using the first four orthonormal Hermite windows. Observe the relation between the order of the window and the number of zeros along the instantaneous frequency curve.
18. Derive formulas for time-frequency estimators of parameters of a signal whose amplitude and phase are given by a fourth-degree polynomial, using time derivatives of the STFT.
19. Generate a linearly frequency-modulated signal. Using the script `run_est_params.m` from the author's repository, estimate its frequency modulation coefficient. Analyze the estimation accuracy for different Gaussian noise levels. Compare the results with those based on Rényi entropy and different window widths.
20. Generate a signal with sinusoidal FM with large oscillations (similar to Figure 2.54a) and add small noise. Using the scripts `run_HSS.m` and `run_VSS.m` from the author's repository, perform horizontal and vertical synchrosqueezing. Compare higher-order synchrosqueezing results and determine which method yields the lowest Rényi entropy. Compare with the reassignment technique.
21. Using the script `run_VSS_decomposition.m` from the author's repository, decompose the signal into components. Analyze how the mask size affects component extraction quality. Apply the same masks to the non-concentrated STFT and evaluate the results.

NOTES

1 These terms have already been used in this book without appropriate explanation, e.g., in (1.22).

2 The radio receiver used to record the signal down-modulated the received wave to sample it in the baseband with a narrower bandwidth than it would be necessary to digitize the signal in the high-frequency range.

3 This configuration is available for passive and bistatic (and multistatic) active radars whenever the transmitter and the receiver are dislocated.

4 In practice, $L_{\min}$ also needs to be settled as it facilitates appropriate triangle merging at the spectrogram boundaries or in the case of several component entanglements.

3 Analysis of Two- and Multidimensional Radar Signals

3.1 INTRODUCTION

Many modern radar systems rely on processing signals that vary in two dimensions: range-Doppler or azimuth-elevation representations. Unlike one-dimensional signals, which are typically functions of time or frequency, two-dimensional signals encapsulate more complex structures and relationships that arise in radar imaging, synthetic aperture radar (SAR), and multidimensional spectral analysis.

One of the fundamental tools for analyzing such signals in the frequency domain is the **two-dimensional Fourier transform**, which extends the conventional Fourier transform described in Chapter 2 to functions of two variables. This transformation plays a crucial role in radar applications by allowing the decomposition of signals into their frequency and spatial-frequency components, facilitating target detection, imaging, and motion estimation.

This chapter explores the mathematical foundations of the 2D Fourier transform and its various modifications relevant to radar signal-processing. Several extensions, including the windowed two-dimensional Fourier transform (W2D-FT) and adaptive Fourier-based techniques, are discussed in the context of improving resolution, mitigating artifacts, and handling nonstationary signals.

Understanding these techniques allows one to analyze and interpret radar data in range-Doppler processing, SAR image formation, and other multidimensional radar applications effectively. The concepts presented here will provide a foundation for advanced radar signal processing methodologies and their practical implementations.

A typical example of a 2D signal is an image, where intensity values are defined over spatial coordinates, e.g., (x,y). Similarly, in radar applications, 2D signals emerge in range-Doppler processing, where the received signal is represented as a function of time (or range) and frequency (or velocity). A 1D signal can be mathematically expressed as $x(t)$, where t is a single independent variable. In contrast, a 2D signal is typically denoted as $x(t,r)$, indicating its dependence on two distinct parameters. This fundamental difference necessitates extending traditional signal-processing tools, such as the Fourier transform, into two dimensions to analyze and interpret these signals effectively.

DOI: 10.1201/9781003765783-3

Two-dimensional signals can be found in radars in the following applications:

- **Range-Doppler (or Range-Velocity) Processing**: a typical approach to radar signal processing is the consecutive pulse organization in a two-dimensional matrix. As such, the first dimension corresponds to each pulse (in particular, the echo from each pulse), while consecutive pulses form the second. Analysis of time and frequency dependencies in such a two-dimensional distribution allows target range and velocity estimation.
- **Range-Time Processing**: similar to range-velocity processing, but without velocity (Doppler) estimation. Instead, one can analyze target range variability over time.
- **Radar Imaging**: SAR systems allow imaging of surfaces, e.g., the Earth or the Moon. Radars are usually mounted on a moving carrier (e.g., satellite, airplane), and the signal illuminates the scene during the flight. Akin to optical images, SAR systems provide a two-dimensional distribution. The difference is that radar images display the reflectivity of each scanned object on the illuminated scene. There is also another type of imaging system called inverse synthetic aperture radar (ISAR). Both SAR and ISAR work similarly, with the only difference being which element of the scene is stationary and which is moving. In ISAR, the radar is stationary, and moving targets are imaged. Notwithstanding, the final image also has a two-dimensional signal distribution, often resulting from a two-dimensional Fourier transform.
- **Multi-Antenna Systems**: when analyzing multi-antenna systems, we usually have spatially distributed antennas, each collecting the signal. Their joint analysis comes down to the distribution of the signal in a two-dimensional space. As such, the first dimension corresponds to time (signal in each antenna), and the second refers to the space (antennas are dislocated).

Through this chapter, we pay attention to Fourier-based techniques for signal and target parameter estimation in such applications. Firstly, we introduce multidimensional radar signals in Section 3.2 using the example of a high-resolution micro-Doppler signature represented in the three-dimensional time-frequency-space domain. Also, the spatial spectrogram is introduced. Next, we incorporate the vertical synchrosqueezing-based mode extraction to filter out noise from ISAR images. The idea is delineated in Section 3.3. In the next step described in Section 3.4, we delve into the two-dimensional Fourier transform theory and examples. We define the basic properties of the two-dimensional Fourier transform and provide some practical examples. Next, in Section 3.5, we derive the explicit formulas for bivariate spectrum processing from the one-dimensional signal, which is widely used in frequency-modulated continuous-wave radars. Sections 3.6 and 3.7 introduce the alternative of the STFT dedicated to two-dimensional signals and its adaptive variants. The resulting distribution is four-dimensional, with the complex values distributed in time, space, frequency, and spatial frequency. Next, Section 3.8 describes the variants of the four-dimensional energy distribution concentration from the previous section. Further, in Section 3.9, we introduce the method for two-dimensional signal decomposition using the windowed two-dimensional Fourier transform. The presented methodologies were validated using different radar data, such as aircraft images. We end up in Section 3.10 with the multitaper spectrum estimation for noise flattening and simultaneous energy concentration for better target visualization and noise suppression.

3.2 SPATIAL MICRO-DOPPLER SIGNATURE

Before introducing the 2D Fourier transform, let us introduce the spatial spectrogram as an extension of the classical one introduced in Chapter 2. In the ordinary STFT, we deem the signal to be one-dimensional such that $x(t) \in \mathbb{C}$. In micro-Doppler analysis, such an implicit assumption sometimes cuts off valuable information about the signal. This statement does not mean that the analysis presented so far was incorrect. Still, it needs to be stressed that we can extract more signal attributes, which, in the micro-Doppler analysis, come down to a precise target movement definition. A typical micro-Doppler extraction algorithm is delineated in Figure 3.1. Note that this scheme presents the pipeline for the frequency-modulated continuous-wave radar. In other radars, the range-compressed signal can be obtained differently (e.g., matched filtering).

The signal processing workflow can be outlined as follows:

1. **Signal Reception**: The transmitted radar signal propagates through space, reflects from objects, and returns to the radar receiver.

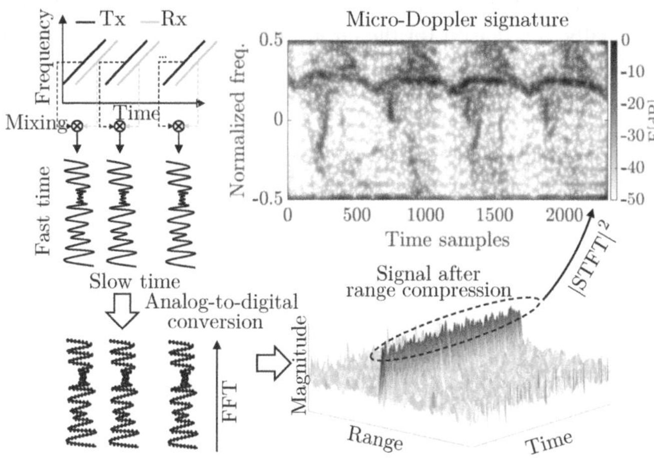

Figure 3.1 Micro-Doppler signature extraction scheme using the frequency-modulated continuous-wave radar.

2. **Analog Mixing**: The received signal is mixed with the transmitted waveform, producing an intermediate frequency signal. This intermediate frequency signal contains essential information about the target's range and Doppler frequency shift.
3. **Intermediate Frequency Signal Digitization**: The intermediate frequency signal undergoes low-pass filtering to get rid of unwanted high-frequency components. It is then sampled at a rate meeting the Nyquist criterion, ensuring no aliasing occurs. The sampled values are subsequently quantized and represented digitally.
4. **Time-Domain Data Organization**: The digitized intermediate frequency signal is structured into a data matrix, where each column corresponds to a single sweep, referred to as fast time. Successive sweeps are aligned along the slow time axis.
5. **Fast Fourier Transform**: A fast Fourier transform is applied to each column of the data matrix (each chirp) along the fast time axis, converting the data from the time domain to the frequency domain. The resulting beat frequency provides information about the target range.
6. **Micro-Doppler Signature Extraction**: A specific range cell (matrix row) is selected for time-frequency analysis to extract the micro-Doppler signature from the range-compressed data matrix. The STFT and spectrogram defined by (2.78) and (2.81), respectively, are used for obtaining the micro-Doppler signature.

The main capability of a frequency-modulated continuous-wave radar system is a range and velocity measurement resulting from the functionality pointed out in Figure 3.2. Typically, the radiated waveform is a continuous linear chirp with a high-frequency carrier (in most cases, a dozen MHz up to hundreds of GHz). The frequency band determines the maximum range (which degrades with increasing frequency) or the penetration ability of some materials (which increases with decreasing frequency). Usually, the reason for the choice of operating frequency is dictated by

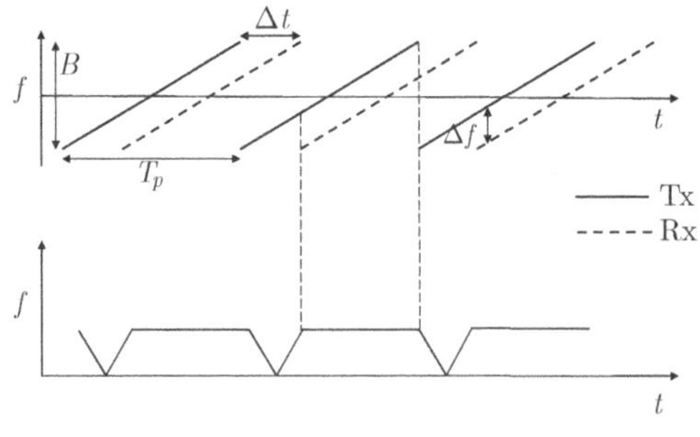

Figure 3.2 A general principle of operation of the frequency-modulated continuous-wave radar. A solid line is the transmitted pulse's instantaneous frequency (Tx), and the dashed line illustrates the instantaneous frequency of the received wave (Rx).

the system's requirements. In fact, the modulation character defines how the target range is measured, which is depicted in Figure 3.2. The received signal (dashed line) is delayed relative to the transmitted (solid line) by a time difference

$$\Delta t = \frac{2R}{c}, \tag{3.1}$$

where R is the range from the radar to the target, and c is the propagation velocity (approximated as the speed of light). The total distance is twice as large because the signal travels from the antenna to the target, reflects off it, and returns along the same path. Since the signal frequency increases linearly, the time delay of the echo relative to the transmitted signal results in a frequency difference between the two waveforms, governed by the following relationship:

$$\Delta f = \frac{2RB}{cT_p}, \tag{3.2}$$

where B stands for the signal bandwidth and T_p its period. In the radar nomenclature, the frequency expressed by (3.2) is known as a beat frequency. In each sweep (each single waveform repetition), the beat signal (provided by mixing the transmitted and received echo and low-pass filtering) is extracted, producing particular information about the range of the target. The radar parameters also determine another important system feature, namely range resolution (1.14).

The range information is directly associated with the beat signal frequency, which, in turn, can be easily estimated using the Fourier transform. In (3.2), B, c, T_p are known, and Δf is estimated, which allows the range to be distinguished using a simple conversion

$$R = \frac{\Delta f c T_p}{2B}. \tag{3.3}$$

This is the main difference between frequency-modulated continuous-wave and pulse radars, namely, the range is computed not through the delay of the received pulse regarding the transmitted one but by using the frequency dependency between these signals. As presented, in the frequency-modulated continuous-wave radar, the range is computed using the Fourier transform based on the (relatively low frequency) beat signal, which does not require fast analog-to-digital converters. The process can be accelerated in the discrete-time domain using its relatively fast implementation – the FFT. Such circumstances allow more bits and higher signal dynamics to be obtained. In pulse radars, the range compression assumes a fast analog-to-digital converter since the received pulse is stored in a full bandwidth. Thus, to fulfill the Nyquist theorem (the sampling rate has to be larger than doubled signal bandwidth), several dozen MSa/s to GSa/s are usually required, which typically results in limiting analog-to-digital converter parameters (resolution, the effective number of bits, etc.) and may prevent the detection of small objects.[1] Now, we take a closer look at the frequency-modulated continuous-wave radar signals.

Let us consider a complex transmitted signal model given by

$$x_T(t) = A_T(t) e^{j\left(\phi_t + \omega_c t + \frac{\alpha t^2}{2}\right)}, \quad 0 < t < T_p, \tag{3.4}$$

where $A_T(t)$ represents the amplitude, modeled as a time-varying function to account for real-world conditions in which the radio front-end affects the envelope of the transmitted signal. The parameter T_p denotes the sweep duration, ϕ_t is the initial phase, $\omega_c = 2\pi f_c$ is the angular frequency corresponding to the initial frequency f_c, and α is the chirp rate. The signal described in (3.4) is referred to as a *sweep* and is periodically transmitted by the radar.

For a stationary single-point target, the received signal is given by

$$x_R(t) = A_R(t-\tau)e^{j\left(\phi_r+\omega_c(t-\tau)+\frac{\alpha(t-\tau)^2}{2}\right)}, \tag{3.5}$$

where $A_R(t)$ represents the envelope of the received signal, delayed according to the target distance and scaled by its radar cross-section [2]. The term ϕ_r denotes a constant phase of the reflected signal, resulting from propagation effects and target reflectivity. The transmitted and received signals, $x_T(t)$ and $x_R(t)$, are mixed in the receiver, yielding

$$\begin{aligned} x_s(t) &= x_T^*(t)\cdot x_R(t) \\ &= A_T^*(t)A_R(t-\tau)e^{j\phi_0 - j\frac{\alpha t^2}{2} + j\frac{\alpha t^2 - 2t\tau\alpha + \alpha\tau^2}{2} - j\omega_c\tau} \\ &= A_T^*(t)A_R(t-\tau)e^{j\left(\phi_0+\frac{\alpha\tau^2}{2}-\frac{2t\tau\alpha}{2}-\omega_c\tau\right)}, \end{aligned} \tag{3.6}$$

where the subscript s stands for *stationary*. The term $\phi_0 = \phi_r - \phi_t$ represents the initial phase difference between the transmitted and received signals.

Assuming that the envelope term $A_T^*(t)A_R(t-\tau)$ in (3.6) is denoted as $A_s(t)$, representing the amplitude of the mixed signal, the whole expression simplifies to

$$x_s(t) = A_s(t)e^{j\left(\phi_0-\omega_c\tau+\omega_b t+\frac{\alpha\tau^2}{2}\right)}, \tag{3.7}$$

where $\omega_b = \tau\alpha = \frac{2R\alpha}{c}$ is the beat angular frequency, and R is the distance between the radar and the target. The Fourier transform of $x_s(t)$ will give a peak at ω_b, and this is how the range is estimated in frequency-modulated continuous-wave radars. It can be written as

$$R = \frac{\hat{\omega}_b f c T_p}{2B}, \tag{3.8}$$

where

$$\hat{\omega}_b = \operatorname{argmax}\left(|\mathscr{F}\{x_s(t)\}|\right). \tag{3.9}$$

When noise or several targets exist (hardly every radar signal is noiseless and contains the echo from a single target), more advanced detection algorithms than the maximum extraction should be applied.

The received signal undergoes a Doppler phase shift for moving targets dependent on the wavelength λ. If the target moves with velocity V, the received signal is modulated according to the Doppler angular frequency

$$\omega_D = 2\pi f_D = -2\pi\frac{2V}{\lambda}. \tag{3.10}$$

Thus, (3.7) can be rewritten as

$$x_m(t) = A_s(t)e^{j\left(\phi_0 - \omega_c\tau + (\omega_D - \omega_b)t + \frac{\alpha\tau^2}{2}\right)}, \tag{3.11}$$

where the subscript m denotes *moving* (we simply assume the amplitude to be unaltered). The signal given in (3.11) is then processed in the receiver.

Remark 3.1. *At first glance, frequency components ω_D and ω_b in (3.11) can be challenging to separate and may bias each other. In practice, the Doppler shift is significantly smaller than the frequency related to the target delay $\omega_D \ll \omega_b$. Thus, the range can be easily estimated using a frequency analysis of a single sweep. At the same time, several pulses allow for velocity estimation, which stems from the phase drift from sweep to sweep. Thus, in practice, ω_D does not affect ω_b significantly in a single sweep.*

Signals analyzed so far were processed under a prerequisite of occupying a single range cell, which cannot always be set to true [166]. If we look at Figure 3.3, we see a monostatic radar receiving the signal from a walking human. Having the radar working with a bandwidth of B, the monostatic range resolution (recall (1.14)) is defined as

$$\Delta R = \frac{c}{2B}. \tag{3.12}$$

Sufficiently high bandwidth allows high-resolution scene projection to be obtained. Namely, the echo spans several range cells if we have a target (e.g., a walking individual) whose spatial spread is higher than the range resolution ΔR. The effect of several range cell occupation by a single target is graphically demonstrated in Figure 3.3. Such a situation occurs for high-resolution radars, and embracing a limb span of 1 m, the monostatic radar bandwidth of $B = 300$ MHz gives the range resolution $\Delta R = 0.5$ m. Such radar parameters are not challenging nowadays.

We must consider the received signal comprising additional dimensions to extract spatial micro-Doppler signatures. As such, we need to analyze the space surrounding the target, which schematically delineates Figure 3.4. The example of a walking person illuminated by a high-resolution radar (which allows the human echo to occupy

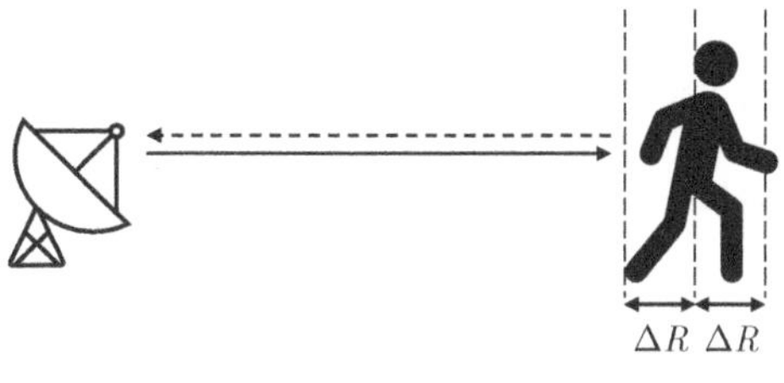

Figure 3.3 Micro-Doppler feature extraction for different range cells. In the displayed example, the radar range resolution permits the extraction of the micro-Doppler signature from each range call. As a result, we obtain a more informative distribution of the target's movement.

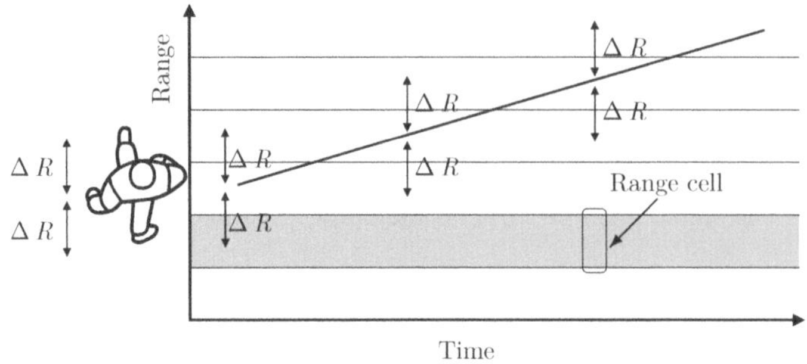

Figure 3.4 Simplistic range-time distribution of a walking human when the radar resolution is smaller than the limb span.

several range cells) demonstrates the range-time distribution. As shown in Figure 3.1, this distribution is obtained by applying the Fourier transform to each chirp.

Remark 3.2. *The range-time distribution can be regarded as the STFT without window overlap for the frequency-modulated continuous-wave radars. It follows from the fact that Δf in (3.2) is proportional to the beat frequency of the mixed echo signal. Its frequency evolution in time can be seen as the time-frequency distribution.*

To capture spatial dependencies in micro-Doppler analysis, let us define the signal as a two-dimensional distribution of complex values $x(t,r) \in \mathbb{C}$, where t is time and r corresponds to the range (space) and embraces spatial coordinates where some parts of the target can be apparent (see Figure 3.4 and range cells surrounding the main target echo). The spatial STFT can be defined as [166]

$$F_x^h(t,r,\omega) = \int_{-\infty}^{\infty} x(\tau,r)h^*(\tau-t)e^{-j\omega(\tau-t)}\mathrm{d}\tau = \int_{-\infty}^{\infty} x(\tau+t,r)h^*(\tau)e^{-j\omega\tau}\mathrm{d}\tau, \quad (3.13)$$

with the spatial spectrogram defined as

$$E_x^h(t,r,\omega) = \left|F_x^h(t,r,\omega)\right|^2. \quad (3.14)$$

As a result, we obtain a spectrogram for each r and, consequently, a 3D energy distribution. It allows for capturing distinctive target features unavailable in a simple time-frequency analysis and hidden in neighboring range cells. Thanks to the additional dimension, we can observe how the signal components vary in time, frequency, and space.

The benefits of using the spatial spectrogram can be demonstrated using a hand gesture obtained using a high-resolution radar. The XY-DemoRad system, described in Chapter 2, has exchangeable front-ends that allow carrier frequency choice. In the experiments presented so far, we used the carrier frequency of $f_c = 24$ GHz. In this experiment, we switched to $f_c = 122$ GHz leading to the bandwidth increase to

$B = 4$ GHz, giving $\Delta R = 3.75$ cm, and the sweeps were sent with the repetition rate of $f_p = 2$ kHz. Such a centimeter resolution admits observation of subtle movements but simultaneously hinders the analysis of movements taken beyond the observed range cell. We can observe the whole gesture if we do not fix r but include the space in our analysis, expanding $x(t)$ to $x(t,r)$.

In the experiment, the person performing the gesture stood approximately 2 m away from the radar antenna. While standing upright, he raised his hand and executed a 'swipe down' motion with his palm. After completing the gesture, he returned his hand to the resting position. In post-processing, 30 range cells were selected for the analysis, corresponding to approximately 1.1 m of the space. The outcomes are presented in Figure 3.5.

The result illustrates different projections of the gesture's spatiotemporal characteristics. Unlike typical spectrograms, the intensity in these plots does not represent energy but the distance of the reflected signal. The visualization has been scaled to suppress noise, showing solely the most prominent elements of the signature. The following observations follow from the visualization:

- **3D Time-Frequency-Space Representation (Figure 3.5a)**: This three-dimensional plot depicts how the micro-Doppler signature evolves across time, frequency, and space. The distinct trajectory reflects the movement of the hand, where the initial upward motion and subsequent downward swipe create characteristic patterns in the time-frequency-space domain.

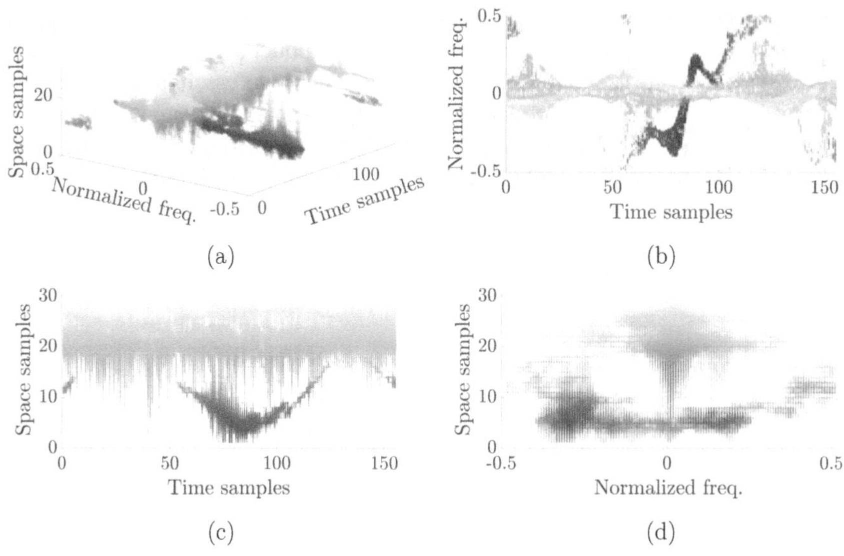

Figure 3.5 Spatial spectrogram as an example of micro-Doppler high-resolution gesture analysis. (a) 3D spatial spectrogram illustrating time-frequency-range dependencies; (b) time-frequency view; (c) time-space view; (d) frequency-space view.

- **Time-Frequency Projection (Figure 3.5b)**: This plot captures the time-frequency evolution of the signal. The curved structure corresponds to the changing velocity of the hand, where the frequency shift indicates acceleration and deceleration phases during the gesture. This view is compatible with the signal's spectrogram, whose r dimension is flattened by integrating the spectrogram for all r.
- **Time-Space Projection (Figure 3.5c)**: The projection onto the time-space plane highlights the spatial movement of the hand throughout the gesture. The downward curvature in the middle of the plot aligns with the uprising of the hand, showing how the reflection moves through space over time. The whole span shows that the gesture occupies approximately 0.5 m.
- **Frequency-Space Projection (Figure 3.5d)**: This plot represents how different spatial components of the reflected signal correspond to specific frequencies. The spread in the frequency axis reflects the Doppler shift introduced by the varying speed of the hand during the gesture.

These results illustrate the utility of spatiotemporal micro-Doppler analysis in capturing fine details of human motion, which can be crucial for gesture recognition in radar-based sensing applications, especially observing large variability in space using high-resolution radars.

Worth noting is that (3.13) is fully invertible and admits utilization of all of the demonstrated STFT and spectrogram readability improvement methods. For example, spatial component decomposition can allow the extraction of the entire hand movement and its separation from stationary obstacles.

So far, we have thoroughly examined the fundamental aspects of micro-Doppler signal analysis, covering key concepts such as time-frequency and time-frequency-space representations. These foundational methods provide essential tools for understanding the micro-Doppler effect and its applications in radar signal processing. Moving forward, we will explore more advanced object detection and imaging challenges. These topics involve sophisticated signal-processing techniques to enhance target recognition, improve resolution, and mitigate interference. Extending our analysis beyond individual micro-Doppler signatures, we will address complex scenarios where robust detection and high-fidelity imaging are crucial.

3.3 ISAR IMAGE FILTERING

Imaging radar systems have been pivotal in remote sensing, surveillance, and military applications since their inception [167]. The earliest developments in radar imaging date back to World War II, with the advancement of SAR techniques in the 1950s and 1960s [8]. These advancements enabled high-resolution imaging of terrain and objects beyond the capabilities of conventional optical systems.

Imaging radars operate by transmitting electromagnetic waves and processing the received echoes to construct a detailed scene representation. Their principle of operation comes down to an inverse problem where, based on the received signal, the

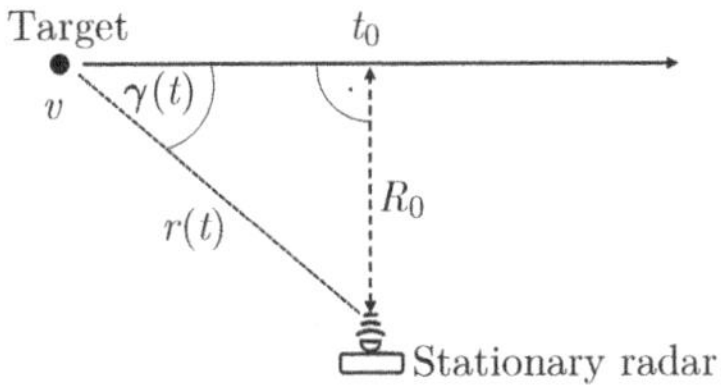

Figure 3.6 A simplified ISAR system geometry.

illuminated terrain is being reconstructed. Unlike traditional radar systems primarily providing range and Doppler information, imaging radars leverage sophisticated signal-processing techniques to achieve satisfactory spatial resolution [168].

A fundamental principle behind imaging radars is the use of aperture synthesis. SAR systems achieve high-resolution by coherently combining echoes received over a moving platform, effectively synthesizing a large aperture [169]. On the other hand, ISAR exploits target motion to generate high-resolution images of moving objects [170].

Signal processing is essential in imaging radar systems. Raw radar echoes undergo several processing stages, including range compression, Doppler processing, and phase correction, to reconstruct a cohèrent image [171]. Advanced techniques, such as phase gradient autofocus and motion compensation, further enhance image clarity and resolution.

This section focuses on ISAR imaging and leverages the STFT-based vertical synchrosqueezing to reduce radar image noise. To begin, we outline the fundamental principles of ISAR and assume a simplified system geometry, as depicted in Figure 3.6. The stationary radar system illuminates the scene in front of the antenna. The target (simplistically presented as a black point) moves perpendicularly to the radiation direction with a velocity v. This geometric model is particularly relevant because the received radar signal consists of two key components

$$x_r(t) = A_x(t) \cdot \exp\left(j2\pi t \left(\underbrace{\frac{2r(t)B}{cT}}_{\substack{\text{range-related} \\ \text{component}}} - \underbrace{\frac{2v}{\lambda}\cos\left(\gamma(t)\right)}_{\text{Doppler shift}} \right) \right), \tag{3.15}$$

where

$$r(t) = \sqrt{R_0^2 + (v(t-t_0))^2}, \quad R_0 \geq 0. \tag{3.16}$$

In this formulation, akin to the previous notation, λ represents the wavelength, and c is the propagation speed (speed of light). The amplitude $A_x(t)$ is influenced primarily by the antenna radiation pattern and the radar cross-section of the target.

Target imaging comes down to trajectory reconstruction and solving the inverse problem, telling what shape can give the signal received in the radar receiver. In practice, (3.15) is too complex to be used directly. Thus, without loss of generality,

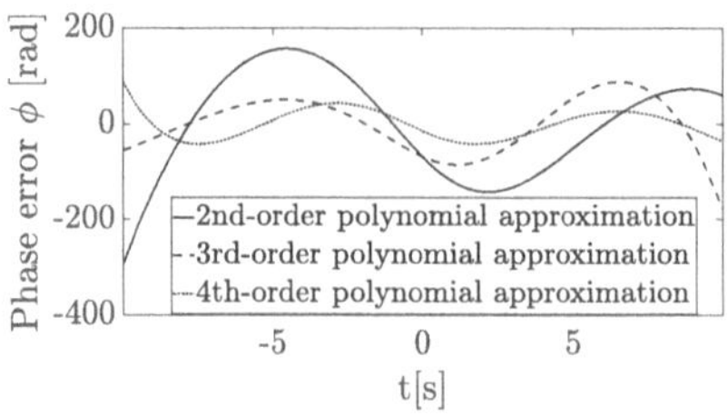

Figure 3.7 Error between the theoretical phase and different order approximations.

the function is approximated using Taylor expansion as a finite-model polynomial. For example, [140] approximated the phase of (3.15) using a second-order Taylor expansion; however, this approach introduces inaccuracies for longer integration times. Increasing the order of nonlinearity in the phase model leads to more precise motion estimation. Some studies, such as [172], propose a cubic (third-order) phase model, and some go further and approximate (3.15) using the fourth-order function [132]

$$\phi_{T,4} = \frac{4\pi BtR_0}{Tc} - \frac{4\pi t^2 v^2}{\lambda R_0} + \frac{2\pi Bt^3 v^2}{TcR_0} + \frac{2\pi t^4 v^4}{\lambda R_0^3}. \tag{3.17}$$

Generally speaking, the higher the approximation order, the better the target motion estimation. The practical choice of Taylor approximation order balances computational efficiency with accuracy in target imaging. Firstly, it enables precise motion parameter estimation over extended integration times. Secondly, it allows for the analysis of strong nonlinearities in motion, which is essential for observing variations in trajectory due to external factors such as wind or turbulence affecting aerial targets.

The advantages of higher-order phase modeling are illustrated in Figure 3.7, where the simulated signal is generated based on (3.15) using the parameters listed in Table 3.1 and the system configuration from Figure 3.6.

The results demonstrate that increasing the phase polynomial order approximation yields the lowest error between the actual and modeled movement, providing a more accurate representation of the theoretical phase model. This improves ISAR image quality, particularly for maneuvering targets, where phase nonlinearities are more pronounced.

Table 3.1
ISAR Simulation Parameters

Parameter	Value
Carrier frequency f_c	5.5 GHz
Bandwidth B	800 MHz
Pulse repetition rate	0.5 kHz
Target range at $t = 0$	100 m
Target velocity	20 m/s

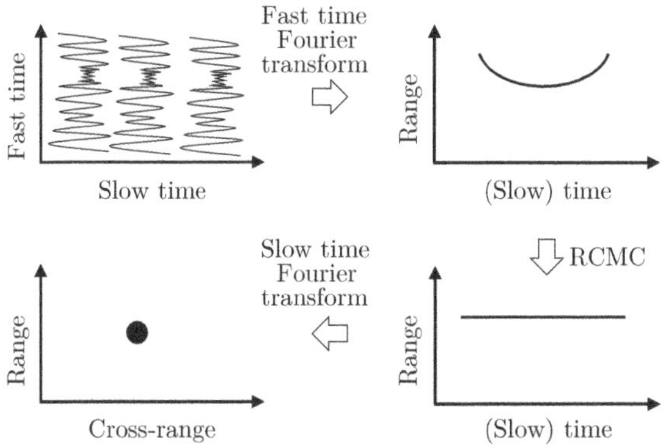

Figure 3.8 Depiction of the general ISAR image processing pipeline for the frequency-modulated continuous-wave radar. The upper left image shows the raw radar signal, where each column represents the beat signal from each sweep. We obtain the range-time distribution after the Fourier transform along the fast time. Target moving perpendicularly to the main radar antenna beam gives the echo of a parabolic shape. Straightening the echo in the range-time domain is called range cell migration correction (RCMC). Next, we obtain the target projection after the Fourier transform along the (slow) time axes.

Remark 3.3. *Radar imaging algorithms constitute a significant branch of radar signal-processing methods. Numerous books, articles, and conference reports have been dedicated to exploring techniques for reconstructing target structures in the ISAR configuration [14, 15, 167, 169, 171, 173–183]. In this book, we do not examine these imaging algorithms in detail. Instead, we focus on how Fourier transforms can be leveraged to enhance existing algorithms and improve radar image quality. Specifically, we do not consider the process of SAR/ISAR image formation itself. When analyzing an image, we assume it has been obtained through some unspecified method, which is beyond the scope of our discussion, and that the resulting image represents a two-dimensional spectrum. Due to the invertibility of the Fourier transform, any two-dimensional spectrum can be transformed back into a spatial-domain signal and subsequently re-transformed into the frequency and spatial-frequency domain. This allows for the application of signal-processing techniques discussed in the subsequent sections of this chapter.*

Signal filtering and component retrieval techniques for one-dimensional signals were discussed in Chapter 2. Now, we can take advantage of their usability by leveraging component extraction to improve ISAR image quality. Radar images, however, arise as the processing of two-dimensional signals, being a collection of range profiles (range-compressed signals forming a range-time distribution). According to Figure 3.6, the most straightforward albeit pivotal (extended and improved by other techniques) algorithm for ISAR image processing is shown in Figure 3.8. Related techniques are based on this methodology with some improvements, but the general sense is basically conserved.

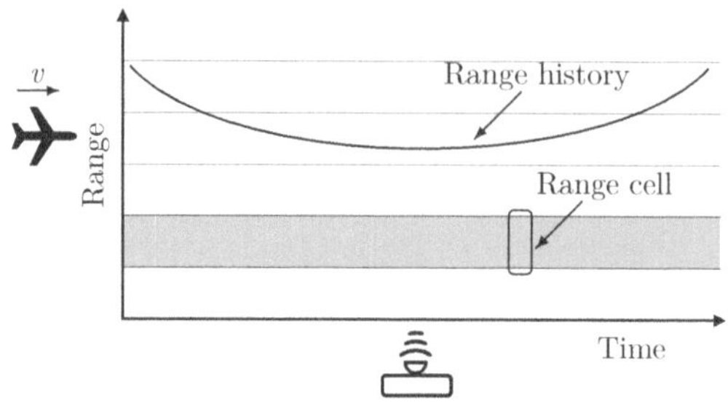

Figure 3.9 A simplified ISAR system geometry.

The general idea is relatively simple. The received signals are firstly range-compressed. For the frequency-modulated continuous-wave radars, it comes down to the Fourier transform computation of each sweep mixing product (see also Figure 3.1). For the simple geometry and trajectory as in Figure 3.6, where the target moves perpendicularly to the radiation direction, the echo on the range-time distribution takes on the parabolic shape, which reflects the decreasing and then increasing target-radar range. This effect is called the range migration and needs to be compensated for precise target shape projection. After this step, the echo forms a straight line, and the Fourier transform along the slow time axis leads to the target image.

Looking at Figure 3.9, we see the range-time formation step in detail. Comparing Figures 3.4 and 3.9, we note a similar mechanism – the target occupies a given range cell (or several range cells in high-resolution radars) and migrates from one range cell to another due to its trajectory. Thus, analyzing the signal as a collection of several range cells admits (3.13) utilization. Moreover, (3.13) preserves other properties belonging to the classical STFT for one-dimensional signals. In the ISAR image example, we can examine the possibility of signal denoising using vertical synchrosqueezing and component retrieval.

Let us define spatial vertical synchrosqueezing for a two-dimensional signal $x(t,r)$

$$V_x^h(t,r,\omega) = \int_{-\infty}^{\infty} F_x^h(t,r,\Omega)\delta\left(\omega - \hat{\omega}_x(t,r,\Omega)\right) \mathrm{d}\Omega, \tag{3.18}$$

where the spatial spectrogram $F_x^h(t,r,\omega)$ is defined as (3.13). The instantaneous frequency estimator $\hat{\omega}_x(t,r,\omega)$ is derived according to the one-dimensional signal with the only difference lying in the space dimension of $x(t,r)$. By analogy to the first-order instantaneous frequency estimator (2.122), we can define its counterpart comprising space r. Thus, we have

$$\hat{\omega}_x(t,r,\omega) = \omega - \Im\left(\frac{F_x^{\mathcal{D}h}(t,r,\omega)}{F_x^h(t,r,\omega)}\right), \tag{3.19}$$

where all symbols keep their meaning as in previous considerations. Higher-order estimators can be defined equivalently; therefore, we omit their replications.

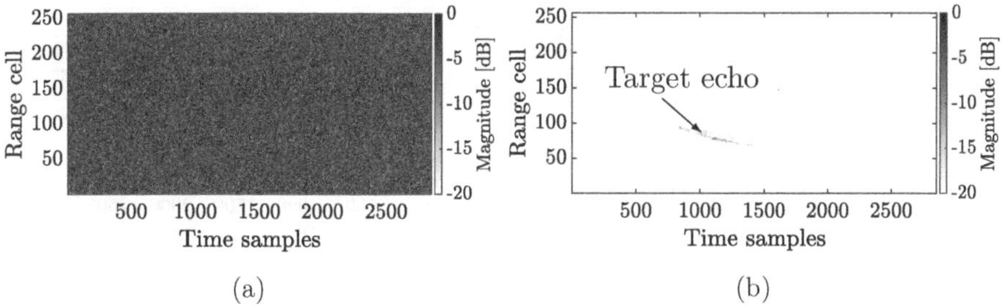

Figure 3.10 Range-time map distributions for the real-life experiment with the Cessna aircraft. (a) Original distribution; (b) denoised distribution. (Results reproduced from [132].)

In [132], the presented approach was used for range-time distribution filtering and denoising of ISAR images. To this end, the frequency-modulated continuous-wave radar working at the carrier frequency of $f_c = 5.5$ GHz, the bandwidth of $B = 1.28$ GHz, and the sweep repetition rate of $f_{\mathrm{SRR}} = 2$ kHz was used. The cooperative target was the Cessna aircraft flying at approximately 50 m/s at approximately 200 meters from the radar. The beat signal from each pulse was used for the range-time distribution creation. Parallelly, this distribution was denoised using the fourth-order (for rationale, see (3.17)) vertical synchrosqueezing method with mode extraction (2.157). The latter step was iteratively performed for each analyzed range cell. The range-time distributions before and after filtering are depicted in Figure 3.10.

The difference clearly shows how noise removal improves the range-time distribution quality. The original map is noisy, disabling target trajectory perception and estimation. After filtering, we clearly see a part of the parabola resulting from the aircraft flight. To determine where this effect follows from, let us look at Figure 3.11, where the spectrogram, concentrated spectrogram, detected signal mode, and filtered mode spectrogram are shown. As already described for vertical synchrosqueezing in Chapter 2, the most prominent signal mode is extracted for each range cell. Such analysis is particularly effective for high-resolution radars, as the signal denoising process is performed iteratively for each range cell. Consequently, even if the target occupies multiple range cells, each of them is independently cleaned of noise, ensuring precise signal reconstruction.

The imaging result before and after signal denoising is compared in Figure 3.12, where signal filtering was applied before motion parameter estimation using the autofocus MapDrift method [132]. The influence of imaging quality results from the range-time distribution, whose noise level was significantly suppressed using vertical synchrosqueezing. Indeed, some noise fractions are still apparent in the filtered signal. They can be removed further to improve the image quality, but the result shows a strong relationship between the range-time distribution and the ISAR image quality. This relationship is evident when analyzing the diagram in Figure 3.8, as the range-time distribution plays a crucial role in ISAR imaging. As shown, signal denoising improves the accuracy of estimating the target's motion parameter, which

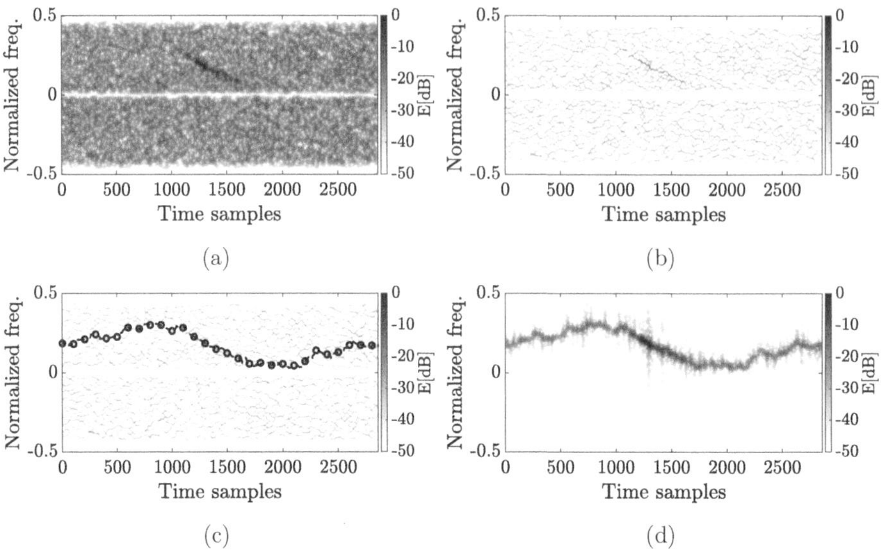

Figure 3.11 Time-frequency distributions for a single range cell. The presented operations are carried out for each range cell to denoise the whole range-time distribution. (a) Initial spectrogram of the selected range cell; (b) vertically synchrosqueezed spectrogram; (c) the same as (b) but with a detected mode superimposed on the spectrogram; (d) spectrogram after filtering.

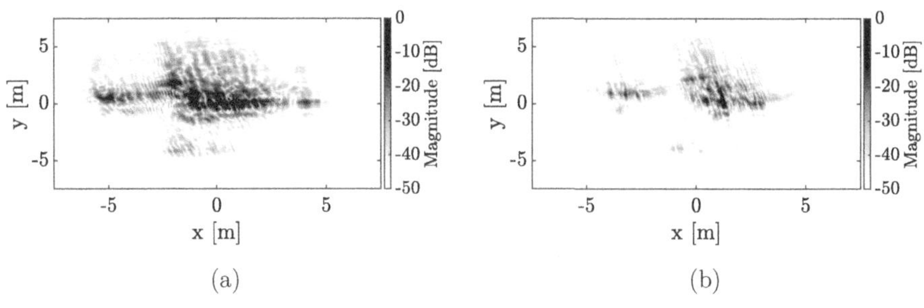

Figure 3.12 Cessna aircraft ISAR image before (a) and after (b) signal filtering. The target projection is more compact and has a lower sidelobe level. For (a), we have SNR = 42.00 dB, contrast equals 2.83. For (b), SNR = 45.19 dB, contrast is 3.22. (Results reproduced from [132].)

is also evident, remembering that these parameters are estimated after signal range compression.

The last example with the PZL Krosno KR-03 Puchatek glider is shown in Figure 3.13. In this case, noise reduction around the target is the most apparent effect. Although the target shape was not as meaningfully improved as in Figure 3.12,

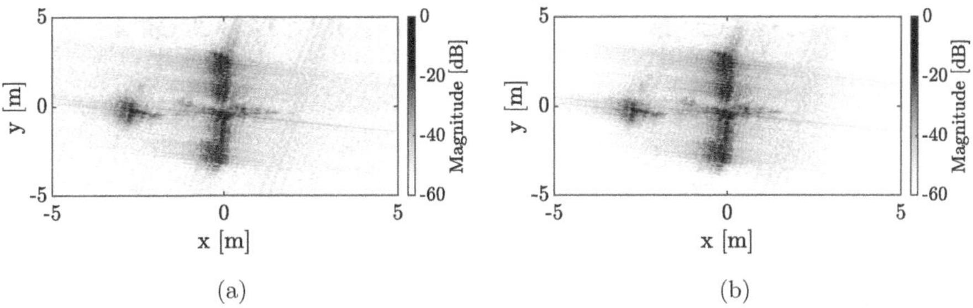

Figure 3.13 Puchatek glider ISAR image. (a) Original image, SNR = 62.48 dB, contrast equals 4.77; (b) filtered image, SNR = 64.11 dB, contrast equals 6.2. (Results reproduced from [132].)

noise filtering is worth appreciating as the whole image exhibits better parameters regarding SNR and contrast.

Although the discussed signal analysis and decomposition techniques were applied to two-dimensional signals, they were processed using a one-dimensional Fourier transform. While this approach provides valuable insights, it does not fully exploit the inherent two-dimensional structure of the data. Moreover, practical applications widely advocate for the two-dimensional Fourier transform as an exceptionally effective tool for analyzing two-dimensional signals. When processed separately, each dimension may overlook important joint frequency components and interdependencies between the two dimensions. To achieve a more comprehensive spectral representation, we now extend our analysis to the two-dimensional Fourier transform, which allows for a more complete characterization of the signal's structure in both dimensions simultaneously.

3.4 TWO-DIMENSIONAL FOURIER TRANSFORM

Our considerations so far have primarily focused on one-dimensional signals and, consequently, one-dimensional Fourier transform. Although the spectrogram resulted in a two-dimensional signal energy distribution for each (t, ω), its obtaining comes down to a one-dimensional transform, and the second dimension reflects the window position. Despite analyzing two-dimensional signals in this chapter, we limited our considerations to the particular space bin, allowing one-dimensional signals to be extracted. Next, we processed the signal for each r. Such a methodology is beneficial, but some readers certainly expect frequency content analysis for t and r simultaneously. For this purpose, a two-dimensional Fourier transform was introduced.

Although Fourier's original formulation dealt with one-dimensional signals, the concept was later extended to multiple dimensions to analyze the functions of two or more variables. The two-dimensional Fourier transform (2D-FT) became particularly important in fields such as image processing, optics, and radar.

3.4.1 CONTINUOUS TWO-DIMENSIONAL FOURIER TRANSFORM

Mathematically, the two-dimensional Fourier transform of a function $x(t,r)$ is defined as

$$F_x(\omega,\eta) = \int_{-\infty}^{\infty}\int_{-\infty}^{\infty} x(t,r)e^{-j(\omega t+\eta r)}\mathrm{d}t\mathrm{d}r. \tag{3.20}$$

The signal is defined in **time** t and **space** r. Equation (3.20) decomposes $x(t,r)$ into its **frequency** ω and **spatial-frequency** η components, where ω and η correspond to variables along the t- and r-axes, respectively. We treat time and space as having a similar meaning, and for the sake of simplicity, we sometimes express both in seconds unless otherwise stated. Similar for frequency and spatial frequency, which we define in Hz. Such an approach is coherent with typical radar nomenclature, where we have fast and slow time for two-dimensional radar signals (see, e.g., Figure 3.7). The inverse transform allows the reconstruction of the original function

$$x(t,r) = \frac{1}{(2\pi)^2}\int_{-\infty}^{\infty}\int_{-\infty}^{\infty} F_x(\omega,\eta)e^{j(\omega t+\eta r)}\mathrm{d}\omega\mathrm{d}\eta. \tag{3.21}$$

In the remainder of this book, we refer to the bivariate energy distribution (or equivalently to the bivariate spectrum) as

$$E_x(\omega,\eta) = |F_x(\omega,\eta)|^2. \tag{3.22}$$

Remark 3.4. *In the literature, bivariate signals are sometimes referred to as complex signals [184], as their real and imaginary components can be treated separately as two distinct real-valued signals or jointly as a single complex-valued signal. In this book, signals are implicitly assumed to be complex by default. The term* ***bivariate*** *indicates that the signal is inherently complex-valued and is defined in both time and space.*

Over the years, the 2D Fourier transform has become a fundamental tool across various scientific and engineering disciplines. Its properties-such as linearity, shifting, convolution, and Parseval's theorem-facilitate efficient signal analysis and manipulation in both theoretical and practical contexts. In the following sections, we systematically present and prove the key properties of the two-dimensional Fourier transform, emphasizing their significance in real-world applications.

The one-dimensional Fourier transform, introduced in Chapter 2, forms the basis of practical signal analysis and processing. In computational applications, it is commonly implemented in three discrete variants: the discrete-time Fourier transform, the discrete Fourier transform, and the fast Fourier transform. These formulations not only enable efficient numerical computations but also extend naturally to two-dimensional signals, playing a crucial role in image and radar signal processing.

While the fundamental properties of these discrete transforms were previously discussed in the context of one-dimensional signals, their two-dimensional counterparts play a crucial role in analyzing spatially distributed data. Now, we formally define the two-dimensional Fourier transform variants, highlighting their significance in practical applications.

3.4.2 TWO-DIMENSIONAL DISCRETE-TIME FOURIER TRANSFORM

The two-dimensional discrete-time Fourier transform of a discrete function $x[m,n]$ is defined as

$$F_x(\omega,\eta) = \sum_{m=-\infty}^{\infty} \sum_{n=-\infty}^{\infty} x[m,n]e^{-j(\omega m+\eta n)}. \tag{3.23}$$

The inverse two-dimensional discrete-time Fourier transform is given by

$$x[m,n] = \frac{1}{(2\pi)^2} \int_{-\pi}^{\pi} \int_{-\pi}^{\pi} F_x(\omega,\eta)e^{j(\omega m+\eta n)}d\omega d\eta. \tag{3.24}$$

3.4.3 TWO-DIMENSIONAL DISCRETE FOURIER TRANSFORM

The two-dimensional discrete Fourier transform of a discrete function $x[m,n]$ of size $M \times N$ is defined as

$$F_x[k,l] = \sum_{m=0}^{M-1} \sum_{n=0}^{N-1} x[m,n]e^{-j2\pi\left(\frac{km}{M}+\frac{ln}{N}\right)}, \tag{3.25}$$

where $k = 0,1,\ldots,M-1$ and $l = 0,1,\ldots,N-1$ are frequency indices. The inverse two-dimensional discrete Fourier transform is given as follows:

$$x[m,n] = \frac{1}{MN} \sum_{k=0}^{M-1} \sum_{l=0}^{N-1} F_x[k,l]e^{j2\pi\left(\frac{km}{M}+\frac{ln}{N}\right)}. \tag{3.26}$$

3.4.4 TWO-DIMENSIONAL FAST FOURIER TRANSFORM

The two-dimensional fast Fourier transform is an efficient algorithm for computing the two-dimensional discrete Fourier transform. It reduces the computational complexity from $\mathcal{O}(M^2N^2)$ to $\mathcal{O}(MN\log_2 M\log_2 N)$ by exploiting the separability property of the discrete Fourier transform.

The two-dimensional fast Fourier transform is computed by applying a one-dimensional fast Fourier transform along each row of the input matrix, followed by a one-dimensional fast Fourier transform along each column. This results in the same output as the discrete Fourier transform but with significantly reduced computational cost.

Table 3.2
Properties of the 2D Fourier Transform

Property	Mathematical Expression
Linearity	$\mathscr{F}\{ax_1(t,r)+bx_2(t,r)\} = aF_{x_1}(\omega,\eta)+bF_{x_2}(\omega,\eta)$
Time/space shifting	$\mathscr{F}\{x(t-t_0,r-r_0)\} = F_x(\omega,\eta)e^{-j(\omega t_0+\eta r_0)}$
Frequency/spatial-frequency shifting	$\mathscr{F}\{x(t,r)e^{j(\omega_0 t+\eta_0 r)}\} = F_x(\omega-\omega_0,\eta-\eta_0)$
Scaling in time/space	$\mathscr{F}\{x(at,br)\} = \frac{1}{\lvert ab\rvert}F_x\left(\frac{\omega}{a},\frac{\eta}{b}\right)$
Duality	$\mathscr{F}\{F_x(t,r)\} = (2\pi)^2 x(-\omega,-\eta)$
Parseval's theorem	$\int\limits_{-\infty}^{\infty}\int\limits_{-\infty}^{\infty} \lvert x(t,r)\rvert^2 \mathrm{d}t\mathrm{d}r = \frac{1}{(2\pi)^2}\int\limits_{-\infty}^{\infty}\int\limits_{-\infty}^{\infty} \lvert F(\omega,\eta)\rvert^2 \mathrm{d}\omega\mathrm{d}\eta$
Convolution theorem	$\mathscr{F}\{x_1(t,r)*x_2(t,r)\} = F_{x_1}(\omega,\eta)F_{x_2}(\omega,\eta)$
Differentiation in time/space	$\mathscr{F}\left\{\frac{\partial x(t,r)}{\partial t}\right\} = (j\omega)F(\omega,\eta)$
Time/space reversal	$\mathscr{F}\{x(-t,-r)\} = F_x(-\omega,-\eta)$
Multiplication in time/space domain	$\mathscr{F}\{x_1(t,r)x_2(t,r)\} = \frac{1}{(2\pi)^2}F_{x_1}(\omega,\eta)*F_{x_2}(\omega,\eta)$

3.4.5 TWO-DIMENSIONAL FOURIER TRANSFORM PROPERTIES

Since the two-dimensional Fourier transform inherits from its one-dimensional counterpart, the latter's properties also hold for the former. The most important properties are listed in Table 3.2. Now, let us prove these.

- **Linearity**: Like its one-dimensional counterpart, the two-dimensional Fourier transform is a linear operation. Given two functions $x_1(t,r)$ and $x_2(t,r)$, and scalars a and b, we have

$$\mathscr{F}\{ax_1(t,r)+bx_2(t,r)\} = aF_{x_1}(\omega,\eta)+bF_{x_2}(\omega,\eta). \tag{3.27}$$

Proof:

$$\begin{aligned} F(\omega,\eta) &= \int\limits_{-\infty}^{\infty}\int\limits_{-\infty}^{\infty} \left[ax_1(t,r)+bx_2(t,r)\right]e^{-j(\omega t+\eta r)}\mathrm{d}t\mathrm{d}r \\ &= a\int\limits_{-\infty}^{\infty}\int\limits_{-\infty}^{\infty} x_1(t,r)e^{-j(\omega t+\eta r)}\mathrm{d}t\mathrm{d}r \\ &\quad + b\int\limits_{-\infty}^{\infty}\int\limits_{-\infty}^{\infty} x_2(t,r)e^{-j(\omega t+\eta r)}\mathrm{d}t\mathrm{d}r \\ &= aF_{x_1}(\omega,\eta)+bF_{x_2}(\omega,\eta). \end{aligned} \tag{3.28}$$

Thus, linearity holds.

- **Time/Space Shifting**: A shift in the spatiotemporal domain results in a phase shift in the frequency domain. If $x(t,r)$ is shifted by (t_0,r_0), then

$$\mathscr{F}\{x(t-t_0,r-r_0)\} = F_x(\omega,\eta)e^{-j(\omega t_0+\eta r_0)}. \tag{3.29}$$

Proof:

$$F_x(\omega,\eta)=\int_{-\infty}^{\infty}\int_{-\infty}^{\infty}x(t-t_0,r-r_0)e^{-j(\omega t+\eta r)}\mathrm{d}t\mathrm{d}r. \tag{3.30}$$

Using the substitution $t'=t-t_0$, $r'=r-r_0$, we obtain

$$\begin{aligned}F_x(\omega,\eta)&=\int_{-\infty}^{\infty}\int_{-\infty}^{\infty}x(t',r')e^{-j[\omega(t'+t_0)+\eta(r'+r_0)]}\mathrm{d}t'\mathrm{d}r'\\&=e^{-j(\omega t_0+\eta r_0)}\int_{-\infty}^{\infty}\int_{-\infty}^{\infty}x(t',r')e^{-j(\omega t'+\eta r')}\mathrm{d}t'\mathrm{d}r'\\&=F_x(\omega,\eta)e^{-j(\omega t_0+\eta r_0)}.\end{aligned} \tag{3.31}$$

Therefore, the time-shifting property holds.

- **Frequency/Spatial-Frequency Shifting**: Multiplication by a complex exponential in the spatiotemporal domain corresponds to a shift in the frequency domain

$$\mathscr{F}\{x(t,r)e^{j(\omega_0 t+\eta_0 r)}\}=F_x(\omega-\omega_0,\eta-\eta_0). \tag{3.32}$$

Proof:

$$\begin{aligned}F_x(\omega,\eta)&=\int_{-\infty}^{\infty}\int_{-\infty}^{\infty}x(t,r)e^{j(\omega_0 t+\eta_0 r)}e^{-j(\omega t+\eta r)}\mathrm{d}t\mathrm{d}r\\&=\int_{-\infty}^{\infty}\int_{-\infty}^{\infty}x(t,r)e^{-j[(\omega-\omega_0)t+(\eta-\eta_0)r]}\mathrm{d}t\mathrm{d}r\\&=F_x(\omega-\omega_0,\eta-\eta_0).\end{aligned} \tag{3.33}$$

From the above, the frequency shifting property is proved.

- **Scaling in Time/Space**: If $x(t,r)$ is scaled in the spatiotemporal domain, its Fourier transform undergoes inverse scaling

$$\mathscr{F}\{x(at,br)\}=\frac{1}{|ab|}F_x\left(\frac{\omega}{a},\frac{\eta}{b}\right). \tag{3.34}$$

Proof: Using the variable change $t'=at, \quad r'=br$, we get

$$F_x(\omega,\eta)=\int_{-\infty}^{\infty}\int_{-\infty}^{\infty}x(at,br)e^{-j(\omega t+\eta r)}\mathrm{d}t\mathrm{d}r. \tag{3.35}$$

Substituting $\mathrm{d}t=\frac{\mathrm{d}t'}{a}, \quad \mathrm{d}r=\frac{\mathrm{d}r'}{b}$, we obtain

$$\begin{aligned}F_x(\omega,\eta)&=\int_{-\infty}^{\infty}\int_{-\infty}^{\infty}x(t',r')e^{-j\left(\omega\frac{t'}{a}+\eta\frac{r'}{b}\right)}\frac{\mathrm{d}t'}{a}\frac{\mathrm{d}r'}{b}\\&=\frac{1}{|ab|}F_x\left(\frac{\omega}{a},\frac{\eta}{b}\right),\end{aligned} \tag{3.36}$$

which ends the proof.

- **Duality**: The duality property states that if $x(t,r)$ has a 2D Fourier transform $F_x(\omega,\eta)$, then the function $F_x(t,r)$ has a Fourier transform related to $x(t,r)$ as follows:

$$\mathscr{F}\{F_x(t,r)\} = (2\pi)^2 x(-\omega,-\eta). \tag{3.37}$$

Proof: By definition, the 2D Fourier transform of $x(t,r)$ is

$$F_x(\omega,\eta) = \int_{-\infty}^{\infty}\int_{-\infty}^{\infty} x(t,r)e^{-j(\omega t+\eta r)}\mathrm{d}t\mathrm{d}r. \tag{3.38}$$

Now, consider taking the Fourier transform of $F_x(t,r)$

$$\mathscr{F}\{F_x(t,r)\} = \int_{-\infty}^{\infty}\int_{-\infty}^{\infty} F_x(t,r)e^{-j(\omega t+\eta r)}\mathrm{d}t\mathrm{d}r. \tag{3.39}$$

Substituting the expression for $F_x(t,r)$

$$\mathscr{F}\{F_x(t,r)\} = \int_{-\infty}^{\infty}\int_{-\infty}^{\infty}\left(\int_{-\infty}^{\infty}\int_{-\infty}^{\infty} x(t',r')e^{-j(tt'+rr')}\mathrm{d}t'\mathrm{d}r'\right)e^{-j(\omega t+\eta r)}\mathrm{d}t\mathrm{d}r. \tag{3.40}$$

Rearranging the order of integration yields

$$\mathscr{F}\{F_x(t,r)\} = \int_{-\infty}^{\infty}\int_{-\infty}^{\infty} x(t',r')\left(\int_{-\infty}^{\infty}\int_{-\infty}^{\infty} e^{-j(tt'+rr')}e^{-j(\omega t+\eta r)}\mathrm{d}t\mathrm{d}r\right)\mathrm{d}t'\mathrm{d}r'. \tag{3.41}$$

Since the exponential terms combine as

$$e^{-j(tt'+rr')}e^{-j(\omega t+\eta r)} = e^{-j[(t'+\omega)t+(r'+\eta)r]}, \tag{3.42}$$

and we know that

$$\int_{-\infty}^{\infty} e^{-j\alpha t}\mathrm{d}t = 2\pi\delta(\alpha), \tag{3.43}$$

the integral simplifies to

$$\mathscr{F}\{F_x(t,r)\} = (2\pi)^2 x(-\omega,-\eta), \tag{3.44}$$

which ends the proof.

- **Parseval's Theorem**: The Parseval's theorem states that the total energy of a function is preserved in both the spatiotemporal and frequency and spatial-frequency domains

$$\int_{-\infty}^{\infty}\int_{-\infty}^{\infty} |x(t,r)|^2\mathrm{d}t\mathrm{d}r = \frac{1}{(2\pi)^2}\int_{-\infty}^{\infty}\int_{-\infty}^{\infty} |F_x(\omega,\eta)|^2\mathrm{d}\omega\mathrm{d}\eta. \tag{3.45}$$

Proof: By definition, the Fourier transform of $x(t,r)$ is given by

$$F_x(\omega,\eta) = \int_{-\infty}^{\infty}\int_{-\infty}^{\infty} x(t,r)e^{-j(\omega t+\eta r)}\mathrm{d}t\mathrm{d}r. \tag{3.46}$$

Let us consider the squared magnitude of $x(t,r)$

$$\int_{-\infty}^{\infty}\int_{-\infty}^{\infty} |x(t,r)|^2\,\mathrm{d}t\mathrm{d}r. \tag{3.47}$$

Substituting $x(t,r)$ in terms of its inverse Fourier transform,

$$x(t,r) = \int_{-\infty}^{\infty}\int_{-\infty}^{\infty} F_x(\omega,\eta)e^{j(\omega t+\eta r)}\mathrm{d}\omega\mathrm{d}\eta, \tag{3.48}$$

we substitute this into $|x(t,r)|^2$ obtaining

$$|x(t,r)|^2 = x(t,r)x^*(t,r) = \left[\int_{-\infty}^{\infty}\int_{-\infty}^{\infty} F_x(\omega,\eta)e^{j(\omega t+\eta r)}\mathrm{d}\omega\mathrm{d}\eta\right] \tag{3.49}$$

$$\times\left[\int_{-\infty}^{\infty}\int_{-\infty}^{\infty} F_x^*(\omega',\eta')e^{-j(\omega' t+\eta' r)}\mathrm{d}\omega'\mathrm{d}\eta'\right]. \tag{3.50}$$

Integrating both sides over t and r yields

$$\int_{-\infty}^{\infty}\int_{-\infty}^{\infty} |x(t,r)|^2\,\mathrm{d}t\mathrm{d}r = \int_{-\infty}^{\infty}\int_{-\infty}^{\infty}\int_{-\infty}^{\infty}\int_{-\infty}^{\infty} F_x(\omega,\eta)F_x^*(\omega',\eta')$$
$$\times\left[\int_{-\infty}^{\infty}\int_{-\infty}^{\infty} e^{j[(\omega-\omega')t+(\eta-\eta')r]}\mathrm{d}t\mathrm{d}r\right]\mathrm{d}\omega\mathrm{d}\eta\mathrm{d}\omega'\mathrm{d}\eta'. \tag{3.51}$$

The integral inside the brackets evaluates to the Dirac delta function

$$\int_{-\infty}^{\infty} e^{j\alpha t}\,\mathrm{d}t = 2\pi\delta(\alpha), \tag{3.52}$$

thus,

$$\int_{-\infty}^{\infty}\int_{-\infty}^{\infty} e^{j[(\omega-\omega')t+(\eta-\eta')r]}\mathrm{d}t\mathrm{d}r = (2\pi)^2\delta(\omega-\omega')\delta(\eta-\eta'). \tag{3.53}$$

Using this property, we obtain

$$\int_{-\infty}^{\infty}\int_{-\infty}^{\infty} |x(t,r)|^2 \mathrm{d}t\mathrm{d}r = \frac{1}{(2\pi)^2}\int_{-\infty}^{\infty}\int_{-\infty}^{\infty} |F_x(\omega,\eta)|^2 \mathrm{d}\omega\mathrm{d}\eta. \tag{3.54}$$

This completes the proof of Parseval's theorem.

- **Convolution Theorem**: The Fourier transform of the convolution of two functions is the product of their Fourier transforms:

$$\mathscr{F}\{x_1(t,r) * x_2(t,r)\} = F_{x_1}(\omega,\eta)F_{x_2}(\omega,\eta). \tag{3.55}$$

Proof: By definition, convolution is given by

$$(x_1 * x_2)(t,r) = \int_{-\infty}^{\infty}\int_{-\infty}^{\infty} x_1(\tau,\rho)x_2(t-\tau,r-\rho)\mathrm{d}\tau\mathrm{d}\rho. \tag{3.56}$$

Taking the Fourier transform

$$F_{x_1 * x_2}(\omega,\eta) = \int_{-\infty}^{\infty}\int_{-\infty}^{\infty}\left[\int_{-\infty}^{\infty}\int_{-\infty}^{\infty} x_1(\tau,\rho)x_2(t-\tau,r-\rho)\mathrm{d}\tau\mathrm{d}\rho\right] e^{-j(\omega t+\eta r)}\mathrm{d}t\mathrm{d}r. \tag{3.57}$$

Changing the order of integration leads to

$$F_{x_1 * x_2}(\omega,\eta) = \int_{-\infty}^{\infty}\int_{-\infty}^{\infty} x_1(\tau,\rho)\left[\int_{-\infty}^{\infty}\int_{-\infty}^{\infty} x_2(t-\tau,r-\rho)e^{-j(\omega t+\eta r)}\mathrm{d}t\mathrm{d}r\right]\mathrm{d}\tau\mathrm{d}\rho. \tag{3.58}$$

Using the substitution $t' = t-\tau$, $r' = r-\rho$, we obtain

$$\begin{aligned}
&\int_{-\infty}^{\infty}\int_{-\infty}^{\infty} x_2(t-\tau,r-\rho)e^{-j(\omega t+\eta r)}\mathrm{d}t\mathrm{d}r \\
&= \int_{-\infty}^{\infty}\int_{-\infty}^{\infty} x_2(t',r')e^{-j[\omega(t'+\tau)+\eta(r'+\rho)]}\mathrm{d}t'\mathrm{d}r' \\
&= e^{-j(\omega\tau+\eta\rho)}\int_{-\infty}^{\infty}\int_{-\infty}^{\infty} x_2(t',r')e^{-j(\omega t'+\eta r')}\mathrm{d}t'\mathrm{d}r' \\
&= e^{-j(\omega\tau+\eta\rho)}F_{x_2}(\omega,\eta).
\end{aligned} \tag{3.59}$$

Substituting back, we get

$$\begin{aligned} F_{x_1 * x_2}(\omega,\eta) &= \int_{-\infty}^{\infty}\int_{-\infty}^{\infty} x_1(\tau,\rho)e^{-j(\omega\tau+\eta\rho)}F_{x_2}(\omega,\eta)\mathrm{d}\tau\mathrm{d}\rho \\ &= F_{x_2}(\omega,\eta)\int_{-\infty}^{\infty}\int_{-\infty}^{\infty} x_1(\tau,\rho)e^{-j(\omega\tau+\eta\rho)}\mathrm{d}\tau\mathrm{d}\rho \\ &= F_{x_1}(\omega,\eta)F_{x_2}(\omega,\eta). \end{aligned} \tag{3.60}$$

- **Differentiation in Time/Space**: Differentiation in the spatiotemporal domain corresponds to multiplication by a frequency term

$$\mathscr{F}\left\{\frac{\partial^m}{\partial t^m}\frac{\partial^n}{\partial r^n}x(t,r)\right\} = (j\omega)^m(j\eta)^n F_x(\omega,\eta). \tag{3.61}$$

Proof:

$$\mathscr{F}\left\{\frac{\partial^m}{\partial t^m}x(t,r)\right\} = \int_{-\infty}^{\infty}\int_{-\infty}^{\infty}\frac{\partial^m}{\partial t^m}x(t,r)e^{-j(\omega t+\eta r)}\mathrm{d}t\mathrm{d}r. \tag{3.62}$$

Integrating by parts m times

$$\begin{aligned} \mathscr{F}\left\{\frac{\partial^m}{\partial t^m}x(t,r)\right\} &= (j\omega)^m\int_{-\infty}^{\infty}\int_{-\infty}^{\infty} x(t,r)e^{-j(\omega t+\eta r)}\mathrm{d}t\mathrm{d}r \\ &= (j\omega)^m F_x(\omega,\eta). \end{aligned} \tag{3.63}$$

Analogously, the operation can be written for the derivative with respect to the second variable r.

- **Time/Space Reversal**: Reversing the function in the spatial domain results in the same transformation with negated frequency arguments:

$$\mathscr{F}\{x(-t,-r)\} = F_x(-\omega,-\eta). \tag{3.64}$$

Proof: For the expression

$$F_x(\omega,\eta) = \int_{-\infty}^{\infty}\int_{-\infty}^{\infty} x(-t,-r)e^{-j(\omega t+\eta r)}\mathrm{d}t\mathrm{d}r, \tag{3.65}$$

we can use the substitution $t' = -t$, $r' = -r$ and change variables

$$\int_{-\infty}^{\infty}\int_{-\infty}^{\infty} x(t',r')e^{-j(-\omega t'-\eta r')}\mathrm{d}t'\mathrm{d}r' = F_x(-\omega,-\eta), \tag{3.66}$$

which ends the proof.

- **Multiplication in Time/Space Domain**: The multiplication property states that the Fourier transform of the product of two functions in the spatiotemporal domain corresponds to the convolution of their Fourier transforms in the frequency and spatial-frequency domains:

$$\mathscr{F}\{x_1(t,r)x_2(t,r)\} = \frac{1}{(2\pi)^2} F_{x_1}(\omega,\eta) * F_{x_2}(\omega,\eta). \tag{3.67}$$

Proof: By definition of the Fourier transform

$$F_{x_1}(\omega,\eta) = \int_{-\infty}^{\infty}\int_{-\infty}^{\infty} x_1(t,r)e^{-j(\omega t+\eta r)}\mathrm{d}t\mathrm{d}r. \tag{3.68}$$

Similarly, for $x_2(t,r)$, we have

$$F_{x_2}(\omega,\eta) = \int_{-\infty}^{\infty}\int_{-\infty}^{\infty} x_2(t,r)e^{-j(\omega t+\eta r)}\mathrm{d}t\mathrm{d}r. \tag{3.69}$$

Taking the Fourier transform of $x_1(t,r)x_2(t,r)$

$$\mathscr{F}\{x_1(t,r)x_2(t,r)\} = \int_{-\infty}^{\infty}\int_{-\infty}^{\infty} x_1(t,r)x_2(t,r)e^{-j(\omega t+\eta r)}\mathrm{d}t\mathrm{d}r$$

$$= \int_{-\infty}^{\infty}\int_{-\infty}^{\infty} x_1(\tau,\rho)e^{-j(\omega\tau+\eta\rho)}\left[\int_{-\infty}^{\infty}\int_{-\infty}^{\infty} x_2(t-\tau,r-\rho)e^{-j(\omega t+\eta r)}\mathrm{d}t\mathrm{d}r\right]\mathrm{d}\tau\mathrm{d}\rho. \tag{3.70}$$

Rewriting the inner integral as a convolution integral, we obtain

$$\mathscr{F}\{x_1(t,r)x_2(t,r)\} = \frac{1}{(2\pi)^2} F_{x_1}(\omega,\eta) * F_{x_2}(\omega,\eta). \tag{3.71}$$

3.4.6 EXAMPLE

In this chapter, we will explore the manifold of 2D signals from different applications. To familiarize ourselves with them, we can start with a generic example showcasing the main principles in bivariate spectrum analysis.

Let us visualize the 2D Fourier transform using a simple, noiseless example. The signal definition reads as

$$x(t,r) = x(t)x(r) = e^{j2\pi 20t}e^{-j2\pi 30r}. \tag{3.72}$$

The signal was sampled over a 1 s interval in both dimensions at a rate of $f_s =$ 100 Sa/s, resulting in a 100×100 sample matrix. The real and imaginary parts in the time-space domain are shown in Figure 3.14a and b, respectively, while its spectrum (using (3.22)) is depicted in Figure 3.14c.

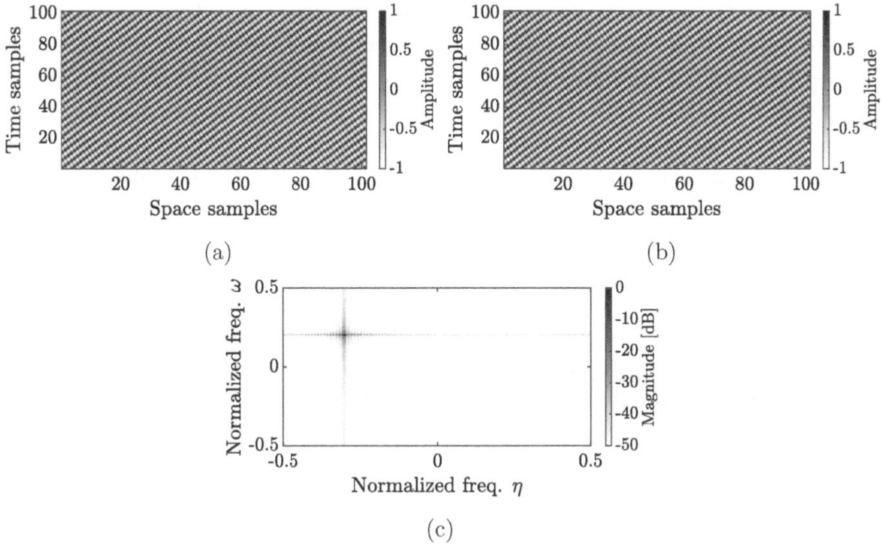

Figure 3.14 Time-space and frequency-spatial-frequency distributions of the simulated harmonic 2D signal. (a) Real and (b) imaginary parts of the signal; (c) two-dimensional spectrum of the signal.

The time-space domain signal reveals a product of two signals. One can notice simple harmonic terms with evolution governed by (3.72) in their cross-sections. In turn, the spectrum has one prominent peak corresponding to $\omega = 0.2$ and $\eta = -0.3$ in the normalized scale. Also, distinct sidelobes are apparent. Their mitigation is a matter of further discussion.

In practical radar applications, two-dimensional signals can be acquired manifold. As mentioned in the introduction to this chapter, they can be received using multi-antenna systems with antennas spatially dislocated. Also, they arise from SAR/ISAR or range-Doppler processing through reorganizing a one-dimensional signal to its bivariate counterpart. The next section explores typical radar signal-processing methods that allow one-dimensional signals to be presented as two-dimensional spectra, which typically take place in active radars.

3.5 FROM ONE-DIMENSIONAL SIGNAL TO BIVARIATE SPECTRUM

Typical radars aim to estimate target range and velocity through the delay and Doppler frequency history analysis. While the former can be considered in different ways, the latter usually comes down to the Fourier analysis of several pulses. In this section, let us focus on the frequency-modulated continuous-wave radar, which can be easily generalized. Now, let us focus on mathematical foundations.

If we contemplate the content of (3.11), we see three main factors. Firstly, we see a constant phase ϕ_0 resulting from the phase shift introduced by the receiver and target. Secondly, we see the frequency-related term $(\omega_D - \omega_b)t$ denoting the (angular) frequency of the Doppler shift and the target range. Ultimately, we can notice a chirp-like term $\frac{\alpha\tau^2}{2}$ resulting from the mixing of the original and delayed signals. The Fourier spectrum of (3.11) should give a peak at the angular frequency $\omega_D - \omega_b$. At this point, we see a necessity to tackle a problem with differentiating between the Doppler frequency and the one related to the signal delay, as only one peak will be apparent (for a single target). The Doppler history, constituted by a collection of several received pulses, allows one to extract the velocity information within a finite observation time. In turn, the Doppler shift is obtainable using the Fourier transform, which suggests that the range-Doppler processing amounts to the two-dimensional Fourier transform on consecutive received and mixed signals. In the receiver, the signal is one-dimensional, so we need to divide it into blocks (batches) so that it forms a two-dimensional distribution. The phase fluctuation between consecutive sweeps enables velocity estimation.

To analyze the phase evolution of the received signal from sweep to sweep in frequency-modulated continuous-wave radars when the target is moving, we can recall the signal definition for the stationary target, which was defined in (3.7) as

$$x_s(t) = A_s(t)e^{j\left(\phi_0 - \omega_c\tau + \omega_b t + \frac{\alpha\tau^2}{2}\right)}. \tag{3.73}$$

For a moving target with velocity V, the round-trip delay changes over successive sweeps as

$$\tau_n = \frac{2(R + nVT_r)}{c}, \tag{3.74}$$

where T_r is the pulse repetition interval. This introduces a Doppler shift (3.10), and the received signal for a moving target becomes as defined by (3.11)

$$x_m(t) = A_s(t)e^{j\left(\phi_0 - \omega_c\tau_n + (\omega_D - \omega_b)t + \frac{\alpha\tau_n^2}{2}\right)}. \tag{3.75}$$

The phase evolution from pulse to pulse can be examined by considering the phase difference between successive sweeps

$$\Delta\phi = \omega_D T_r. \tag{3.76}$$

This shows that each transmitted pulse experiences an additional phase shift proportional to the Doppler frequency, resulting in a linear phase progression across pulses. This phase evolution is the basis for Doppler processing in a frequency-modulated continuous-wave radar, enabling velocity estimation through phase differences between successive sweeps.

To allow the target velocity to be estimated, we can introduce the Zak transform. Let $\mathscr{Z}(t,\eta)$ denote the Zak transform [185, 186], which, although originally introduced in 1967 in the context of quantum mechanics, has also been utilized in signal-processing applications [187, 188]. The Zak transform of a signal $x(t)$ is given by

$$\mathscr{Z}_x(t,\eta) = \sqrt{T}\sum_{k=\infty}^{\infty} x(t+kT)e^{-j\eta kT}, \quad k \in \mathbb{Z}, T \in \mathbb{R}_+. \tag{3.77}$$

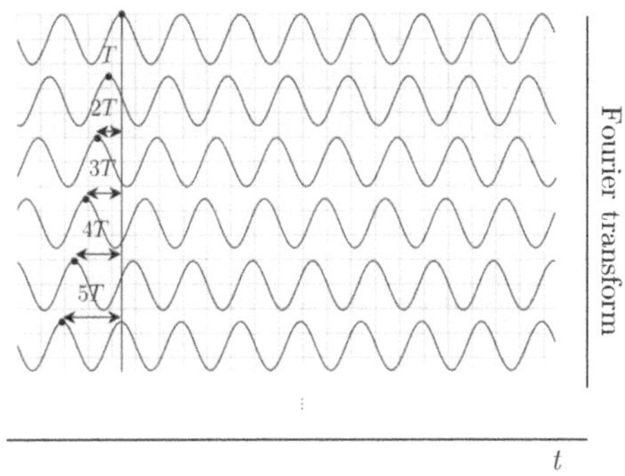

Figure 3.15 Simplified depiction of the Zak transform. (Reproduced from [189], licensed under CC BY 4.0.)

Since the scaling factor $\sqrt{T}$ does not influence the detection performance in radar systems, it will be omitted in the subsequent analysis for simplicity. A conceptual representation of the Zak transform is shown in Figure 3.15.

The Zak transform $\mathscr{Z}_x(t,\eta)$ reorganizes a signal $x(t)$ in a way that reveals its periodic and frequency-related properties. It can be thought of as a transformation that arranges signals along two-dimensional time-frequency axes. To have the Zak transform computed, the signal $x(t)$ is first shifted by integer multiples of a fixed period $T \in \mathbb{R}_+$. These shifted versions of the signal are then weighted by complex exponentials $e^{-j\eta kT}$, which is the discrete-time Fourier transform kernel. The discrete nature stems from shifting a non-infinite small fraction of t defined by kT. The Fourier transform result can be continuous according to the discrete-time Fourier transform rule. The result, $\mathscr{Z}_x(t,\eta)$, is a function of both time t and spatial frequency η. This two-dimensional representation allows for a more structured analysis of signals, which is particularly useful in radar and communication systems. In elementary words, the Zak transform can be viewed as the Fourier transform on equally spaced signal samples (see Figure 3.15). Its key property is that it organizes the signal into a periodic structure in both time and spatial frequency. This means that analyzing a single period in each dimension provides complete information about the original signal.

At this point, one can wonder how (3.77) is related to frequency-modulated continuous-wave radars. Having $T = T_p$ (with T_p being the sweep duration – see Figure 3.2) allows for pulse alignment and the Fourier transform along the so-called slow time. The result lacks the range compression step, which aims to extract the frequency components related to the target range (see Figure 3.11). This step amounts to the Fourier transform computation along the fast time. The whole process takes on the following form

$$M_x(\omega,\eta) = \int_0^{T_p} \sum_{k=-K/2}^{K/2} x(t+kT_p)e^{-j\eta kT_p}e^{-j\omega t}\mathrm{d}t, \quad k \in \mathbb{Z}, T_p \in \mathbb{R}_+, \tag{3.78}$$

where $K = T/T_b$ with T being the whole signal length, is the integer number of sweeps in the received signal. Equation (3.78), as any Fourier transform, is a complex-valued function whose squared magnitude is usually displayed in practice

$$\mathscr{M}_x(\omega, \eta) = |M_x(\omega, \eta)|^2 . \tag{3.79}$$

We have modified the original Zak transform (3.77) in the following ways:

- **Removal of the Scaling Factor**: The $\sqrt{T}$ term in (3.77) has been omitted, as it does not impact detection performance in radar applications. This simplification makes the formulation more practical for signal-processing.
- **Adaptation to Frequency-Modulated Continuous-Wave Radar Context**: Instead of a general period T, we have set $T = T_p$, where T_p represents the sweep duration of a frequency-modulated continuous-wave radar signal. This choice aligns the transformation with the radar system's timing structure.
- **Fourier Transform Along Slow Time**: The sum over integer shifts of the signal is retained, but an additional Fourier transform is introduced along the slow-time axis. This step enables efficient extraction of Doppler-related frequency components.
- **Introduction of Range Compression**: The integral in (3.78) represents the Fourier transform along the fast time. This transformation isolates frequency components corresponding to different target ranges, linking the modified equation directly to the range-Doppler processing step in radar systems.

While the spectral analysis of the single sweep (3.11) allows for range estimation, several pulses processed by (3.78) provide a comprehensive insight into the signal nature leading to both range and velocity estimation simultaneously. In particular, (3.78) leads to the phase content analysis between several pulses according to (3.76).

By incorporating these modifications, the equation now better reflects the processing steps used in frequency-modulated continuous-wave radar analysis while preserving the essential properties of the Zak transform. We associate the spatial (angular) frequency η with the target velocity and the (angular) frequency ω with the target range. The simplistic graphical interpretation of (3.78) is shown in Figure 3.16. In general, we divide the whole received signal into single sweep-related batches. Sub-signals form the two-dimensional distribution, whose bivariate Fourier transform gives the spectrum. This dual-processing step results in a range-velocity map, where each point represents a specific combination of distance and radial velocity. Ultimately, we obtain the two-dimensional spectrum from the one-dimensional input signal. After appropriate scaling, the spectrum can be represented in m and m/s, respectively. These values are defined as follows:

$$R = \frac{f_b c T_p}{2B}, \tag{3.80}$$

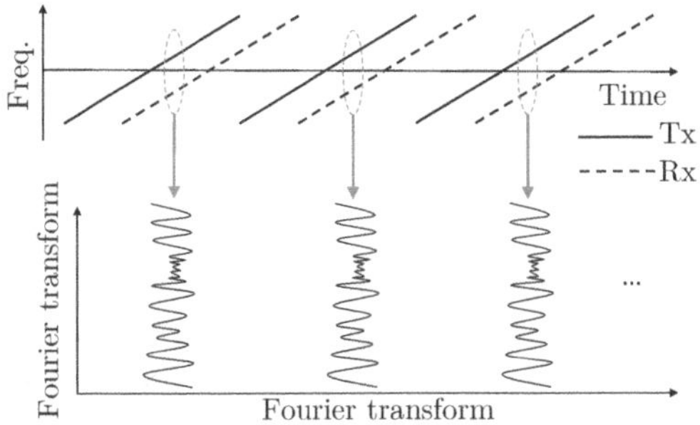

Figure 3.16 Depiction of how the range-velocity map is created as a two-dimensional spectrum from the one-dimensional signal. The input signal is divided into batches. They form a two-dimensional signal whose bivariate Fourier transform leads to the range-velocity map (after appropriate axis scaling).

and

$$V = -\frac{f_D \lambda}{2} = -\frac{c f_D}{2 f_c}, \tag{3.81}$$

where all the parameters are as defined previously, except for two variables $f_b = \frac{\omega_b}{2\pi}$ and $f_D = \frac{\omega_D}{2\pi}$ (see (3.11)).

Remark 3.5. *We have just made a specific simplification, deriving the bivariate Fourier transform using a one-dimensional signal. Unlike conventional spectral analysis, which typically maps a one-dimensional signal to a one-dimensional frequency representation, this approach decomposes the signal along two distinct time scales: fast time (within a single sweep) and slow time (across multiple sweeps).*

In radar systems, several challenges arise, with weak target echoes being one of the primary obstacles. Targets with low magnitude echo may remain undetectable as they can be masked by ambient noise in range-velocity maps. As a companion argument, strong targets can produce high sidelobes, which may also obscure weaker echo detectability. As discussed in Sections 2.2 (see Figure 2.8) and 2.8 (see Figure 2.37), applying a window function to the signal before performing the Fourier transform helps reduce sidelobes at the cost of widening the spectral peaks. This trade-off is often acceptable in radar applications, as it enhances the detection of weak targets, even though their echoes become more spread out. A slight degradation of the range-velocity target signature in exchange for potentially more detections ultimately improves the radar's effectiveness and performance.

Equation (3.78) can be extended to its windowed version. Assuming the Gaussian window as an example (but any window is admissible), we can define the bivariate Gaussian bell-shaped function as

$$h(t, kT_p) = h(t)h(kT_p) = \underbrace{\pi^{-1/4} e^{-t^2/2}}_{h(t)} \underbrace{\pi^{-1/4} e^{-kT_p^2/2}}_{h(kT_p)}. \tag{3.82}$$

Putting (3.82) into (3.78) yields

$$M_x^h(\eta,\omega) = \int_0^{T_p} \sum_{k=-K/2}^{K/2} x(t+kT_p)h(t,kT_p)e^{-j\eta kT_p}e^{-j\omega t}\mathrm{d}t. \tag{3.83}$$

Remark 3.6. *If we compare (3.83) and the STFT definition (2.78), we can notice a link between these two equations. In particular, for vanishing terms related to η, we obtain the STFT with the window stride equal to T_p. Therefore, (3.83) can be viewed as the distribution created from the STFT with the additional Fourier transform along the time axes.*

The window plays a pivotal role in sidelobe suppression and affects the radar target detection performance. The effect of signal windowing is shown in Figure 3.17. The results were simulated assuming the total observation time $T_i = 0.5$ s, the sweep duration $T_p = 1$ ms, and the sweep repetition rate is 1 kHz. The carrier frequency was $f_c = 24$ GHz, and the system bandwidth was set to $B = 1$ GHz. Two moving targets are considered in this scenario. Their initial conditions are:

- **Target 1**: Initial range $R_0 = 50$ m and velocity $v_0 = 1$ m/s.
- **Target 2**: Initial range $R_0 = 50$ m and velocity $v_0 = 1.5$ m/s.

The final received signal is obtained by summing the two echoes, with the second target having a $\times 20$ lower amplitude. Therefore, the second target occupies the same range but is significantly weaker than the first one.

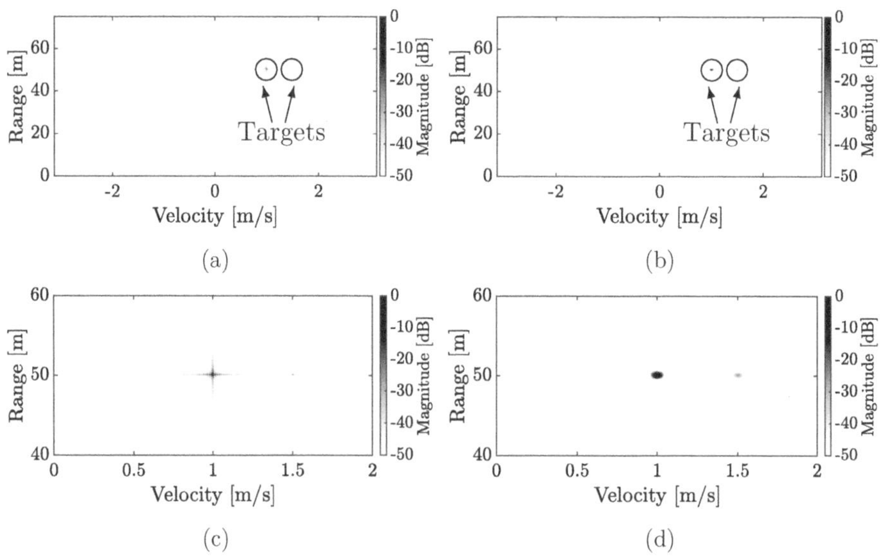

Figure 3.17 Simulated range-velocity maps for the noiseless case. (a) Range-velocity map; (b) windowed range-velocity map; (c) close-up on the range-velocity map around the target; (d) close-up on the windowed range-velocity map around the target.

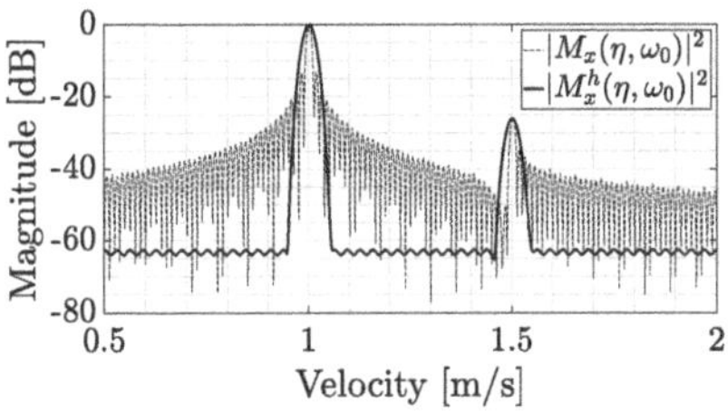

Figure 3.18 Cross-sections of the range-velocity maps for the target range. The solid line corresponds to the windowed distribution, and the dotted line corresponds to no window.

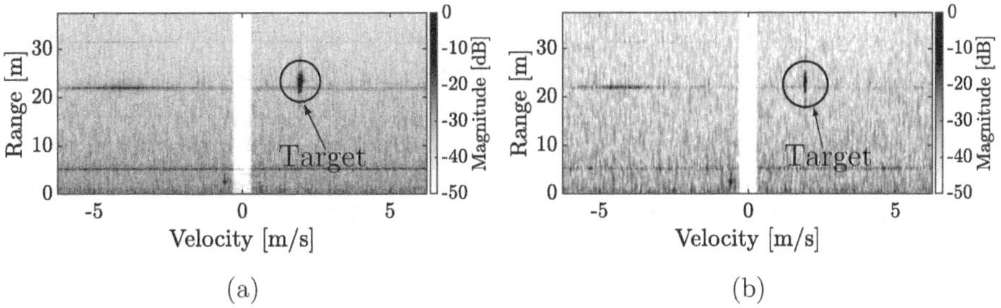

Figure 3.19 Range-velocity maps for the real-life signal with a drone's echo. The example displays a processing result comparison when the window was not used prior to the Fourier transform (a) and when the window was applied (b).

From Figure 3.17 it is apparent that the weaker echo (that at 50 m and 1.5 m/s) is barely visible at Figure 3.17a and c. The sidelobes of the first target meaningfully cover the echo of the second target, and such a result in practice would lead to the echo not being detected. Windowing, however, spreads the target echo but allows two targets to be clearly visible and possibly detected. The effect is also depicted in Figure 3.18 where these two distributions (windowed and without window) are superimposed. While the range-velocity map without a window has a weaker target that is hardly visible (its magnitude is at the sidelobes' level), the windowed one allows unequivocal result interpretation.

The importance of signal windowing is stressed in this book because of its practical usage. In Figure 3.19, we see a real-life range-velocity map recorded using the XY-DemoRad sensor at the carrier frequency of $f_c = 24$ GHz, with the bandwidth of $B = 1$ GHz and the sweep repetition rate of $f_p = 1$ kHz. The signal was filtered so that the echoes from stationary obstacles at $V = 0$ m/s were removed, and this effect manifested itself as a white region around the zero velocity. The target was a drone, whose echo is apparent at $R = 22.76$ m and the velocity $V = 1.95$ m/s. Since windowing allows for sidelobe influence alleviation at the expense of the main peak widening, the range-velocity map in Figure 3.19b has the echo better visible in the background of noise. Moreover, the influence of interference (horizontal line at the range of approximately 22 m) is reduced as its sidelobes impact less after windowing.

Presented examples, although practical and informative, do not exhaust two-dimensional signal analysis, notably in the frequency domain. So far, we deemed the signals stationary in time and space so that single peaks could describe their frequency and spatial-frequency components after spectrum analysis. Similar to one-dimensional counterparts, two-dimensional signals can be nonstationary and they widely exist in practice. In the next section, we extend the bivariate spectrum to a four-dimensional distribution of time, space, frequency, and spatial frequency.

3.6 STFT FOR TWO-DIMENSIONAL SIGNAL – WINDOWED TWO-DIMENSIONAL FOURIER TRANSFORM

In the STFT context described in Section 2.4, we assumed a one-dimensional signal to be multiplied by the moving window, and for each position, we computed the Fourier transform from the window-signal product. The result was the two-dimensional distribution of the signal spectrum for each block defined by the window position. Now, we extend the approach to bivariate signals, introducing the windowed two-dimensional Fourier transform (W2D-FT).

Consider a general two-dimensional signal model defined as a function of t and r

$$x(t,r) = \mathscr{A}_x e^{-\frac{(t-t_x)^2}{2T_x^2}} e^{-\frac{(r-r_x)^2}{2R_x^2}} e^{j\left(\phi_x+\omega_x t+\frac{\alpha_x t^2}{2}\right)} e^{j\left(\psi_x+\eta_x r+\frac{\beta_x r^2}{2}\right)}, \quad -T < t < T, -R < r < R, \tag{3.84}$$

where $\mathscr{A}_x$ is the real-valued amplitude, ϕ_x and ψ_x denote the initial phases, and ω_x and η_x represent the frequency and spatial frequency, respectively, and α_x and β_x denote the frequency and spatial-frequency rate factors. The term $\mathscr{A}_x e^{-\frac{(t-t_x)^2}{2T_x^2}} e^{-\frac{(r-r_x)^2}{2R_x^2}}$ forms a bivariate Gaussian envelope characterized by time spread parameters T_x and R_x and centered at (t_x, r_x). This signal representation will be utilized in subsequent sections.

The bivariate spectrum of $x(t,r)$, as introduced in (3.20), represents its two-dimensional Fourier transform. It is defined as follows:

$$F_x(\omega,\eta) = \int_{-\infty}^{\infty}\int_{-\infty}^{\infty} x(t,r)e^{-j(\omega t+\eta r)}\mathrm{d}t\mathrm{d}r. \tag{3.85}$$

An extension of (3.85) is its windowed counterpart, known as the W2D-FT. As described in [190], the W2D-FT of the complex signal $x(t,r)$ is given by

$$\begin{aligned} F_x^h(t,r,\omega,\eta) &= \int_{-\infty}^{\infty}\int_{-\infty}^{\infty} x(\tau,\rho)h^*(\tau-t,\rho-r)e^{-j(\omega(\tau-t)+\eta(\rho-r))}\mathrm{d}\tau\mathrm{d}\rho \\ &= \int_{-\infty}^{\infty}\int_{-\infty}^{\infty} x(\tau+t,\rho+r)h^*(\tau,\rho)e^{-j(\omega\tau+\eta\rho)}\mathrm{d}\tau\mathrm{d}\rho, \end{aligned} \tag{3.86}$$

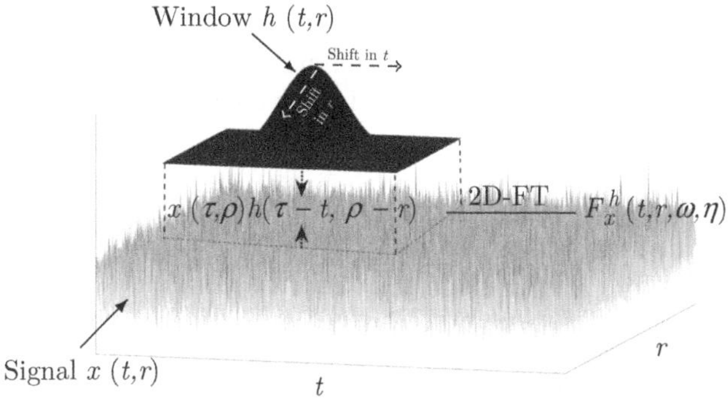

Figure 3.20 Illustration of the W2D-FT process. The window $h(t,r)$ slides along time t and space r. For each window location, we compute the window-signal product and then the two-dimensional Fourier transform. The resulting distribution is four-dimensional as we have the bivariate Fourier transform for each t and r coordinate.

where $h(t,r)$ is a continuous and differentiable two-dimensional analysis window, which moves along t and r and can be factorized as

$$h(t,r) = h(t)h(r). \tag{3.87}$$

Unless otherwise stated, the window is deemed to have real values; thus, the notation without complex conjugation $h(t,r)$ instead of $h^*(t,r)$ is used in the further part of this book.

Similarly to the STFT, the transform (3.86) can be interpreted twofold. In the first formulation of (3.86), the signal has a fixed position, and the window is sliding along time t and space r. In the right-hand side equation, the window is fixed in time and space, but the signal moves. For the graphical visualization of this property in one dimension, go back to Figure 2.7.

Akin to the STFT, the window time and space spread determine the resolution. By the one-dimensional window, its short-time spread allows for obtaining good time but poor frequency resolution, and vice versa – time spread extension leads to the time resolution deterioration at the cost of frequency resolution improvement. In the space domain, these properties are conserved.

Equation (3.86) corresponds to the two-dimensional Fourier transform of a signal portion weighted by the analysis window, both being two-dimensional functions. The key distinction between (3.85) and (3.86) lies in their dimensionality: the former evaluates the spectrum once for the entire signal, while the latter involves sliding the window across spatial dimensions, generating localized spectral estimates at each position. The conceptual framework of this transform is illustrated in Figure 3.20.

The four-dimensional energy distribution is determined as the squared modulus of (3.86) and is given by

$$E_x^h(t,r,\omega,\eta) = \left|F_x^h(t,r,\omega,\eta)\right|^2. \tag{3.88}$$

The W2D-FT distribution is invertible through the integration of $F_x^h(t,r,\omega,\eta)$ in the frequency and spatial-frequency domains for each t and r. The inverse formula can then be formulated as

$$x(t,r) = \frac{1}{(2\pi)^2 h(0,0)} \int_{-\infty}^{\infty}\int_{-\infty}^{\infty} F_x^h(t,r,\omega,\eta)\mathrm{d}\omega\mathrm{d}\eta, \tag{3.89}$$

where $h(0,0)$ is the window value at its center. The signal is recovered to its initial form after applying (3.89).

The four-dimensional distribution cannot be presented concisely and efficiently interpreted. Therefore, the results provided in this chapter focus on cross-sections, mostly (but not always) for $t = 0$ and $r = 0$, which correspond to a two-dimensional Fourier transform with a window centered on the signal. This choice is justified because when the window is positioned at the signal's center and maintains the same size, it fully overlaps with the signal, covering the same domain. Consequently, the Fourier transform of the windowed signal simplifies to a straightforward two-dimensional Fourier transform of the entire signal-window product (see Figure 3.20).

If the window is placed at different locations, it becomes possible to capture nonstationary characteristics of the signal. For example, to detect a transient at the beginning of the signal, one should examine the cross-section where the window encompasses the portion containing the transient. This feature will then be observable across all ω and η parameters. In other words, analyzing different (t,r) window positions allows the identification of various signal characteristics, such as discontinuities or rapid changes.

Now, we compare the classical bivariate spectrum to its windowed counterpart. The simulated signal is based on a continuous-time model defined as follows:

$$x(t) = e^{j2\pi\frac{100}{2}t^2} e^{j2\pi\frac{100}{2}r^2}, t \in [-50,50], r \in [-50,50]. \tag{3.90}$$

As can be seen, the signal is a product of two chirps. Next, it was sampled with the rate of $f_s = 101$ Sa/s in time and space. An additive white Gaussian noise component was added to the signal to model the effects of noise in a real-world scenario, resulting in the SNR of 20 dB. Figure 3.21 displays the classical two-dimensional Fourier transform without windowing and the cross-section of the W2D-FT in $t = 0$ and $r = 0$. The signal is designed to fully occupy the available bandwidth in both dimensions, ensuring that its frequency content extends up to the Nyquist limit in each direction. This effect is apparent as the spectrum in Figure 3.21a occupies the whole plane. In Figure 3.21b, the effect of windowing is apparent as the spectrum takes on the window's shape.

Although the presented windowed spectrum is not impressive, we must remember that this energy distribution is four-dimensional. Thus, we can plot several cross-sections, which is not the case with the classical two-dimensional Fourier transform. The W2D-FT's performance is shown in Figures 3.22–3.24 where three selected cross-sections in different domains are delineated.

The results correspond to three cross-sections for time and range bins $t_c = \{20,50,80\}$, $r_c = \{20,50,80\}$, and frequency bins $\omega_c = \{20,128,220\}$ and $\eta_c =$

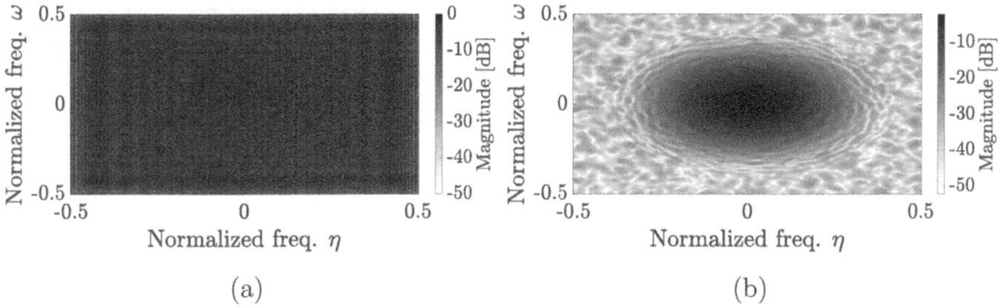

Figure 3.21 Comparison of the classical two-dimensional Fourier transform (a) and the cross-section in $t = 0$ and $r = 0$ for the W2D-FT.

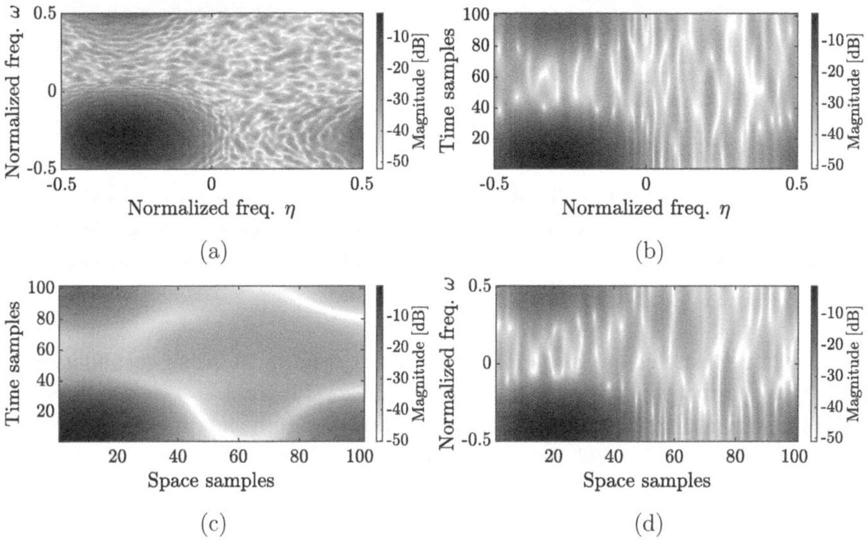

Figure 3.22 Cross-sections of the W2D-FT energy distribution for the first set of indices i.e. $t_c = 20$, $r_c = 20$, $\omega_c = 20$, and $\eta_c = 20$. (a) Spatial frequency-frequency distribution; (b) spatial frequency-time distribution; (c) space-time distribution; (d) space-frequency distribution.

$\{20, 128, 220\}$. The distribution consists of 101 samples in t and r, with 256 points in the Fourier transform along ω and η. The abovementioned values of t_c, r_c, ω_c and η_c are given in the sample indices.

For comparison, the results are organized into three groups of four, corresponding to the same cross-section indices—specifically, i.e., the first index for t, the first index for r, the first index for ω, and the first index for η. For example, the first set in Figure 3.22 corresponds to $t_c = 20$, $r_c = 20$, $\omega_c = 20$, and $\eta_c = 20$. Subsequent comparison in Figure 3.23 follows the second index ($t_c = 50$, $r_c = 50$, $\omega_c = 128$, $\eta_c = 128$) and the third in Figure 3.24 is displayed for the last index ($t_c = 80$, $r_c = 80$, $\omega_c = 220$, $\eta_c = 220$).

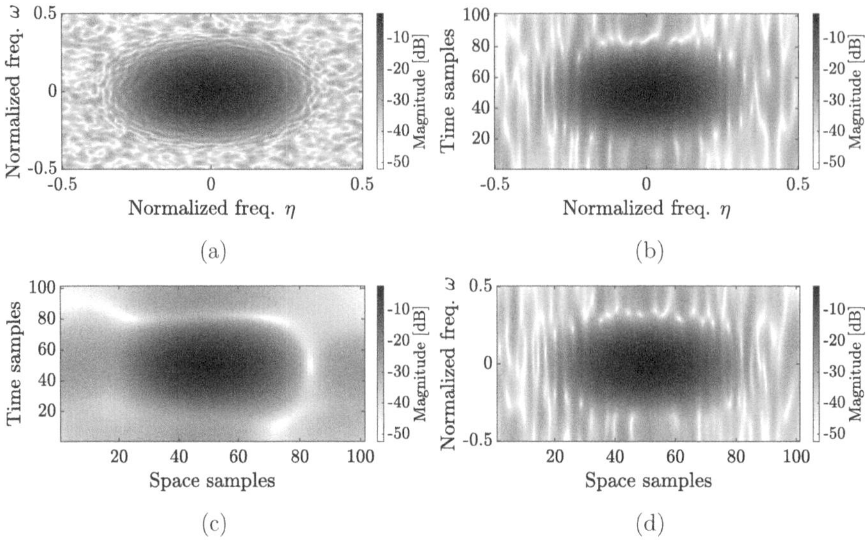

Figure 3.23 Cross-sections of the W2D-FT energy distribution for the second set of indices i.e. $t_c = 50$, $r_c = 50$, $\omega_c = 128$, and $\eta_c = 128$. (a) Spatial frequency-frequency distribution; (b) spatial frequency-time distribution; (c) space-time distribution; (d) space-frequency distribution.

From the illustrated cross-sections, we see the evolution of the signal and its spectrum in two dimensions. For the first pair of indices (Figure 3.22), we see the signal in its initial phase when the peaks occupy a lower spectrum and time fragments. The signal evolves, and the peak moves toward higher values of time, space, frequency, and spatial frequency. In each cut, we can assess the coordinates of the peak in each dimension, allowing for its characterization.

The constraints of the STFT naturally extend to the W2D-FT, as the latter generalizes the former to higher dimensions. Since W2D-FT is a direct extension of STFT principles, it inherits the same fundamental trade-offs and limitations.

One of the most significant inherited constraints is the Heisenberg-Gabor uncertainty principle obeying the inequality for temporal (Δ_t), spatial (Δ_r), frequency (Δ_ω), and spatial frequency (Δ_η)

$$\Delta_t \Delta_\omega \geq 1/4\pi \text{ and } \Delta_r \Delta_\eta \geq 1/4\pi. \tag{3.91}$$

This condition states that a function cannot be simultaneously well-localized in both the time (space) and frequency (spatial frequency) domains. This limitation remains valid in four-dimensional analysis, where the localization in the spatial domain inherently constrains resolution in the spatial-frequency domain. Consequently, just as in STFT, a trade-off must be made between achieving high spatial resolution and precise frequency characterization.

These limitations impact the interpretation and practical applications of W2D-FT, as they impose a fundamental bound on the precision with which spatial and

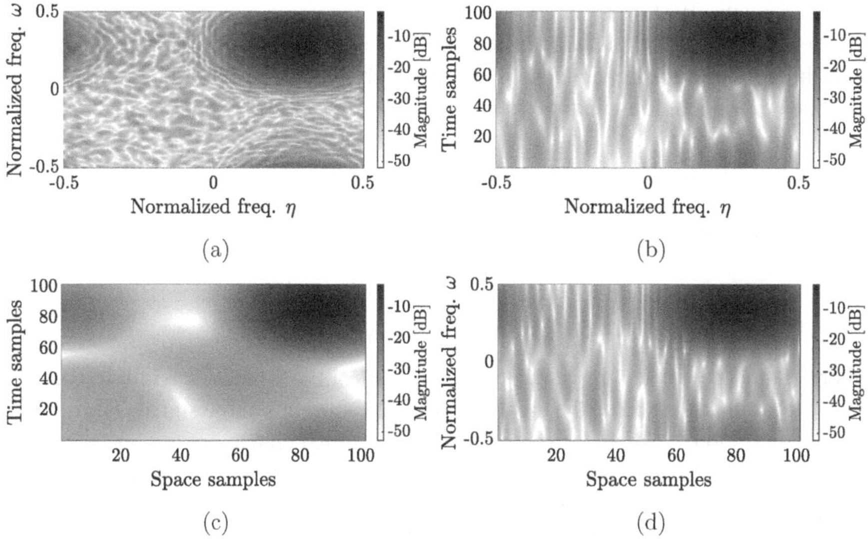

Figure 3.24 Cross-sections of the W2D-FT energy distribution for the third set of indices i.e. $t_c = 80$, $r_c = 80$, $\omega_c = 220$, and $\eta_c = 220$. (a) Spatial frequency-frequency distribution; (b) spatial frequency-time distribution; (c) space-time distribution; (d) space-frequency distribution.

frequency information can be simultaneously represented. In the next sections, we will convey the STFT framework, especially the resolution-improving tricks, to the W2D-FT.

3.7 IMPROVING W2D-FT READABILITY

In this section, we introduce extensions of the W2D-FT inspired by modifications applied to the STFT. These extensions focus on adapting the analysis window, thereby enhancing the energy concentration of the resulting four-dimensional signal representation. By carefully selecting and modifying the window function, a more precise localization of signal components in both spatial and frequency domains can be achieved, analogous to the resolution considerations discussed for the classical STFT in Chapter 2.

For a more comprehensive exploration of the analyzed extensions, we encourage the reader to conduct independent experiments using the provided toolbox. This allows for an in-depth analysis of the various windowing approaches and their effects on representing two-dimensional signals in the frequency domain.

3.7.1 ADAPTIVE W2D-FT

The process of adaptive STFT described in Section 2.5 aims to optimize the analysis window time spread based on the signal frequency modulation rate and can be extended to the W2D-FT. Since W2D-FT is a generalization of the STFT to two additional dimensions, the adaptation must consider the two-dimensional nature of both the window and the signal. As a result, each dimension should be treated separately, ensuring that the window's shape and size are optimized independently along both t and r axes.

In the adaptive W2D-FT, the window function is no longer fixed (compared to (3.86)) but varies depending on the local characteristics of the signal. The two-dimensional Gaussian function can be written as

$$h(t,r) = \frac{1}{\sqrt{2\pi}\sigma_t} e^{-\frac{t^2}{2\sigma_t^2}} \frac{1}{\sqrt{2\pi}\sigma_r} e^{-\frac{r^2}{2\sigma_r^2}} = \frac{1}{2\pi\sigma_t\sigma_r} e^{-\frac{t^2}{2\sigma_t^2}-\frac{r^2}{2\sigma_r^2}}, \tag{3.92}$$

where σ_t and σ_r are scaling parameters that control the window's width in each dimension, adjusted based on the local modulation properties of the signal. The optimal values of σ_t and σ_r can be derived from the local signal characteristics, ensuring that time-frequency and spatial-frequency resolutions are properly balanced.

Thus, unlike the one-dimensional case, the adaptation process must independently optimize the window along both t and r, allowing for more precise localization of the signal's features in the four-dimensional representation space.

In the case of a one-dimensional window, the optimal window width is determined based on the second-order derivative of the phase function $\phi(t)$, as given by (see (2.85)) [51]

$$\text{Optimal window width} \sim \frac{1}{2\left|\phi''(t)\right|}. \tag{3.93}$$

For the two-dimensional case, where the signal $x(t,r)$ has a phase function $\phi(t,r)$, the window must be adapted independently in both dimensions. Thus, the optimal window widths along t and r are determined by the second-order derivatives of the phase function in each respective direction

$$\text{Optimal window time width} \sim \frac{1}{2\left|\frac{\partial^2\phi(t,r)}{\partial t^2}\right|}, \tag{3.94}$$

$$\text{Optimal window space width} \sim \frac{1}{2\left|\frac{\partial^2\phi(t,r)}{\partial r^2}\right|}. \tag{3.95}$$

This ensures that the resolution is optimized separately in each dimension, allowing for a more accurate representation of the signal in the four-dimensional domain of the W2-DFT. For the signal given by (3.84) and the Gaussian window, we have the optimal time spread as

$$\sigma_t^o = \sqrt{\frac{1}{|\alpha_x|}}, \tag{3.96}$$

and

$$\sigma_r^o = \sqrt{\frac{1}{|\beta_x|}}, \tag{3.97}$$

where $\alpha_x \neq 0$ and $\beta_x \neq 0$ is the frequency modulation rate and spatial-frequency modulation rate, respectively.

If the signal exhibits mixed frequency components, the cross-dependencies between the dimensions should also be considered. This can be achieved by incorporating the Hessian matrix of the phase function

$$H_\phi = \begin{bmatrix} \frac{\partial^2 \phi(t,r)}{\partial t^2} & \frac{\partial^2 \phi(t,r)}{\partial t \partial r} \\ \frac{\partial^2 \phi(t,r)}{\partial r \partial t} & \frac{\partial^2 \phi(t,r)}{\partial r^2} \end{bmatrix}. \tag{3.98}$$

In this case, the optimal window widths can be derived based on the eigenvalues of H_ϕ, ensuring an adaptive finetuning of the window shape according to the local modulation characteristics of the signal. However, matrix (3.98) is 2×2 thus having two eigenvalues. It may happen that one needs to process the signal with two different windows to optimize the energy concentration for t and r separately.

When α_x or β_x evolve in time and space, the obvious time- and space-dependent relation arises. When $\alpha_x \neq$ const. and $\beta_x \neq$ const., we may write the following adaptive W2D-FT relation (h_a in the superscript denotes the adaptive window)

$$F_x^{h_a}(t,r,\omega,\eta) = \int_{-\infty}^{\infty}\int_{-\infty}^{\infty} x(\tau,\rho) h_a(\tau - t, \rho - r) e^{-j(\omega(\tau-t)+\eta(\rho-r))} \mathrm{d}\tau \mathrm{d}\rho, \tag{3.99}$$

which, for the adaptive Gaussian window, takes on the following form

$$F_x^{h_a}(t,r,\omega,\eta) = \int_{-\infty}^{\infty}\int_{-\infty}^{\infty} x(\tau,\rho) \underbrace{\frac{1}{2\pi\sigma_t(\tau-t)\sigma_r(\rho-r)} e^{-\frac{(\tau-t)^2}{2(\sigma_t(\tau-t))^2} - \frac{(\rho-r)^2}{2(\sigma_r(\rho-r))^2}}}_{\text{Adaptive bivariate window}} \times e^{-j(\omega(\tau-t)+\eta(\rho-r))} \mathrm{d}\tau \mathrm{d}\rho. \tag{3.100}$$

The corresponding energy distribution, whose cross-sections are depicted in the further part of this chapter, reads as

$$E_x^{h_a}(t,r,\omega,\eta) = \left| F_x^{h_a}(t,r,\omega,\eta) \right|^2. \tag{3.101}$$

Using a simple bivariate signal given by the model (3.90) with disjoint time and space relations, we can analyze the performance of the adaptive W2D-FT in practice. If we deem the frequency and spatial-frequency rate to be known (in practice, it should be estimated if unknown), we have

$$\sigma_t^o = \sigma_r^o = \sqrt{\frac{1}{|2\pi 100|}} \approx 0.04, \tag{3.102}$$

which, considering the sampling rate in t and r equal to 101 Sa/s, gives $\sigma_t^o = \sigma_r^o \approx 4$. Having common relations with entangled t and r in the frequency modulation would require (3.98) to consider all frequency modulation dependencies.

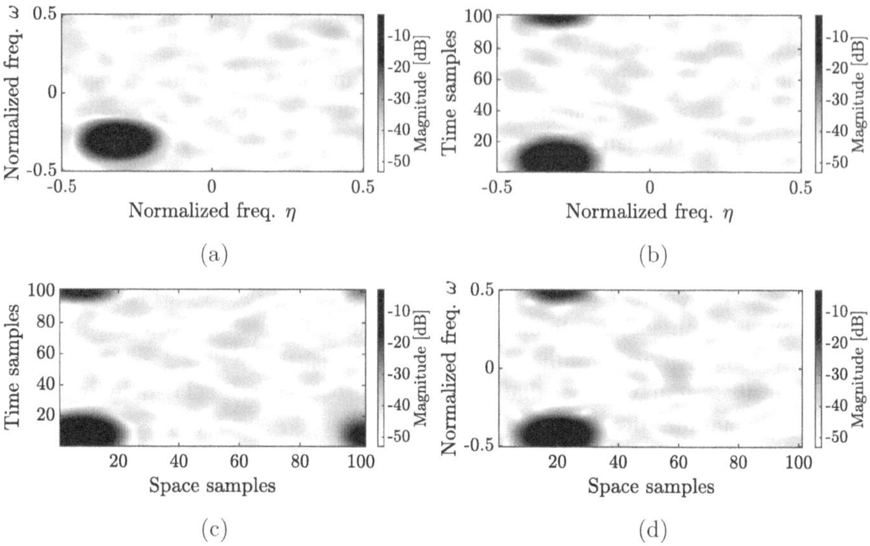

Figure 3.25 Cross-sections of the adaptive W2D-FT energy distribution for the first set of indices i.e. $t_c = 20$, $r_c = 20$, $\omega_c = 20$, and $\eta_c = 20$. (a) Spatial frequency-frequency distribution; (b) spatial frequency-time distribution; (c) space-time distribution; (d) space-frequency distribution.

Figures 3.25–3.27 present the cross-sections corresponding to those already delineated in Figures 3.22–3.24. The illustrated planes were obtained for the same time, space, frequency, and spatial-frequency indices so that we can directly compare the influence of the window width on the energy concentration.

When we look at the outcomes, we see the signal peak significantly narrower than in the initial results with the default window in Figures 3.22–3.24. The signal component occupies a significantly smaller area in each cross-section, leading to a compact representation and more precise parameter estimation.

Perception of the illustrated cross-sections allows for visual insight without a numerical measure of the improvement of energy concentration. To do so, let us extend the Rényi entropy (2.92) comprising the additional space and spatial-frequency dimensions. The new information content measure, but for the four-dimensional W2D-FT energy distribution, reads as

$$H_R^\gamma\left(E_x^h(t,r,\omega,\eta)\right) = \frac{1}{1-\gamma}\log_2\left(\frac{\int\limits_{-\infty}^{\infty}\int\limits_{-\infty}^{\infty}\int\limits_{-\infty}^{\infty}\int\limits_{-\infty}^{\infty}\left(E_x^h(t,r,\omega,\eta)\right)^\gamma \mathrm{d}t\mathrm{d}r\mathrm{d}\omega\mathrm{d}\eta}{\int\limits_{-\infty}^{\infty}\int\limits_{-\infty}^{\infty}\int\limits_{-\infty}^{\infty}\int\limits_{-\infty}^{\infty} E_x^h(t,r,\omega,\eta)\mathrm{d}t\mathrm{d}r\mathrm{d}\omega\mathrm{d}\eta}\right), \tag{3.103}$$

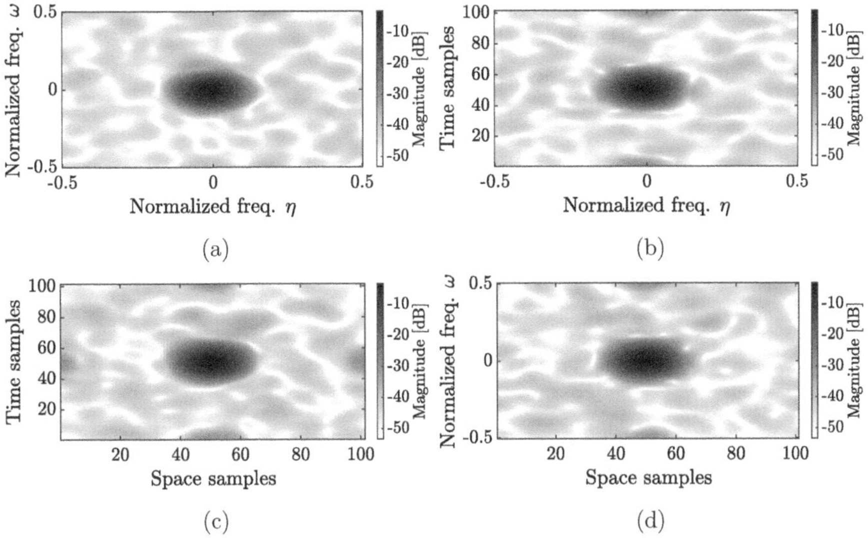

Figure 3.26 Cross-sections of the adaptive W2D-FT energy distribution for the second set of indices i.e. $t_c = 50$, $r_c = 50$, $\omega_c = 128$, and $\eta_c = 128$. (a) Spatial frequency-frequency distribution; (b) spatial frequency-time distribution; (c) space-time distribution; (d) space-frequency distribution.

with the entropy order γ. For $\gamma = 1$, Rényi entropy reduces to Shannon entropy, providing a standard measure of uncertainty. For higher values of γ (e.g., $\alpha = 2, 3...$), the measure emphasizes regions where the probability distribution is high, making it more sensitive to concentrated peaks in the energy distribution. Akin to the spectrogram-like approach, we assume $\gamma = 3$ in the further experiments.

Equation (3.103) describes how well the signal energy is localized in the four-dimensional representation. Its character is generally identical to the previously introduced Rényi entropy for the spectrogram. Lower entropy indicates that the energy is highly concentrated in a small region, meaning that the signal has a well-defined structure with negligible dispersion. This corresponds to sharp, well-resolved features in the W2D-FT energy representation. In turn, higher entropy suggests a more uniform energy distribution across the domain, indicating more significant uncertainty in the signal's localization in both time and frequency dimensions.

Figure 3.28 illustrates the third-order Rényi entropy for the W2D-FT energy distribution of the analyzed signal (3.90). For simplicity, the experiments fixed $\sigma_t = \sigma_r$ and spanned their normalized values in $\sigma_t \& \sigma_r \in [1, 20]$.

The results in Figure 3.28 exhibit a cohesion between our considerations at the beginning of this section (in particular, the condition on (3.94) and (3.95)) and the Rényi entropy for the four-dimensional energy distribution. We observe a minimum for $\sigma_t \& \sigma_r$ precisely for the optimal Gaussian window, where the time spread parameter exacerbates to attain its minimum at 4.

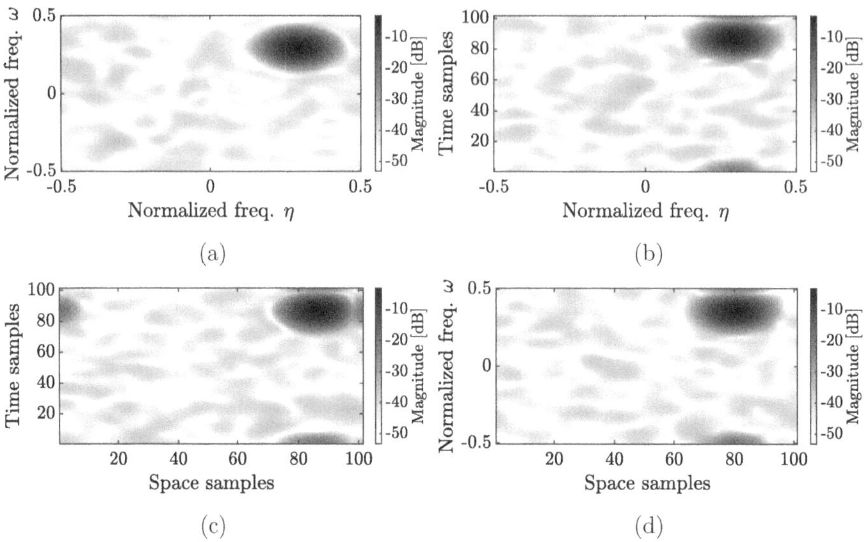

Figure 3.27 Cross-sections of the adaptive W2D-FT energy distribution for the third set of indices i.e. $t_c = 80$, $r_c = 80$, $\omega_c = 220$, and $\eta_c = 220$. (a) Spatial frequency-frequency distribution; (b) spatial frequency-time distribution; (c) space-time distribution; (d) space-frequency distribution.

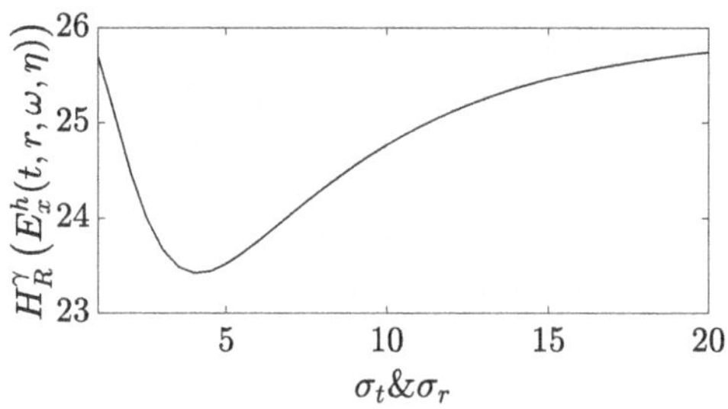

Figure 3.28 Depiction of the third-order ($\gamma = 3$) Rényi entropy of the four-dimensional energy distribution for different window widths in time and space dimensions. In the experiment, the signal model (3.90) disturbed by the noise giving SNR $=$ 20 dB was used.

In this section, we have demonstrated the application of adaptive W2D-FT, incorporating simulation examples and introducing a measure of energy concentration in the four-dimensional domain of time-space-frequency-spatial frequency. The presented results illustrate how adaptive window selection outperforms the localization of signal components, leading to enhanced resolution in both temporal and spatial-frequency dimensions.

As shown in Chapter 2, modifications of the analysis window can also include modulation by a complex chirp signal. This approach enabled further refinement of time-frequency representations by adapting to the local signal characteristics. In the next section, we extend the concept of chirplet transform to higher dimensions, achieving even greater energy concentration than in the adaptive W2D-FT. This extension leverages the benefits of chirp-adaptive analysis, providing a more precise representation of multidimensional nonstationary signals.

3.7.2 ADAPTIVE WINDOWED TWO-DIMENSIONAL CHIRPLET TRANSFORM

Bivariate windows have properties corresponding to their one-dimensional counterparts. Therefore, accompanying the time and space spread, we can define the rotation as modulation by a chirp-like signal. In this section, we introduce the adaptive windowed two-dimensional chirplet transform (W2D-CT), an extension of the previously discussed W2D-FT. While W2D-FT employs a fixed analysis window (or its adaptively varying time and space spread), W2D-CT enhances time-space-frequency-spatial-frequency resolution by utilizing a complex window modulated by a chirp signal in two dimensions. The key advantage of this approach lies in its ability to concentrate signal energy within the four-dimensional representation further, thereby outperforming localization in both time-frequency and space-spatial-frequency domains. This is particularly useful for signals exhibiting frequency and/or spatial-frequency modulation, where conventional windowed transforms may suffer from energy dispersion. By adapting the chirp parameters to the local characteristics of the signal, W2D-CT allows for a more accurate estimation of its instantaneous properties. The following section details the mathematical formulation of W2D-CT, its adaptive parameter selection, and its performance evaluation through simulation examples.

To define the W2D-CT, let us introduce the two-dimensional complex-valued window with a chirp. For the example of the Gaussian window (but any window is admissible), we may write it as follows

$$h_x(t,r) = \frac{1}{2\pi\sigma_t\sigma_r} e^{-\frac{t^2}{2\sigma_t^2} - \frac{r^2}{2\sigma_r^2}} e^{j\left(\frac{\alpha t^2}{2} + \frac{\beta r^2}{2}\right)}, \tag{3.104}$$

where α and β denote the frequency and spatial-frequency modulation rate, respectively, and the last exponent carries chirps in t and r domains. Substituting (3.104) into (3.86) gives the W2D-CT definition. We can also incorporate it with (3.100), giving the adaptive window width and the frequency rate (in t and r), and such a general definition reads as

$$F_x^{h_{\mathrm{dac}}}(t,r,\omega,\eta) = \int_{-\infty}^{\infty}\int_{-\infty}^{\infty} x(\tau,\rho) h^*_{\mathrm{dac}}(\tau - t, \rho - r) e^{-j(\omega(\tau-t)+\eta(\rho-r))} d\tau d\rho, \tag{3.105}$$

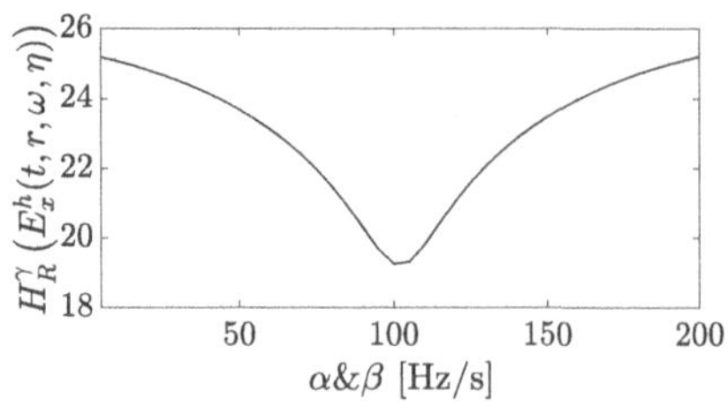

Figure 3.29 Depiction of the third-order ($\gamma = 3$) Rényi entropy of the four-dimensional energy distribution for different window frequency and spatial-frequency modulation rates. In the experiment, the signal model (3.90) disturbed by the noise giving SNR = 20 dB was used.

and for the adaptive Gaussian chirplet window, we have

$$F_x^{h_{\mathrm{dac}}}(t,r,\omega,\eta) = \int_{-\infty}^{\infty}\int_{-\infty}^{\infty} x(\tau,\rho) \underbrace{\frac{1}{2\pi\sigma_t(\tau-t)\sigma_r(\rho-r)} e^{-\frac{(\tau-t)^2}{2(\sigma_t(\tau-t))^2}-\frac{(\rho-r)^2}{2(\sigma_r(\rho-r))^2}}}_{\text{Adaptive bivariate window}} \\ \times \underbrace{e^{-j\frac{\alpha(\tau-t)(\tau-t)^2}{2}-j\frac{\beta(\rho-r)(\rho-r)^2}{2}}}_{\text{Adaptive modulation}} e^{-j(\omega(\tau-t)+\eta(\rho-r))} \mathrm{d}\tau \mathrm{d}\rho. \tag{3.106}$$

Remark 3.7. *In (3.105) and (3.106), we directly use the complex-valued window.*

In (3.106), we generalize the window so that it can have varying time and space width and chirp rate in two dimensions. Therefore $\sigma_t(t), \sigma_r(r), \alpha(t)$ and $\beta(r)$ are time/space-varying functions. Naturally, we define the energy distribution of (3.106) as

$$E_x^{h_{\mathrm{dac}}}(t,r,\omega,\eta) = \left| F_x^{h_{\mathrm{dac}}}(t,r,\omega,\eta) \right|^2. \tag{3.107}$$

We analyze the performance of (3.106) using the same simulated example as in previous cases, where the model defined by (3.90) was processed. By employing Rényi entropy to measure information content, we estimate the frequency and spatial-frequency modulation rates while treating the signal as unknown.

Figure 3.29 presents the Rényi entropy for different values of the modulation rates α and β. For simplicity, the simulations assume $\alpha = \beta$. The estimated modulation rate corresponds to the minimum of the Rényi entropy and is 100 Hz/s for both analyzed parameters.

The value estimated thanks to the plot in Figure 3.29 was used for processing the signal with the optimal window parameters. The cross-sections of the W2D-CT energy distribution are shown in Figures 3.30–3.32. The resolution improvement is significant. The signal peak is more focused in each analyzed cut. To numerically compare the improvement, let us look at the minima of the Rényi entropy in Figures 3.28 and 3.29. The minimum value for analysis comprising only the window time and space width attained 23.42 bits, while for the W2D-CT, we obtained 19.26

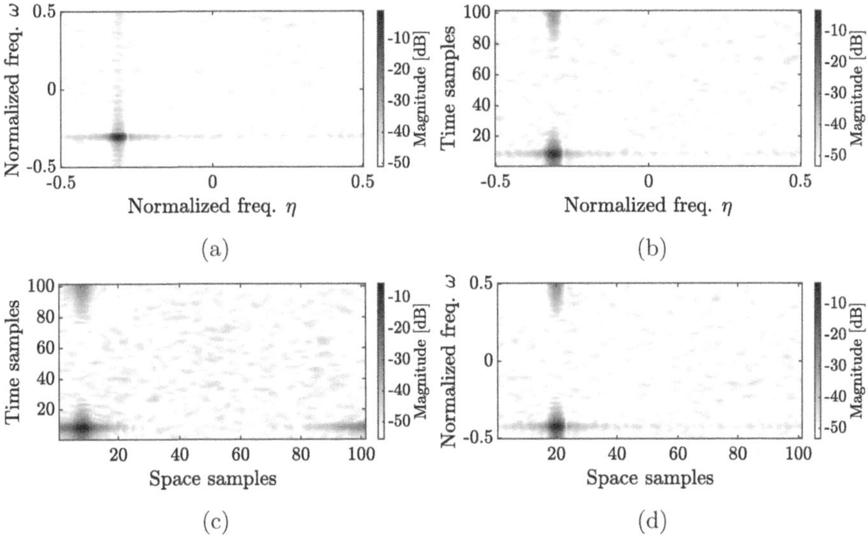

Figure 3.30 Cross-sections of the adaptive W2D-CT energy distribution for the first set of indices i.e. $t_c = 20$, $r_c = 20$, $\omega_c = 20$, and $\eta_c = 20$. (a) Spatial frequency-frequency distribution; (b) spatial frequency-time distribution; (c) space-time distribution; (d) space-frequency distribution.

bits. The difference of 4 bits is meaningful and shows the possibility of precise signal representations.

The results presented in Figures 3.28 and 3.29 highlight another practical aspect of the Rényi entropy. Both plots demonstrate that the frequency and spatial-frequency modulation rates can be estimated relatively straightforwardly. Although this approach is iterative and computationally demanding, it enables accurate signal parameter estimation even in the presence of high-noise levels.

To numerically illustrate the performance and feasibility of estimating the modulation rate in a high-noise environment, we plot the Rényi entropy for SNR $= -10$ dB. Figure 3.33 presents a distinct minimum for the two compared methods: W2D-FT and W2D-CT. Despite the presence of significant noise, the minimum remains clearly identifiable, corresponding to the correct frequency and spatial-frequency modulation rates. Although the Rényi entropy extreme is higher than for higher SNR, we can still differentiate it from the ambient values.

In this section, we have introduced the adaptive W2D-CT as an extension of the previously discussed W2D-FT. Incorporating a chirp-modulated analysis window enables improved energy concentration in the four-dimensional representation, particularly for signals exhibiting frequency and spatial-frequency modulation, and using Rényi entropy as a measure of information content allowed for the estimation of modulation rates, demonstrating the method's effectiveness even in high-noise environments.

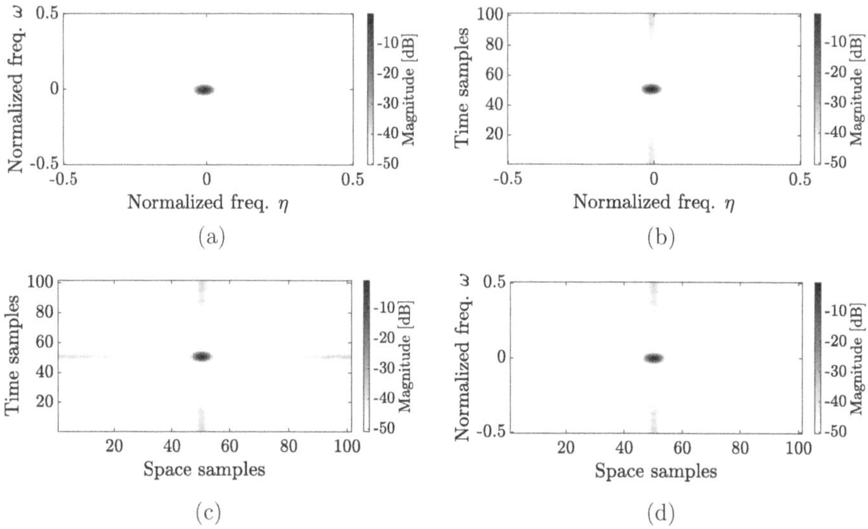

Figure 3.31 Cross-sections of the adaptive W2D-CT energy distribution for the second set of indices i.e. $t_c = 50$, $r_c = 50$, $\omega_c = 128$, and $\eta_c = 128$. (a) Spatial frequency-frequency distribution; (b) spatial frequency-time distribution; (c) space-time distribution; (d) space-frequency distribution.

The numerical simulations confirmed that the modulation rate could be reliably estimated by identifying the entropy minimum. The comparison of W2D-FT and W2D-CT showed that the latter provides better localization and robustness, as illustrated in Figures 3.28, 3.29, and 3.33. Despite the computational cost associated with the iterative nature of the approach, its ability to extract key signal parameters makes it a powerful tool for bivariate signal analysis.

In the next section, we analyze a real-world radar signal acquired from a frequency-modulated continuous-wave radar system. This case study will assess the practical feasibility of the examined techniques and validate their feasibility for real measurement data.

3.7.3 REAL-LIFE EXAMPLE

As an illustrative example of the presented methods' applicability, we analyze a frequency-modulated continuous-wave radar signal, previously discussed and depicted in Figure 3.17.

Remark 3.8. *If we compare the Zak-based definition of the range-velocity radar signal processing (3.78) and the W2D-FT (3.86) or any of its adaptive versions, we can notice a similarity between these formulas. The first equation expresses a transformation where a one-dimensional signal $x(t)$ is extended into a two-dimensional representation by incorporating a summation over an index k. This index effectively*

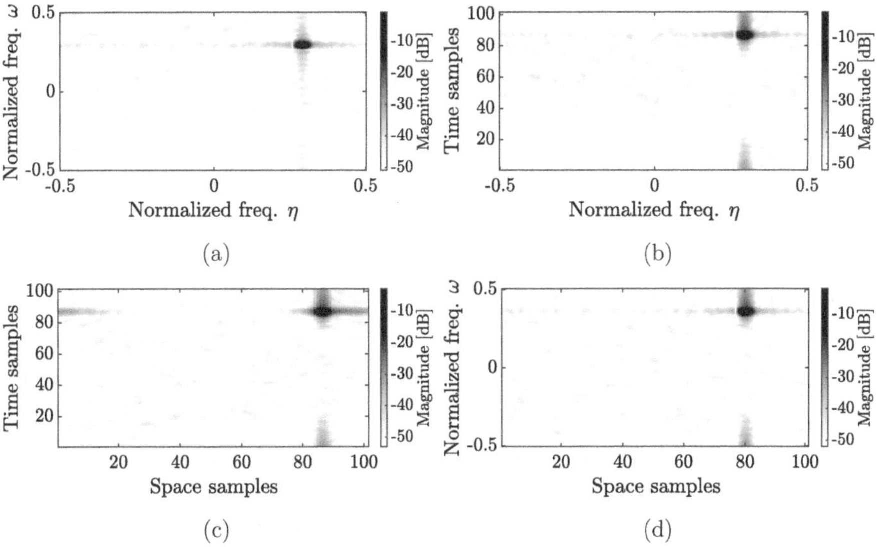

Figure 3.32 Cross-sections of the adaptive W2D-CT energy distribution for the third set of indices i.e. $t_c = 80$, $r_c = 80$, $\omega_c = 220$, and $\eta_c = 220$. (a) Spatial frequency-frequency distribution; (b) spatial frequency-time distribution; (c) space-time distribution; (d) space-frequency distribution.

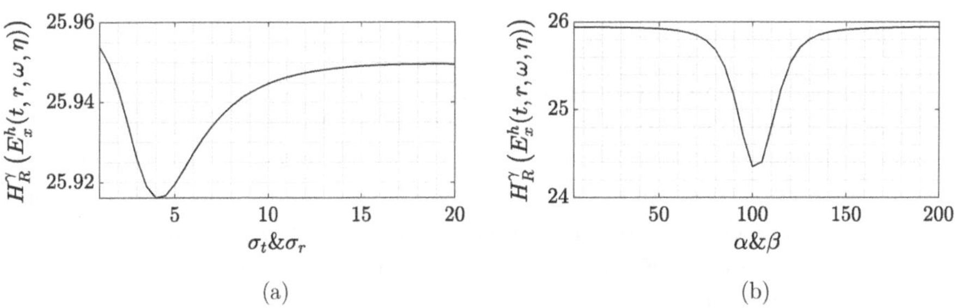

Figure 3.33 Rényi entropy for the W2D-FT (a) and the W2D-CT (b) for SNR $= -10$ dB. Two presented plots show the possibility of estimating the frequency and spatial frequency for a high-noise environment.

introduces a second dimension into the analysis. The window function $h(t,kT_p)$ emphasizes specific signal components before they are transformed via an exponential term that includes both η and ω. The integral then ensures that the transformation is computed over time, preserving temporal dependencies

$$M_x^h(\omega,\eta) = \int_0^{T_p} \sum_{k=-K/2}^{K/2} x(t+kT_p)h(t,kT_p)e^{-j\eta kT_p}e^{-j\omega t}\mathrm{d}t. \tag{3.108}$$

The second equation, on the other hand, explicitly sets up the prerequisite of a two-dimensional signal representation $x(t,r)$, making the two-dimensional nature of the operation explicit from the start. The transformation involves integration over both dimensions, leading to a function that describes joint time and frequency characteristics

$$F_x^h(t,r,\omega,\eta) = \int_{-\infty}^{\infty}\int_{-\infty}^{\infty} x(\tau,\rho)h(\tau-t,\rho-r)e^{-j(\omega(\tau-t)+\eta(\rho-r))}\mathrm{d}\tau\mathrm{d}\rho. \qquad (3.109)$$

Rewriting the equation, we obtain

$$F_x^h(t,r,\omega,\eta) = \int_{-\infty}^{\infty}\int_{-\infty}^{\infty} x(\tau+t,\rho+r)h(\tau,\rho)e^{-j(\omega\tau+\eta\rho)}\mathrm{d}\tau\mathrm{d}\rho. \qquad (3.110)$$

While the structure appears different, it fundamentally achieves the same effect as the first equation, as the additional dimension in the first equation is merely introduced through the summation over k. In contrast, in the second equation, it is inherently present.

From an analytical perspective, both formulations can be used interchangeably depending on the context of the problem. The first equation is valid when transitioning from a one-dimensional signal to a two-dimensional representation. In contrast, the second equation assumes the existence of a two-dimensional function from the outset. Both approaches lead to similar conclusions regarding the signal's temporal and spectral properties.

A segment containing the drone echo and its surroundings was extracted and processed to determine the optimal window parameters. The signal was modulated to center the main echo around $\omega = 0$ and $\eta = 0$. The original echo, obtained using the default window, is shown in Figure 3.34, where the main target appears in the center, with sidelobes predominantly spread along the η axis shaping a crest.

Next, the third-order Rényi entropy was computed for $\sigma_t \in [0.1, 5]$ and $\sigma_r \in [2, 12]$. The minimum value in Figure 3.35a indicates the optimal window parameters, ensuring the best range-velocity distribution in terms of information content. The optimized target echo, obtained using $\sigma_t = 1.3$ and $\sigma_r = 6.5$, is shown in

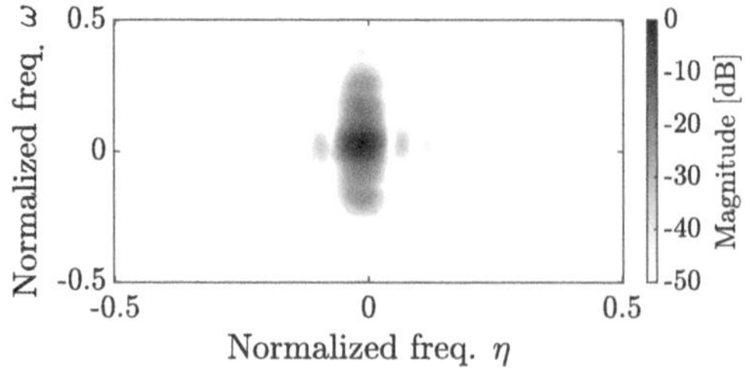

Figure 3.34 Range-velocity map fragment with a drone echo.

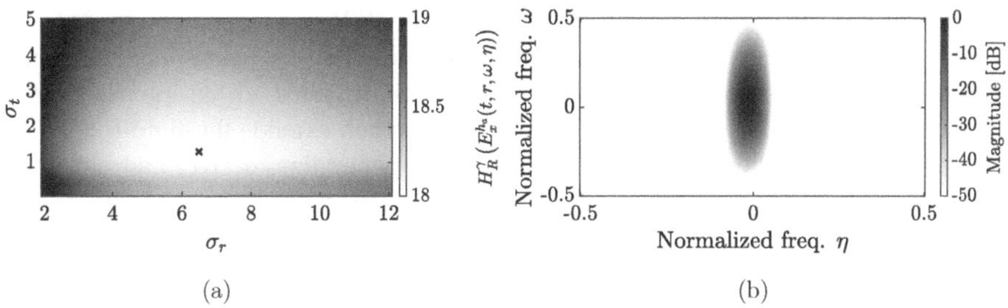

Figure 3.35 Optimized range-velocity map with a drone echo. (a) Rényi entropy optimization with the minimum value (with the superimposed cross) indicating optimal window parameters; (b) optimized range-velocity map. The outcomes were obtained using the W2D-FT with the optimized window time and space spread.

Figure 3.35b, demonstrating a well-defined peak with enhanced resolution and reduced sidelobes.

The same procedure was applied to the adaptive W2D-CT, where the window was optimized not only for temporal and spatial spread but also for frequency and spatial-frequency modulation. The results, illustrated in Figure 3.36, follow a similar optimization approach to that shown in Figure 3.35a. However, the modulation rate yielded only a marginal improvement in the Rényi entropy, suggesting that energy concentration is primarily influenced by the window's temporal and spatial spread. The estimated modulation rates are $\alpha = 65.31$ and $\beta = 6.33$ Hz/s, marked with a cross in Figure 3.36a.

The effectiveness of the adaptive W2D-CT was demonstrated using a frequency-modulated continuous-wave radar signal containing a drone echo. Optimal window parameters were determined by extracting and analyzing a relevant segment of the signal to improve the range-velocity distribution. The third-order Rényi entropy was employed to identify the best settings, leading to an optimized echo with enhanced

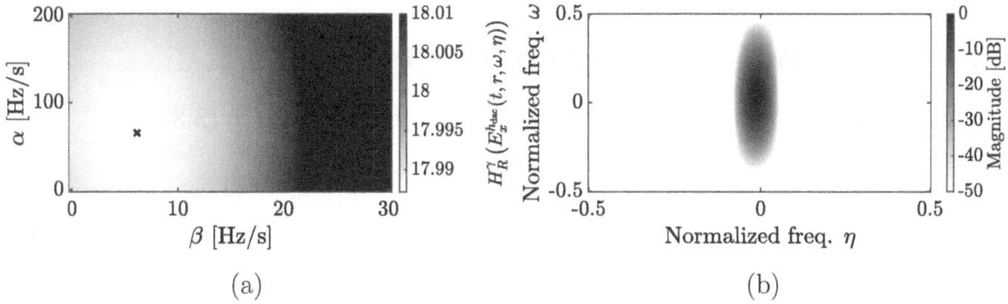

Figure 3.36 Optimized range-velocity map with a drone echo. (a) Rényi entropy optimization with the minimum value (with the superimposed cross) indicating optimal window parameters; (b) optimized range-velocity map. The outcomes were obtained using the W2D-CT with the optimized window time and space chirp rate.

resolution and reduced sidelobes. Additionally, the W2D-CT approach was investigated, incorporating both window width optimization and modulation adjustments. In the next section, we focus on the more advanced method for an intense energy concentration of the W2D-FT. The presented approach extends the time-frequency reassignment described in Section 2.7. Now, we generalize, derive, and elucidate the fundamentals for the W2D-FT concentration for higher-dimensional signals and apply this technique to simulated and real-life radar signals.

3.8 W2D-FT CONCENTRATION

So far, we have discussed methods for enhancing the W2D-FT concentration in four-dimensional space. While the presented techniques have tangibly improved the readability of the W2D-FT, particularly in its cross-sections, one may anticipate that reassignment and synchrosqueezing methods could further refine its performance beyond the previously described outcomes. In this section, we delve deeper into the problem of W2D-FT energy concentration by extending the aforementioned techniques known from the time-frequency analysis to higher-dimensional signal representations.

Inherited from time-frequency analysis, the W2D-FT can be generally relocated in all or selected dimensions. However, due to the four-dimensional nature of the distribution, the problem of computing the gradient extends to additional dimensions related to space and spatial frequency. To address this, we introduce time and space group delay estimators for time/space relocation and instantaneous frequency and spatial-frequency estimators to facilitate frequency and spatial-frequency reassignment. By leveraging these estimators, we establish a refined framework that enhances the energy concentration of the W2D-FT.

The theoretical analysis is further validated through practical radar applications, with a particular focus on ISAR radar imaging. Here, the demonstrated techniques are applied to improve image clarity and feature extraction, proving their effectiveness in real-world scenarios. The following subsections detail the mathematical formulation, computational aspects, and simulation results supporting this approach.

3.8.1 FOUR-DIMENSIONAL REASSIGNMENT

Let us recall the generic two-dimensional signal model defined in (3.84)

$$x(t,r) = \mathscr{A}_x e^{-\frac{(t-t_x)^2}{2T_x^2}} e^{-\frac{(r-r_x)^2}{2R_x^2}} e^{j\left(\phi_x+\omega_x t+\frac{\alpha_x t^2}{2}\right)} e^{j\left(\psi_x+\eta_x r+\frac{\beta_x r^2}{2}\right)}, \tag{3.111}$$

with all symbols preserving their meaning. In the first step, let us assume $T_x \to \infty$ and $R_x \to \infty$ so that the amplitude is constant in two dimensions. Also, we can temporarily set $\alpha_x = 0$ and $\beta_x = 0$ discarding the frequency and spatial-frequency deviation. Then, one has

$$x(t,r) = \mathscr{A}_x e^{j(\phi_x+\omega_x t)} e^{j(\psi_x+\eta_x r)}. \tag{3.112}$$

Differentiating (3.112) with respect to t yields

$$\frac{\partial x(t,r)}{\partial t} = j\omega_x x(t,r), \tag{3.113}$$

and differentiating $x(t,r)$ with respect to r gives

$$\frac{\partial x(t,r)}{\partial r} = j\eta_x x(t,r). \tag{3.114}$$

A partial derivative of $F_x^h(t,r,\omega,\eta)$ with respect to t leads to

$$\begin{aligned}\frac{\partial F_x^h(t,r,\omega,\eta)}{\partial t} &= \int_{-\infty}^{\infty}\int_{-\infty}^{\infty} x(\tau,\rho)\frac{\partial h(\tau-t,\rho-r)e^{-j(\omega(\tau-t)+\eta(\rho-r))}}{\partial t}\mathrm{d}\tau\mathrm{d}\rho \\ &= \int_{-\infty}^{\infty}\int_{-\infty}^{\infty} \frac{\partial x(\tau+t,\rho+r)}{\partial t}h(\tau,\rho)e^{-j(\omega\tau+\eta\rho)}\mathrm{d}\tau\mathrm{d}\rho,\end{aligned} \tag{3.115}$$

which can be written as

$$\begin{aligned}&\int_{-\infty}^{\infty}\int_{-\infty}^{\infty} x(\tau,\rho)\frac{\partial h(\tau-t,\rho-r)}{\partial t}e^{-j(\omega(\tau-t)+\eta(\rho-r))}\mathrm{d}\tau\mathrm{d}\rho + j\omega F_x^h(t,r,\omega,\eta) \\ &= \int_{-\infty}^{\infty}\int_{-\infty}^{\infty} \frac{\partial x(\tau+t,\rho+r)}{\partial t}h(\tau,\rho)e^{-j(\omega\tau+\eta\rho)}\mathrm{d}\tau\mathrm{d}\rho.\end{aligned} \tag{3.116}$$

Identifying the first term in (3.116) as the W2D-FT with the window derivative with respect to time, we may write

$$j\omega F_x^h(t,r,\omega,\eta) - F_x^{\partial_t h}(t,r,\omega,\eta) = \int_{-\infty}^{\infty}\int_{-\infty}^{\infty} \frac{\partial x(\tau+t,\rho+r)}{\partial t}h(\tau,\rho)e^{-j(\omega\tau+\eta\rho)}\mathrm{d}\tau\mathrm{d}\rho, \tag{3.117}$$

where the superscript $\partial_t h$ denotes the window derivative with respect to time.

Analogously, for r, one has

$$\begin{aligned}\frac{\partial F_x^h(t,r,\omega,\eta)}{\partial r} &= \int_{-\infty}^{\infty}\int_{-\infty}^{\infty} x(\tau,\rho)\frac{\partial h(\tau-t,\rho-r)e^{-j(\omega(\tau-t)+\eta(\rho-r))}}{\partial r}\mathrm{d}\tau\mathrm{d}\rho \\ &= \int_{-\infty}^{\infty}\int_{-\infty}^{\infty} \frac{\partial x(\tau+t,\rho+r)}{\partial r}h(\tau,\rho)e^{-j(\omega\tau+\eta\rho)}\mathrm{d}\tau\mathrm{d}\rho,\end{aligned} \tag{3.118}$$

which simplifies to

$$j\eta F_x^h(t,r,\omega,\eta) - F_x^{\partial_r h}(t,r,\omega,\eta) = \int_{-\infty}^{\infty}\int_{-\infty}^{\infty} \frac{\partial x(\tau+t,\rho+r)}{\partial r}h(\tau,\rho)e^{-j(\omega\tau+\eta\rho)}\mathrm{d}\tau\mathrm{d}\rho, \tag{3.119}$$

where $\partial_r h$ denote the window derivative with respect to r. From (3.115) and (3.118), we see a relation between differentiating the signal and the window in $F_x^h(t,r,\omega,\eta)$.

Next, we can substitute (3.113) into the last expression of (3.117) obtaining

$$j\omega F_x^h(t,r,\omega,\eta) - F_x^{\partial_t h}(t,r,\omega,\eta) = j\omega_x F_x^h(t,r,\omega,\eta), \tag{3.120}$$

Dividing both sides of (3.120) by $F_x^h(t,r,\omega,\eta)$ leads to the instantaneous frequency estimator

$$\hat{\omega}_x^{[1]}(t,r,\omega,\eta) = \omega - \Im\left(\frac{F_x^{\partial_t h}(t,r,\omega,\eta)}{F_x^h(t,r,\omega,\eta)}\right). \tag{3.121}$$

The same framework can be leveraged for the spatial-frequency estimator. Doing so, we can combine (3.114) and the last term of (3.119) obtaining

$$j\eta F_x^h(t,r,\omega,\eta) - F_x^{\partial_r h}(t,r,\omega,\eta) = j\eta_x F_x^h(t,r,\omega,\eta). \tag{3.122}$$

We end up with the spatial-frequency estimator in the following form:

$$\hat{\eta}_x^{[1]}(t,r,\omega,\eta) = \eta - \Im\left(\frac{F_x^{\partial_r h}(t,r,\omega,\eta)}{F_x^h(t,r,\omega,\eta)}\right). \tag{3.123}$$

Since we operate in four dimensions, we need to define two more reassignment operators responsible for the energy dislocation in the time and space domains. The time group delay estimator can be derived conventionally using differentiating (3.86) with respect to ω [89, 191] (see Section 2.7):

$$\begin{aligned}\frac{\partial F_x^h(t,r,\omega,\eta)}{\partial\omega} &= \\ &\int_{-\infty}^{\infty}\int_{-\infty}^{\infty} x(\tau,\rho)h(\tau-t,\rho-r)\frac{\partial e^{-j(\omega(\tau-t)+\eta(\rho-r))}}{\partial\omega}\mathrm{d}\tau\mathrm{d}\rho \\ &= jtF_x^h(t,r,\omega,\eta) - j\tau F_x^h(t,r,\omega,\eta).\end{aligned} \tag{3.124}$$

Assuming $\tau F_x^h(t,r,\omega,\eta) = F_x^{\mathcal{T}h}(t,r,\omega,\eta)$, where $\mathcal{T}h = th(t,r)$ is the time-weighted window, one has

$$\frac{\partial F_x^h(t,r,\omega,\eta)}{\partial\omega} = j\left(tF_x^h(t,r,\omega,\eta) - F_x^{\mathcal{T}h}(t,r,\omega,\eta)\right), \tag{3.125}$$

which leads to

$$\hat{t}_x(t,r,\omega,\eta) = t + \Re\left(\frac{F_x^{\mathcal{T}h}(t,r,\omega,\eta)}{F_x^h(t,r,\omega,\eta)}\right). \tag{3.126}$$

In a similar fashion, the space group delay can be derived as

$$\begin{aligned}\frac{\partial F_x^h(t,r,\omega,\eta)}{\partial\eta} &= \\ &\int_{-\infty}^{\infty}\int_{-\infty}^{\infty} x(\tau,\rho)h(\tau-t,\rho-r)\frac{\partial e^{-j(\omega(\tau-t)+\eta(\rho-r))}}{\partial\eta}\mathrm{d}\tau\mathrm{d}\rho \\ &= jrF_x^h(t,r,\omega,\eta) - j\rho F_x^h(t,r,\omega,\eta),\end{aligned} \tag{3.127}$$

giving (under the assumption on $rh(t,r) = \mathscr{R}h$)

$$\hat{r}_x(t,r,\omega,\eta) = r + \Re\left(\frac{F_x^{\mathscr{R}h}(t,r,\omega,\eta)}{F_x^h(t,r,\omega,\eta)}\right). \tag{3.128}$$

Equations (3.126) and (3.128) correspond to the first-order group delay estimator defined in Chapter 2 by (2.129). Precisely, (3.126) corresponds to time, whereas (3.128) is associated with space.

The four-dimensional reassignment can be built upon the reassignment operators (3.121), (3.123), (3.126), and (3.128), leading to its ultimate form

$$\begin{aligned} R_x^h(t,r,\omega,\eta) = &\int_{-\infty}^{\infty}\int_{-\infty}^{\infty}\int_{-\infty}^{\infty}\int_{-\infty}^{\infty} \left|F_x^h(\tau,\rho,\Omega,\zeta)\right|^2 \\ &\times\delta(t-\hat{t}_x(\tau,\rho,\Omega,\zeta), r-\hat{r}_x(\tau,\rho,\Omega,\zeta), \omega-\hat{\omega}_x(\tau,\rho,\Omega,\zeta), \\ &\eta-\hat{\eta}_x(\tau,\rho,\Omega,\zeta))\mathrm{d}\tau\mathrm{d}\rho\mathrm{d}\Omega\mathrm{d}\zeta. \end{aligned} \tag{3.129}$$

Simply put, (3.129) moves energy from each (t,r,ω,η) coordinate to its point of maximum contribution $(t,r,\omega,\eta) \mapsto (\hat{t}_x(t,r,\omega,\eta),\ \hat{r}_x(t,r,\omega,\eta),\ \hat{\omega}_x(t,r,\omega,\eta),\ \hat{\eta}_x(t,r,\omega,\eta))$ sometimes called a center of gravity. Note that each estimator works locally regardless of number of signal component.

To comprehensively illustrate the effectiveness of four-dimensional reassignment, we apply it to the same signal model (3.90) as in the previous analyses. The signal, corrupted by noise with a signal-to-noise ratio of SNR = 20 dB, was processed to achieve concentration across all four dimensions. As a result, the third-order Rényi entropy was reduced to 15.67 bits. This represents a significant drop of approximately 10 bits compared to the entropy obtained using adaptive transforms, highlighting a substantial improvement in the clarity and interpretability of the four-dimensional energy distribution.

To further illustrate this enhancement, Figures 3.37–3.39 present cross-sectional views of the reassigned energy distribution at selected bins along the time, space, frequency, and spatial-frequency dimensions. The signal characteristics are clearly discernible in each of these visualizations, demonstrating the improved localization and concentration of energy across all dimensions.

Our benchmark signal, defined by (3.90), is a bivariate linear chirp in two dimensions. While useful, this model imposes limitations, as each cross-section is interpreted as a single-component waveform. To demonstrate the capability of the W2D-FT to concentrate energy across all four dimensions simultaneously at each (t,r,ω,η) coordinate, we construct a temporary two-component signal by introducing a delay of 20 samples (0.2 s) in both t and r to (3.90) to simulate e.g., multipath propagation. Figures 3.40 and 3.41 illustrate the W2D-FT energy distribution before and after four-dimensional concentration for a selected set of indices.

Firstly, we observe two main components with improved separability after concentration. When relocating the energy, we operate on local centers of gravity, so each local salient point attracts the energy individually, thus concentrating the

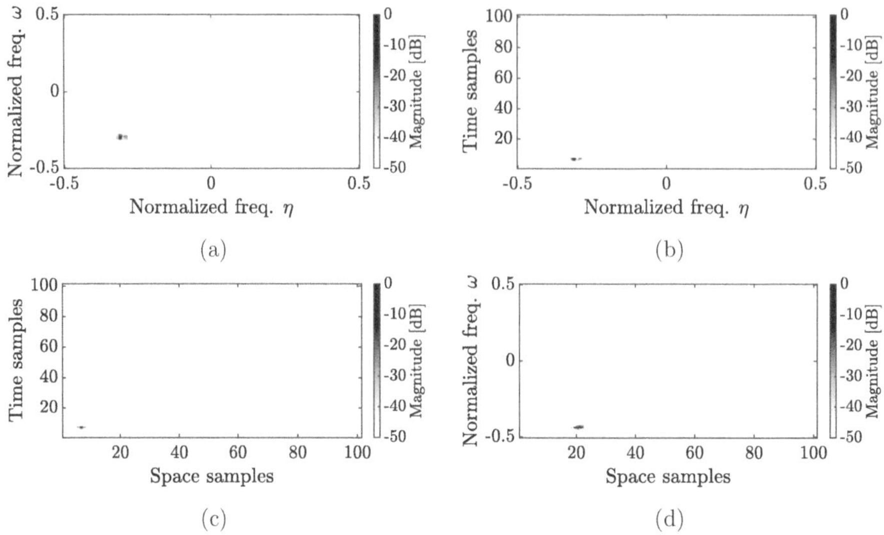

Figure 3.37 Cross-sections after four-dimensional reassignment for the first set of indices, i.e., $t_c = 20$, $r_c = 20$, $\omega_c = 20$, and $\eta_c = 20$. (a) Spatial frequency-frequency distribution; (b) spatial frequency-time distribution; (c) space-time distribution; (d) space-frequency distribution.

distribution. Secondly, we identify artifacts appearing between signal peaks in each cross-section, indicating that similar to time-frequency reassignment, four-dimensional reassignment is not a super-resolution method. When components are closely located in any two analyzed domains, interference may arise between them due to overlapping in the time-range-frequency-spatial-frequency space, leading to a mixing of reassigned energy. This effect results from the nonlinearity of the reassignment vectors, which depend on local gradient estimates of the phase (see Figure 2.36 for a comparison with time-frequency analysis).

Having validated the efficiency of four-dimensional reassignment using synthetic data, we now transition to analyzing real-world signals. Figure 3.42 illustrates the effect of four-dimensional reassignment on the concentration of the drone echo. The figure presents two sets of cross-sectional views, each comparing the original W2D-FT energy distribution with its reassigned counterpart. The first set, shown in Figure 3.42a and b, corresponds to cross-sections taken at $t = 0$ and $r = 0$. The second set, depicted in Figure 3.42c and d, represents cross-sections at $\omega = 0$ and $\eta = 0$.

In the original W2D-FT representation, the energy of the drone echo is more dispersed, hindering precise localization. For this case, the calculated third-order Rényi entropy is 20.02 bits, indicating a relatively high degree of energy spread. After the relocation of W2D-FT values, the energy is effectively concentrated, so the Rényi entropy is reduced to 9.78 bits, reflecting the improved localization and compactness of the signal.

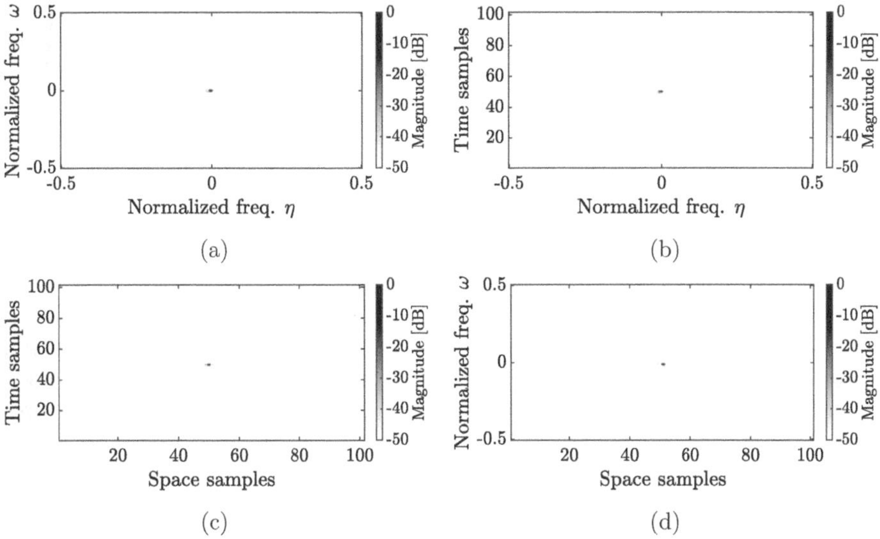

Figure 3.38 Cross-sections after four-dimensional reassignment for the second set of indices, i.e., $t_c = 50$, $r_c = 50$, $\omega_c = 128$, and $\eta_c = 128$. (a) Spatial frequency-frequency distribution; (b) spatial frequency-time distribution; (c) space-time distribution; (d) space-frequency distribution.

The concentrated distributions clearly show that the four-dimensional reassignment method effectively reduces the signal energy dispersion related to the Heisenberg-Gabor uncertainty principle and improves target visibility in time, space, frequency, and spatial-frequency domains. This enhanced energy localization facilitates a more accurate determination of the drone echo across multiple dimensions, demonstrating the W2D-FT concentration performance in readability improvement. Note that (3.129) works for each (t, r, ω, η) coordinate, thus leading to better component separability.

3.8.2 GENERALIZATION

The strategy for signal concentration in bivariate signals can be extended to any number of dimensions. Let us consider an N-dimensional signal, where $N \in \mathbb{N}^+$, defined as a product of monochromatic components that depend on time $t_x^{[n]}$ and angular frequency $\omega_x^{[n]}$. The subscript notation specifies the dimension index of a given component. The signal model can be expressed as

$$x(t^{[1\ldots N]}) = \mathscr{A}_x \prod_{n=1}^{N} e^{j\omega_x^{[n]} t^{[n]}}, \tag{3.130}$$

where the superscript $[n]$ denotes the nth dimension.

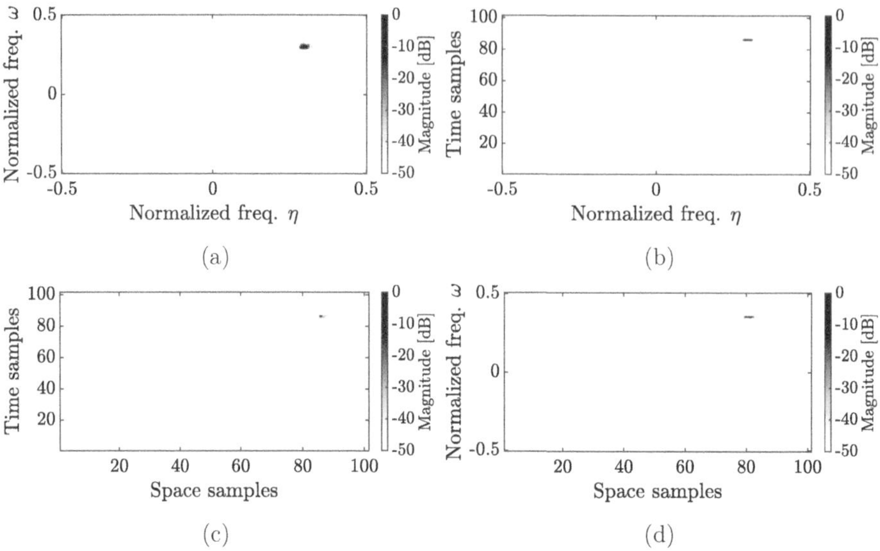

Figure 3.39 Cross-sections after four-dimensional reassignment for the third set of indices, i.e., $t_c = 50$, $r_c = 50$, $\omega_c = 128$, and $\eta_c = 128$. (a) Spatial frequency-frequency distribution; (b) spatial frequency-time distribution; (c) space-time distribution; (d) space-frequency distribution.

Following the same approach as in the bivariate case, the windowed N-dimensional Fourier transform can be formulated as

$$F_x^h(t^{[1\dots N]}, \omega^{[1\dots N]}) = \int_{-\infty}^{\infty} \cdots \int_{-\infty}^{\infty} x(\tau^{[1\dots N]} - t^{[1\dots N]}) h(\tau^{[1\dots N]}) \times e^{-j\omega^{[1\dots N]}(\tau^{[1\dots N]} - t^{[1\dots N]})} \mathrm{d}\tau^{[1\dots N]}, \tag{3.131}$$

where $h(t^{[1\dots N]})$ represents the N-dimensional window function.

The reassignment operators can be categorized into two groups: those associated with time (interpreted broadly as time, space, and additional dimensions) and those related to frequency (such as frequency, spatial frequency, etc.). These are denoted generally as $\hat{t}_x^{[1\dots N]}(\omega^{[1\dots N]}, t^{[1\dots N]})$ and $\hat{\omega}_x^{[1\dots N]}(\omega^{[1\dots N]}, t^{[1\dots N]})$, respectively. The nth frequency reassignment operator is given by [192]

$$\hat{\omega}_x^{[n]}(t^{[1\dots N]}, \omega^{[1\dots N]}) = \omega_x^{[n]} - \Im\left(\frac{F_x^{\partial_{t^{[n]}} h}(t^{[1\dots N]}, \omega^{[1\dots N]})}{F_x^h(t^{[1\dots N]}, \omega^{[1\dots N]})}\right), \tag{3.132}$$

where $\partial_{t^{[n]}} h = \frac{\partial h(t^{[1\dots N]})}{\partial t^{[n]}}$. Similarly, the nth time reassignment operator is given by

$$\hat{t}_x^{[n]}(t^{[1\dots N]}, \omega^{[1\dots N]}) = t^{[n]} + \Re\left(\frac{F_x^{\mathscr{T}^{[n]} h}(t^{[1\dots N]}, \omega^{[1\dots N]})}{F_x^h(t^{[1\dots N]}, \omega^{[1\dots N]})}\right), \tag{3.133}$$

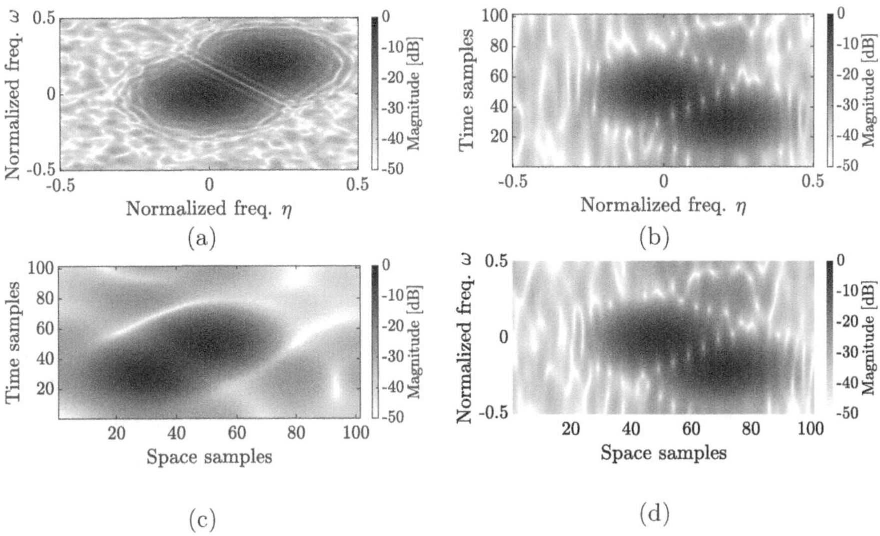

Figure 3.40 Cross-sections of the W2D-FT energy distribution for the second set of indices, i.e., $t_c = 50$, $r_c = 50$, $\omega_c = 128$, and $\eta_c = 128$. (a) Spatial frequency-frequency distribution; (b) spatial frequency-time distribution; (c) space-time distribution; (d) space-frequency distribution.

with $\mathscr{T}^{[n]}h = t^{[n]}h(t^{[1\ldots N]})$. Finally, the reassigned N-dimensional representation is given by

$$\begin{aligned} R_x^h(t^{[1\ldots N]}, \omega^{[1\ldots N]}) = \int_{-\infty}^{\infty} \cdots \int_{-\infty}^{\infty} & \left| F_x^h(t^{[1\ldots N]}, \omega^{[1\ldots N]}) \right|^2 \\ & \times \delta\big(t^{[1\ldots N]} - \hat{t}_x(t^{[1\ldots N]}, \omega^{[1\ldots N]}), \\ & \omega^{[1\ldots N]} - \hat{\omega}_x(t^{[1\ldots N]}, \omega^{[1\ldots N]})\big)\, \mathrm{d}\tau^{[1\ldots N]}\, \mathrm{d}\Omega^{[1\ldots N]}. \end{aligned} \tag{3.134}$$

We omit visualization of the N-dimensional distributions for obvious reasons. Nevertheless, (3.134) can find practical utilization for high-dimensional signal analysis, e.g., in 3D radar imaging, where the signal is represented in the three-dimensional space corresponding to the three-dimensional frequency space. We leave these considerations for the readers, who will undoubtedly find their applications of the N-dimensional reassignment. Now, let us go back to more practical problems.

3.8.3 DIMENSION-SELECTIVE REASSIGNMENT OF W2D-FT

Four-dimensional processing is time and resource-consuming. As can be seen in the open-source code provided within this book, four-dimensional reassignment needs four nested `for` loops iterating for time, space, frequency, and spatial-frequency bins, making the process relatively demanding. Not every single problem requires

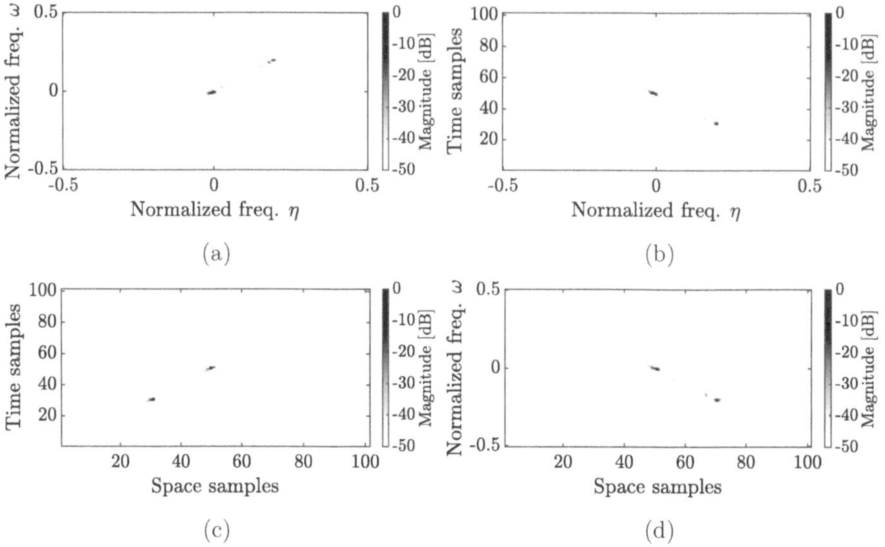

Figure 3.41 Cross-sections of the W2D-FT energy distribution for the second set of indices after four-dimensional concentration, i.e., $t_c = 50$, $r_c = 50$, $\omega_c = 128$, and $\eta_c = 128$. (a) Spatial frequency-frequency distribution; (b) spatial frequency-time distribution; (c) space-time distribution; (d) space-frequency distribution.

all these dimensions to have concentrated energy, especially in radar applications. A perfect example is an ISAR image, where we ultimately want the target projection to be precisely distributed in ω and η. ISAR imaging comes down to frequency and spatial-frequency analysis, and time and space relations can be neglected (of course, this statement is a simplification that can be made when discussing the final ISAR image, not the whole processing pipeline). In contrast, some applications can skip the frequency and spatial-frequency analysis and focus more on time and space for better signal synchronization. On the other hand, only time-frequency reassignment (or equivalently, space-spatial frequency) comes down to the discussion presented in Chapter 2, as the two remaining dimensions remain unaltered. Now, we can look closer at some exceptional cases of four-dimensional reassignment in which we concentrate the energy in one of two dimensions.

3.8.3.1 Frequency and Spatial–Frequency Reassignment

In Section 2.9, we introduced the time-frequency reassignment-like technique, which aims to shift the STFT values toward their local maxima. However, in this method, the transformation points are relocated exclusively along one axis. In the time-frequency context, synchrosqueezing differs from reassignment in the following respects:

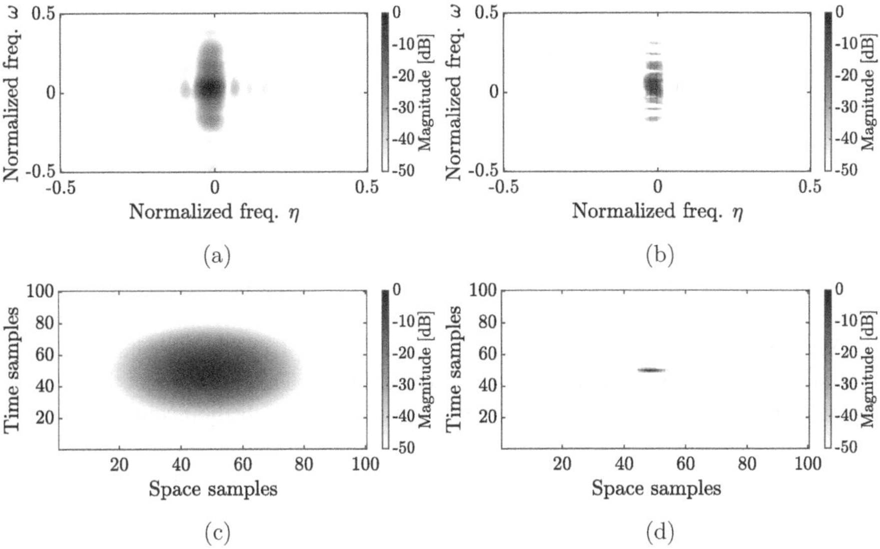

Figure 3.42 Real-life example of the drone echo concentration. (a) Original W2D-FT energy distribution in $t = 0$ and $r = 0$; (b) concentrated distribution corresponding to (a); (c) original W2D-FT energy distribution in $\omega = 0$ and $\eta = 0$; (d) concentrated (c).

- Reassignment operates both in the time and frequency domains while synchrosqueezing shifts the STFT values solely in one domain. In particular, vertical synchrosqueezing relocates the STFT points along the time (horizontal synchrosqueezing) or frequency (vertical synchrosqueezing) axes.
- Reassignment works with the squared modulus of the STFT (energy) while synchrosqueezing with the complex values.
- Synchrosqueezing (horizontal and vertical) is invertible. Firstly, we relocate the complex values instead of their squared magnitude. Secondly, the phase content is preserved when only one dimension is relocated. These factors make synchrosqueezing a versatile technique for signal retrieval and mode extraction.

In the four-dimensional space, we can define the synchrosqueezing counterparts operating in the selected domains. To keep the structured train of thought, we successively expand the W2D-FT with consecutive modifications in the following part of this chapter.

The W2D-FT can be, in general, concentrated in one, two, three, and four domains. The latter case was described in the previous section as the most comprehensive example corresponding to time-frequency reassignment. Starting with dimension-selective reassignment, we can identify two simple cases associated with frequency and spatial-frequency concentration. Let them be defined respectively

$$W_x^h(t,r,\omega,\eta) = \int_{-\infty}^{\infty} \left|F_x^h(t,r,\Omega,\eta)\right|^2 \delta(\omega - \hat{\omega}_x(t,r,\Omega,\eta))\mathrm{d}\Omega, \tag{3.135}$$

$$N_x^h(t,r,\omega,\eta) = \int_{-\infty}^{\infty} \left|F_x^h(t,r,\omega,\zeta)\right|^2 \delta(\eta - \hat{\eta}_x(t,r,\omega,\zeta))\mathrm{d}\zeta, \tag{3.136}$$

where $\hat{\omega}_x(t,r,\omega,\eta)$ and $\hat{\eta}_x(t,r,\omega,\eta)$ are given by (3.121) and (3.123). Let us disregard the practical utilization of (3.135) and (3.136), as their benefits are significantly lower compared to their combined application [192].

With the estimators (3.121) and (3.123), the W2D-FT can be simultaneously focused in two frequency-related dimensions: ω and η. This joint relocation in both frequency and spatial-frequency domains is particularly advantageous for applications where time and space relations are secondary, while frequency content and spatial-frequency distribution are of primary importance. By incorporating (3.135) and (3.136), we obtain

$$S_x^h(t,r,\omega,\eta) = \int_{-\infty}^{\infty}\int_{-\infty}^{\infty} \left|F_x^h(t,r,\Omega,\zeta)\right|^2 \delta(\omega - \hat{\omega}_x(t,r,\Omega,\zeta), \eta - \hat{\eta}_x(t,r,\Omega,\zeta))\mathrm{d}\Omega\mathrm{d}\zeta. \tag{3.137}$$

The fundamental concept behind (3.137) is to shift the squared magnitude of transform values within the frequency and spatial-frequency domains toward their most representative locations. This is accomplished by reallocating the distribution energy from its original position in the spatial-frequency domain to a new location that more accurately reflects the actual characteristics of the signal: $(t,r,\omega,\eta) \mapsto (t,r,\hat{\omega}_x(t,r,\omega,\eta),\hat{\eta}_x(t,r,\omega,\eta))$. The concentrated W2D-FT representation is computed by summing the squared modulus of transform values across all (t,r,ω,η) coordinates in the original representation, aligning them with the actual spectral distribution of the signal. This leads to an improved spectral representation with enhanced spatial-frequency localization, facilitating a more precise identification of signal components.

Equations (3.135), (3.136), and (3.137) draw the following conclusions. We relocate the transform energy in each variant, disabling the inverse transform. Moreover, we do not affect the remaining domains, so the precise localization in time and space can still be hampered as in the original W2D-FT. Nevertheless, not every application requires distribution readability enhancement in each domain. Instead, we can focus on the selected domains, e.g., frequency and spatial frequency.

If we look at the ISAR images (analyzed in Section 3.5) illustrated in Figures 3.12 and 3.13, we can deduce that the primary focus is on the frequency and spatial-frequency coordinates (scaled appropriately in meters). Indeed, time and range play a crucial role in ensemble processing; however, the final image is a two-dimensional spectrum representing the target projection on the ω and η plane. Therefore, we can use the technique in question to enhance the ISAR image readability.

The motivation for employing dimension-selective W2D-FT concentration in ISAR imaging can be outlined as follows. Increasing the signal bandwidth is one of the most direct methods to enhance ISAR image resolution. Numerous high-resolution radar systems have already been developed, as demonstrated in [193–195]. However, bandwidth expansion has inherent constraints, making it impractical to extend without repercussions indefinitely.

Firstly, a broader system bandwidth directly translates to increased system costs. Expanding the bandwidth necessitates leveraging high-speed digital-to-analog converters and wideband analog front-ends, leading to significant financial implications, especially for ultra-high-resolution imaging systems. Secondly, processing wideband signals introduces additional challenges. Conventional ISAR processing assumes a relatively constant Doppler shift across the observation bandwidth, yet in reality, the Doppler shift is frequency-dependent. This assumption becomes invalid when the signal bandwidth exceeds approximately 10% of the carrier frequency. Consequently, the entire processing chain must incorporate Doppler shift corrections that account for instantaneous frequency variations. This discrepancy arises because the Doppler shift at one end of the bandwidth differs significantly from that at the other end. Traditional narrowband processing methods become inadequate, necessitating Doppler frequency scaling, which increases computational complexity. Lastly, bandwidth availability is a critical limiting factor, as regulatory and technological constraints impose strict frequency allocation, restricting further enhancements in resolution.

Given these limitations, radar system designers increasingly shift the focus from hardware-based solutions to advanced signal-processing techniques. This approach allows cost-effective wideband systems to be used while leveraging sophisticated algorithms to improve ISAR image quality. Manifold strategies for ISAR imaging have been explored in the literature [196], including sparse aperture techniques [77, 197]. Additionally, ISAR imaging leveraging orbital angular momentum modes has been investigated in [198], demonstrating azimuth resolution enhancement. Other research efforts focus on compressive sensing methods, which assume that the ISAR image consists of a limited number of dominant components within a more extensive set [199–204]. However, compressive sensing techniques require prior knowledge, such as the number of signal components, which, if unknown, must be estimated, thereby increasing computational demands. Furthermore, as demonstrated in [132], compressive sensing does not inherently enhance ISAR resolution but rather prioritizes specific components within the image.

Analyzing the ISAR image as a bivariate spectrum enables the application of alternative methods, such as multiple signal classification (MUSIC) [205]. Recently, an ISAR image concentration technique has been introduced, tangibly enhancing the interpretability of ISAR images [132]. This method offers a promising direction for enhancing ISAR image quality without relying on bandwidth expansion, making it an attractive solution for high-resolution radar imaging.

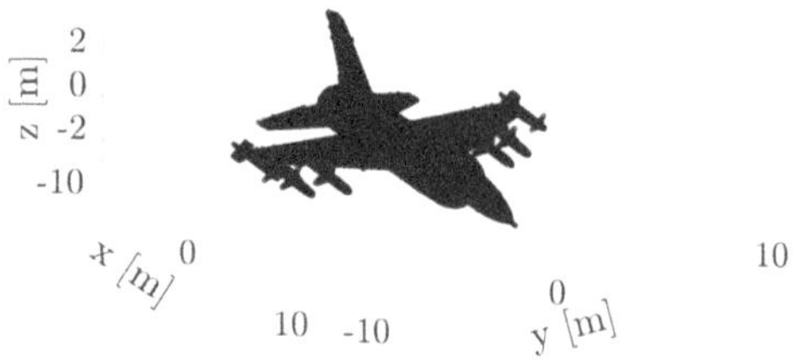

Figure 3.43 Jet fighter model used in the simulation.

In view of the fact that an ISAR image can be reduced to a two-dimensional spectrum while neglecting spatiotemporal dependencies, the dimension-selective reassignment of W2D-FT is particularly relevant in this context. By applying this technique, we enhance the distinguishability of radar signal scatterers while maintaining a lower computational burden. As demonstrated in (3.137), energy focusing without considering the dimensions t and r requires fewer computational resources, making it a justified approach for ISAR image concentration. Therefore, we present the practical application of frequency and spatial-frequency W2D-FT reassignment using ISAR imaging as an example.

Remark 3.9. *In the examples related to ISAR imaging, we focus solely on the frequency ω and spatial-frequency η domains. Henceforth, to easily visualize the result as a target projection on the plane, the outcomes presented in this section are limited to the cross-sections for $t = 0$ and $r = 0$ (with a window in the signal center) unless otherwise stated. It is also assumed that the window size equals the signal size; thus, the presented ISAR images can be interpreted as the bivariate windowed spectrum. Such an assumption permits focusing on the ISAR image readability improvement without analysis of time and range relations as they do not play a crucial role in the presented ISAR image analysis.*

The W2D-FT frequency and spatial-frequency reassignment can be visualized using simple ISAR imaging with a geometry illustrated in Figure 3.6, where the target flights perpendicularly to the monostatic radar beam. Let us examine the simulated ISAR images of the same target (jet fighter, see Figure 3.43), illuminated by signals with different bandwidths. During the 0.5-second simulation, the transmitter generated narrowband Gaussian noise with a flat amplitude spectrum at a carrier frequency of $f_c = 10$ GHz. Two bandwidths were used: $B_1 = 50$ MHz and $B_2 = 100$ MHz, with a signal repetition rate of $f_r = 1$ kHz (the signal was continuous, but the same fragment was repeated). The original image was processed using the polar format algorithm [206–208], and the resulting data was utilized to visualize the W2D-FT dimension-selective concentration. Before image processing, the signal was corrupted with weak additive white Gaussian noise, resulting in an SNR of 50 dB.

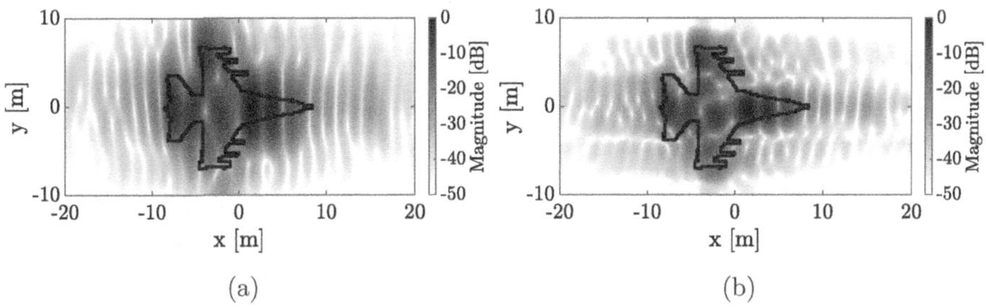

Figure 3.44 Original ISAR images of the simulated jet fighter. (a) Image for $B_1 = 50$ MHz; (b) image for $B_2 = 100$ MHz.

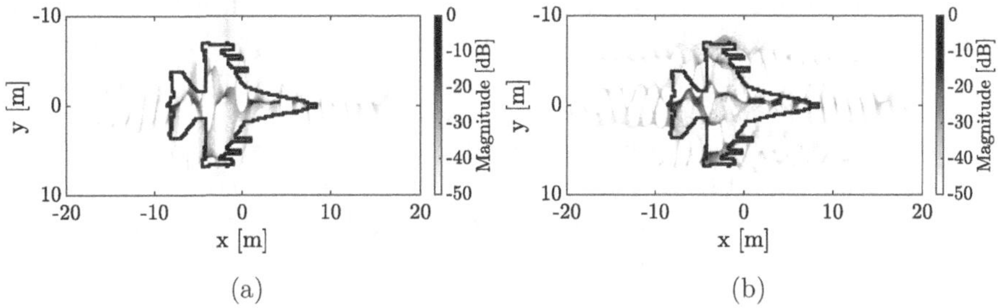

Figure 3.45 Concentrated ISAR images of the simulated jet fighter. (a) Image for $B_1 = 50$ MHz; (b) image for $B_2 = 100$ MHz.

The original images for two signal bandwidths were processed and are shown in Figure 3.44. Also, the target contour is superimposed to alleviate its shape perception. The images can be interpreted as cross-sections of the W2D-FT energy distribution for $t = 0$ and $r = 0$. This comparison shows the obvious relation between the signal bandwidth and the range resolution. We can extract more target features in the range (y) dimension when the signal bandwidth increases. Also, the cross-range resolution remains unaltered as it is associated with the integration time and the carrier frequency (see (1.19)). Therefore, when comparing the ISAR images from Figure 3.44, we see a range resolution improvement for $B_2 - 100$ MHz, manifesting as more distinctive and prominent scattering points. Despite the different resolutions, we can extract the jet fighter's shape from both images, especially its hull and wings.

Image concentration, allowing for better scatterer echo separation, can be obtained using the dimension-selective W2D-FT concentration according to (3.137). The results are shown in Figure 3.45.

The comparison between the original ISAR images (Figure 3.44) and the concentrated ISAR images (Figure 3.45) demonstrates the benefits of applying the concentration method. In the original images, the target contours appear blurred, with noticeable sidelobes and spread out scattering effects, hindering the ability to distinguish individual reflectors. Additionally, the background interference is less prominent, especially for the lower bandwidth case ($B_1 = 50$ MHz).

In contrast, the concentrated ISAR images exhibit significant outperformance in the sharpness and clarity of the target shape. The salient points are more localized, reducing the effect of energy smearing across the image. This enhancement is especially noticeable in the case of $B_2 = 100$ MHz, where finer details of the target are more clearly visible. The reduction of sidelobes and unwanted artifacts also enhances the interpretability of the images, making the concentrated ISAR images more suitable for automatic target recognition and classification applications.

However, energy is still spread in the concentrated images. If we look at the phase approximation (3.17), we see several nonlinear terms in the signal received by the imaging radar. Although the target flights with a constant velocity, its movement affects geometry (in particular, γ angle in Figure 3.6), thereby making the frequency of the received signal vary. The consequence is that the frequency and spatial-frequency estimators (3.121) and (3.123), which build upon a prerequisite of the constant amplitude and frequency, are biased.

Similar to the vertical synchrosqueezing profoundly examined in Section 2.9, we can derive the bias of (3.121) and (3.123) for the quadratic phase. In this case, if α_x and β_x can no longer be neglected and T_x and R_x do not tend to infinity, (3.111) retains its original form (whereas previously, we simplified it to (3.112)). Therefore, differentiating (3.111) with respect to t leads to

$$\frac{\partial x(t,r)}{\partial t} = (a_x t + b_x)x(t,r), \tag{3.138}$$

where $a_x = -1/T_x^2 + j\alpha_x$, $b_x = t_x/T_x^2 + j\omega_x$. In the same way, one obtains the derivative of $x(t,r)$ with respect to r:

$$\frac{\partial x(t,r)}{\partial r} = (c_x r + d_x)x(t,r), \tag{3.139}$$

where $c_x = -1/R_x^2 + j\beta_x$, $d_x = r_x/R_x^2 + j\eta_x$. Terms $(a_x t + b_x)$ and $(c_x r + d_x)$ are the complex instantaneous frequency in the ω and η domain, respectively.

Substituting the signal derivative (3.138) into (3.117) yields

$$j\omega F_x^h(t,r,\omega,\eta) - F_x^{\partial_t h}(t,r,\omega,\eta) = (a_x t + b_x)F_x^h(t,r,\omega,\eta) + a_x F_x^{\mathcal{T}h}(t,r,\omega,\eta). \tag{3.140}$$

Remembering that $a_x = -1/T_x^2 + j\alpha_x$, $b_x = t_x/T_x^2 + j\omega_x$, we can write

$$\begin{aligned} j\omega F_x^h(t,r,\omega,\eta) - F_x^{\partial_t h}(t,r,\omega,\eta) =& (-t/T_x^2 + j\alpha_x t + t_x/T_x^2 + j\omega_x)F_x^h(t,r,\omega,\eta) \\ & - (-1/T_x^2 + j\alpha_x)F_x^{\mathcal{T}h}(t,r,\omega,\eta). \end{aligned} \tag{3.141}$$

Dividing by $F_x^h(t,r,\omega,\eta)$ yields

$$\begin{aligned} j\omega - \frac{F_x^{\partial_t h}(t,r,\omega,\eta)}{F_x^h(t,r,\omega,\eta)} =& (-t/T_x^2 + j\alpha_x t + t_x/T_x^2 + j\omega_x) \\ & - (-1/T_x^2 + j\alpha_x)\frac{F_x^{\mathcal{T}h}(t,r,\omega,\eta)}{F_x^h(t,r,\omega,\eta)}. \end{aligned} \tag{3.142}$$

The imaginary part of (3.141) should give a pure ω_x when unbiased, but the frequency modulation term affects the equation. The conclusion is that the simple frequency estimator is biased when the linear (and generally nonlinear) frequency modulation occurs. The same holds for $\hat{\eta}_x^{[1]}(t,r,\omega,\eta)$ with analogous derivations.

The bias effect is visible in Figure 3.45, where the scatterers, although focused, do not exhibit strong concentration. To alleviate the effect of frequency modulation, let us derive the second-order estimators $\hat{\omega}_x(t,r,\omega,\eta)$ and $\hat{\eta}_x(t,r,\omega,\eta)$. From (3.124), we know that

$$\begin{aligned}\frac{\partial F_x^h(t,r,\omega,\eta)}{\partial\omega} &= \iint x(\tau+t,\rho+r)h(\tau,\rho)\frac{\partial e^{-j(\omega\tau+\eta\rho)}}{\partial\omega}\mathrm{d}\tau\mathrm{d}\rho \\ &= -j\iint \tau x(\tau+t,\rho+r)h(\tau,\rho)e^{-j(\omega\tau+\eta\rho)}\mathrm{d}\tau\mathrm{d}\rho \\ &= -jF_x^{\mathcal{T}h}(t,r,\omega,\eta),\end{aligned} \tag{3.143}$$

and we can differentiate

$$F_x^{\partial_t h}(t,r,\omega,\eta) = -(a_xt+b_x-j\omega)F_x^h(t,r,\omega,\eta)-a_xF_x^{\mathcal{T}h}(t,r,\omega,\eta), \tag{3.144}$$

with respect to ω obtaining

$$\begin{aligned}-jF_x^{\partial_t\mathcal{T}h}(t,r,\omega,\eta) =& j(a_xt+b_x-j\omega)F_x^{\mathcal{T}h}(t,r,\omega,\eta) \\ &-jF_x^h(t,r,\omega,\eta)+ja_xF_x^{\mathcal{T}^2h}(t,r,\omega,\eta),\end{aligned} \tag{3.145}$$

which simplifies to

$$\begin{aligned}&F_x^{\partial_t\mathcal{T}h}(t,r,\omega,\eta)+F_x^h(t,r,\omega,\eta) = \\ &\qquad -(a_xt+b_x-j\omega)F_x^{\mathcal{T}h}(t,r,\omega,\eta)-a_xF_x^{\mathcal{T}^2h}(t,r,\omega,\eta).\end{aligned} \tag{3.146}$$

Combining (3.144) and (3.146) into the linear system leads to

$$\begin{cases} F_x^{\partial_t h}(t,r,\omega,\eta) = -(a_xt+b_x-j\omega)F_x^h(t,r,\omega,\eta) \\ \qquad -a_xF_x^{\mathcal{T}h}(t,r,\omega,\eta), \\ F_x^{\partial_t\mathcal{T}h}(t,r,\omega,\eta)+F_x^h(t,r,\omega,\eta) = -(a_xt+b_x-j\omega)F_x^{\mathcal{T}h}(t,r,\omega,\eta) \\ \qquad -a_xF_x^{\mathcal{T}^2h}(t,r,\omega,\eta).\end{cases} \tag{3.147}$$

Its solution leads to two estimators

$$\hat{\omega}_x^{[\omega]}(t,r,\omega,\eta) = \Im\left(\frac{\begin{array}{l}F_x^{\partial_t\mathcal{T}h}(t,r,\omega,\eta)F_x^{\mathcal{T}h}(t,r,\omega,\eta) \\ \quad -F_x^{\mathcal{T}^2h}(t,r,\omega,\eta)F_x^{\partial_t h}(t,r,\omega,\eta) \\ \quad\quad +F_x^{\mathcal{T}h}(t,r,\omega,\eta)F_x^h(t,r,\omega,\eta)\end{array}}{\begin{array}{l}\left(F_x^{\mathcal{T}h}(t,r,\omega,\eta)\right)^2 \\ \quad -F_x^{\mathcal{T}^2h}(t,r,\omega,\eta)F_x^h(t,r,\omega,\eta)\end{array}}\right), \tag{3.148}$$

$$\hat{\alpha}_x^{[\omega]}(t,r,\omega,\eta) = \Im\left(\frac{\begin{array}{l} F_x^h(t,r,\omega,\eta)^2 \\ \quad +F_x^h(t,r,\omega,\eta)F_x^{\partial_t \mathscr{T}h}(t,r,\omega,\eta) \\ \qquad -F_x^{\partial_t h}(t,r,\omega,\eta)F_x^{\mathscr{T}h}(t,r,\omega,\eta) \end{array}}{\begin{array}{l} \left(F_x^{\mathscr{T}h}(t,r,\omega,\eta)\right)^2 \\ \quad -F_x^{\mathscr{T}^2h}(t,r,\omega,\eta)F_x^h(t,r,\omega,\eta) \end{array}}\right), \tag{3.149}$$

where $\mathscr{T}^n h = t^n h(t,r)$ and $\partial_t^m \mathscr{T}^n h = \frac{\partial^m}{\partial t^m} t^n h(t,r)$.

We can also derive the estimators using derivatives with respect to time. Differentiating (3.144) once yields

$$F_x^{\partial_t^2 h}(t,r,\omega,\eta) = -(a_x t + b_x - j\omega)F_x^{\partial_t h}(t,r,\omega,\eta) - a_x F_x^{\partial_t \mathscr{T}h}(t,r,\omega,\eta), \tag{3.150}$$

where $\partial_t^m h = \frac{\partial^m}{\partial t^m} h(t,r)$. Combining (3.150) with (3.144) into the linear system allows us to end up with the following estimators:

$$\hat{\omega}_x^{[t]}(t,r,\omega,\eta) = \omega - \Im\left(\frac{F_x^{\partial_t^2 h}(t,r,\omega,\eta)F_x^{\mathscr{T}h}(t,r,\omega,\eta) - F_x^{\partial_t \mathscr{T}h}(t,r,\omega,\eta)F_x^{\partial_t h}(t,r,\omega,\eta)}{F_x^{\partial_t h}(t,r,\omega,\eta)F_x^{\mathscr{T}h}(t,r,\omega,\eta) - F_x^{\partial_t \mathscr{T}h}(t,r,\omega,\eta)F_x^h(t,r,\omega,\eta)}\right), \tag{3.151}$$

$$\hat{\alpha}_x^{[t]}(t,r,\omega,\eta) = \Im\left(\frac{F_x^{\partial_t^2 h}(t,r,\omega,\eta)F_x^h(t,r,\omega,\eta) - \left(F_x^{\partial_t h}(t,r,\omega,\eta)\right)^2}{F_x^{\partial_t h}(t,r,\omega,\eta)F_x^{\mathscr{T}h}(t,r,\omega,\eta) - F_x^{\partial_t \mathscr{T}h}(t,r,\omega,\eta)F_x^h(t,r,\omega,\eta)}\right). \tag{3.152}$$

Analogously, we can derive the estimators for the η domain. Since the derivations are very same as for ω, we let them as a practical exercise for the readers, finishing with their final forms as follows:

$$\hat{\eta}_x^{[\eta]}(t,r,\omega,\eta) = \Im\left(\frac{\begin{array}{l} F_x^{\partial_r \mathscr{R}h}(t,r,\omega,\eta)F_x^{\mathscr{R}h}(t,r,\omega,\eta) \\ \quad -F_x^{\mathscr{R}^2h}(t,r,\omega,\eta)F_x^{\partial_r h}(t,r,\omega,\eta) \\ \qquad +F_x^{\mathscr{R}h}(t,r,\omega,\eta)F_x^h(t,r,\omega,\eta) \end{array}}{\begin{array}{l} \left(F_x^{\mathscr{R}h}(t,r,\omega,\eta)\right)^2 \\ \quad -F_x^{\mathscr{R}^2h}(t,r,\omega,\eta)F_x^h(t,r,\omega,\eta) \end{array}}\right), \tag{3.153}$$

$$\hat{\beta}_x^{[\eta]}(t,r,\omega,\eta) = \Im\left(\frac{\begin{array}{l} F_x^h(t,r,\omega,\eta)^2 \\ \quad +F_x^h(t,r,\omega,\eta)F_x^{\partial_r \mathscr{R}h}(t,r,\omega,\eta) \\ \qquad -F_x^{\partial_r h}(t,r,\omega,\eta)F_x^{\mathscr{R}h}(t,r,\omega,\eta) \end{array}}{\begin{array}{l} \left(F_x^{\mathscr{R}h}(t,r,\omega,\eta)\right)^2 \\ \quad -F_x^{\mathscr{R}^2h}(t,r,\omega,\eta)F_x^h(t,r,\omega,\eta) \end{array}}\right), \tag{3.154}$$

$$\hat{\eta}_x^{[r]}(t,r,\omega,\eta) = \eta - \Im\left(\frac{F_x^{\partial_r^2 h}(t,r,\omega,\eta)F_x^{\mathscr{R}h}(t,r,\omega,\eta) - F_x^{\partial_r \mathscr{R}h}(t,r,\omega,\eta)F_x^{\partial_r h}(t,r,\omega,\eta)}{F_x^{\partial_r h}(t,r,\omega,\eta)F_x^{\mathscr{R}h}(t,r,\omega,\eta) - F_x^{\partial_r \mathscr{R}h}(t,r,\omega,\eta)F_x^{h}(t,r,\omega,\eta)}\right), \tag{3.155}$$

$$\hat{\beta}_x^{[r]}(t,r,\omega,\eta) = \Im\left(\frac{F_x^{\partial_r^2 h}(t,r,\omega,\eta)F_x^{h}(t,r,\omega,\eta) - \left(F_x^{\partial_r h}(t,r,\omega,\eta)\right)^2}{F_x^{\partial_r h}(t,r,\omega,\eta)F_x^{\mathscr{R}h}(t,r,\omega,\eta) - F_x^{\partial_r \mathscr{R}h}(t,r,\omega,\eta)F_x^{h}(t,r,\omega,\eta)}\right). \tag{3.156}$$

where $\mathscr{R}^n h = r^n h(t,r)$ and $\partial_r^m \mathscr{R}^n h = \frac{\partial^m}{\partial r^m} r^n h(t,r)$.

We can see a clear analogy between (3.148) and (3.151), comparing them to (2.153) and (2.154) derived in Chapter 2. Indeed, as an extension of the STFT to higher dimensions, the W2D-FT demonstrates cohesion and natural enrichment of the former with a spatial dimension while maintaining consistent relationships. From (3.148)–(3.152), let us take advantage of the instantaneous frequency and spatial-frequency estimators. Robust (second-order) instantaneous frequency and spatial-frequency estimators substituted to (3.137) allow intense energy concentration for signal components with frequency modulation [192].

Of course, the estimators can be combined, specifically the first-order ones (3.121) with either (3.153) or (3.155), as well as (3.123) with either (3.148) or (3.151). As in the time-frequency domain, the difference between the estimators derived as derivatives of the W2D-FT with respect to time (space) or frequency (spatial frequency) is negligible, and they can be used interchangeably, just like the modulation rate estimators. Table 3.3 presents possible combinations for several typical cases, considering the signal characteristics and the required accuracy.

When optimizing computational efficiency, each estimator's complexity should be considered. The first-order estimators (3.121) and (3.123) are particularly efficient for harmonic spatial frequencies. Their computation requires three W2D-FTs: the first one utilizing a window function $h(t,r)$, while the remaining two employ its

Table 3.3
Suggested Estimators' Combinations for Different Cases

Case	(3.121)	(3.123)	(3.148) or (3.151)	(3.153) or (3.155)
Frequency harmonic in t, harmonic in r	Yes	Yes	No	No
Frequency harmonic in t, modulated in r	Yes	No	No	Yes
Frequency modulated in t, harmonic in r	No	Yes	Yes	No
Frequency modulated in t, modulated in r	No	No	Yes	Yes
Higher precision	No	No	Yes	Yes
Faster processing	Yes	Yes	No	No

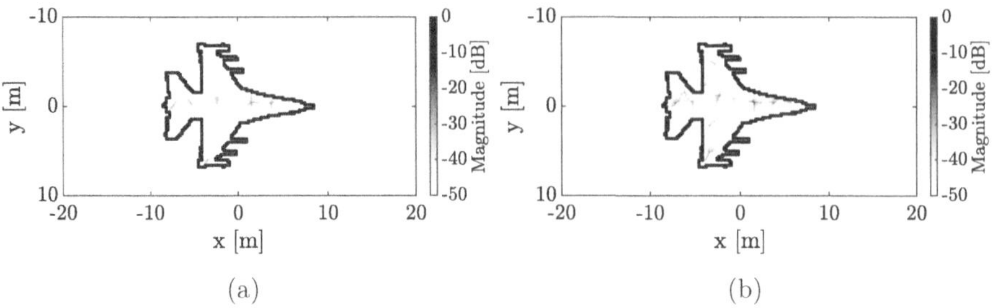

Figure 3.46 Strongly concentrated ISAR images of the simulated jet fighter. (a) Image for $B_1 = 50$ MHz; (b) image for $B_2 = 100$ MHz.

derivatives with respect to t and r. For second-order estimators (both derived from the W2D-FT derivative with respect to time/space or frequency/spatial frequency), five W2D-FTs are needed for each $\hat{\omega}(t,r,\omega,\eta)$ and $\hat{\eta}(t,r,\omega,\eta)$ case.

Ignoring computational overhead, let us process the ISAR images from Figure 3.44 using (3.148) and (3.153) - the frequency and spatial-frequency estimators derived from the W2D-FT derivatives with respect to ω and η as an example. The outcomes can be seen in Figure 3.46.

Compared to the ISAR images in Figure 3.45, the images in Figure 3.46 exhibit significantly improved scatterer localization and reduced artifacts. The enhanced concentration is a direct result of employing the second-order instantaneous frequency and instantaneous spatial-frequency estimators. While the previous results from Figure 3.45 already provided better focus than the original ISAR images, some blurring and unwanted distortions remained, particularly in regions with weaker reflections. In contrast, the refined estimation method used in Figure 3.46 further suppresses these distortions, leading to well-defined reflecting points of the jet fighter. The scatterers appear sharper, with significantly attenuated side lobes, making the dominant reflectors more distinguishable. This enhancement clearly demonstrates the effectiveness of second-order frequency and spatial-frequency estimators in ISAR imaging, particularly when imaging complex objects with fine structural details and acceleration in both time and space domains. Moreover, the focused images for $B = 50$ and 100 MHz remain comparable, showing no significant differences even at bandwidths lower than twice the original. This approach represents a step toward super-resolution radar imaging,[2] enabling high-quality image reconstruction despite reduced system bandwidth and associated costs.

Remark 3.10. *ISAR images are usually compared through their contrast and entropy as typical images. Since we analyze only one cross-section of the W2D-FT for $t = 0$ and $r = 0$, this kind of analysis is reasonable as the Rényi entropy takes all dimensions into account. Neglecting t and r relationships allows for simplifying the image analysis process. Image contrast is defined as*

$$C = \frac{1}{MN} \sum_{m=1}^{M} \sum_{n=1}^{N} (I(m,n) - \mu)^2, \tag{3.157}$$

Table 3.4
Quantitative Comparison of Image Metrics for Different Processing Methods for the Simulated ISAR Image for $B_1 = 50$ MHz

Image	Entropy	Contrast
Original ISAR image – Figure 3.44a	1.52	5.40
Simply concentrated ISAR image – Figure 3.45a	0.02	60.61
Strongly concentrated ISAR image – Figure 3.46a	0.002	204.61

Table 3.5
Quantitative Comparison of Image Metrics for Different Processing Methods for the Simulated ISAR Image for $B_1 = 100$ MHz

Image	Entropy	Contrast
Original ISAR image – Figure 3.44b	1.27	6.09
Simply concentrated ISAR image – Figure 3.45b	0.05	34.66
Strongly concentrated ISAR image – Figure 3.46b	0.002	209.01

where $I(m,n)$ represents the pixel intensity at location (m,n), μ is the mean intensity of the image, and $M \times N$ denotes the image dimensions. A higher contrast value signifies a sharper image with well-defined features, which is crucial for accurate ISAR target recognition. The image entropy reads as

$$H = -\sum_{n=1}^{N} p_n \log_2 p_n, \tag{3.158}$$

where p_n represents the normalized histogram distribution of pixel intensities in the ISAR image, and N is the total number of intensity levels. A lower entropy value typically indicates a more concentrated ISAR image, as energy is focused on fewer pixels, leading to higher interpretability. These metrics are used for ISAR image comparison. Note that they are calculated for images in the linear scale after normalization, although they are depicted in decibels for visualization purposes.

Tables 3.4 and 3.5 summarize metrics (contrast and entropy) of the presented images from Figures 3.44–3.46. The results clearly present reinforcement of image concentration and the enhancement of their readability. We see the drop (for entropy) and growth (for contrast) of approximately two degrees of magnitude between the original images and those concentrated using the second-order estimators, ensuring strong concentration.

The real-life signal-processing results can be seen in Figure 3.47. The target was the KR-03 Puchatek glider, already analyzed in Section 3.3.

The comparison from Figure 3.47 presents the differences between the original and post-processed images. The top image (a) represents the original ISAR reconstruction obtained using conventional processing techniques (in this case, it was the phase gradient algorithm). The image exhibits noticeable smearing and artifacts, which obscure the target's structural details and reduce interpretability. The

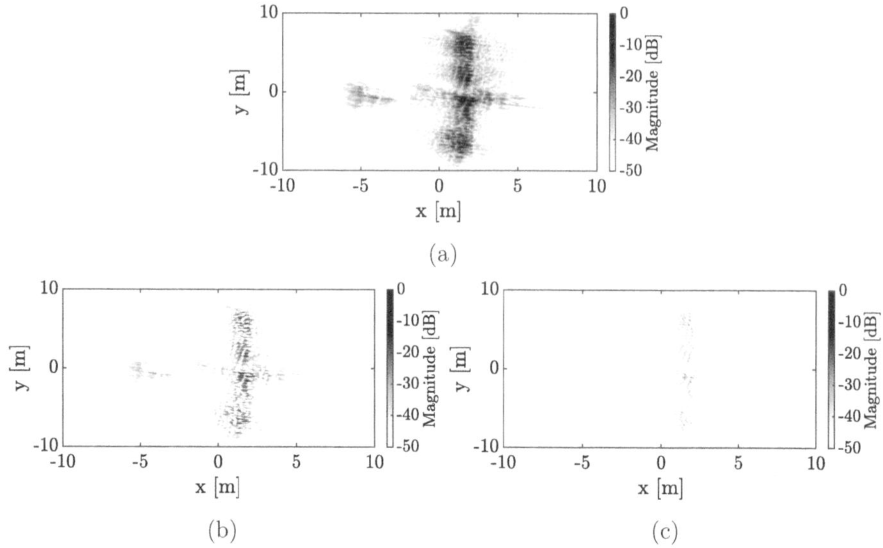

Figure 3.47 Real-life ISAR image after the original processing (a), concentrated using simple estimators (b) and robust estimators (c).

bottom-left image (b) results from applying simple frequency and spatial-frequency estimators for ISAR image concentration. The target's structure appears slightly more defined than the original image, but residual smearing effects remain. While the energy distribution is somewhat more localized, the image still lacks sufficient contrast for precise target recognition. The bottom-right image (c) demonstrates the effect of applying robust estimators (3.151) and (3.155) for frequency and instantaneous frequency concentration (those comprising the signal phase acceleration). This approach leads to a significantly improved ISAR image with higher energy localization. The target features are more clearly defined, with reduced noise and sidelobe artifacts. The concentration of energy results in an enhanced contrast, making this processing method particularly useful for practical ISAR applications.

Overall, comparing these three images highlights the effectiveness of advanced concentration techniques. The robust estimators lead to a more focused ISAR image, improving readability and potential target identification accuracy. The same conclusion follows from Table 3.6.

Table 3.6

Quantitative Comparison of Image Metrics for Different Processing Methods for the Real-Life ISAR Image in Figure 3.47

Image	Entropy	Contrast
Original ISAR image – Figure 3.47a	16.23	2.88
Simply concentrated ISAR image – Figure 3.47b	13.05	18.70
Strongly concentrated ISAR image – Figure 3.47c	11.91	51.20

Applying the W2D-FT concentration to the range-velocity map in the frequency-modulated continuous-wave radar, we can quickly obtain a concentrated target echo (generally, the whole range-velocity map). The possibility stems from the fact that according to (3.78), the one-dimensional signal is chopped into batches and forms the bivariate signal, whose two-dimensional spectrum gives the frequency and spatial-frequency distribution, corresponding to the range and velocity of the target. In the same way as for the dimension-selective W2D-FT concentration given by (3.137), we can apply the procedure directly for the frequency-modulated continuous-wave radar.

Figure 3.48 demonstrates the feasibility of range-velocity map concentration with the real-life target. This example encompasses the KR-03 Puchatek glider without specific ISAR processing. Since the radar was high-resolution ($\Delta R = 11.7$ cm), the wing span embraces approximately 15 m and was much larger than the single range cell, and therefore, the target shape, despite the lack of a dedicated ISAR imaging algorithm, appeared with a rough shape on the range-velocity distribution. However, due to the echo spread in the range-velocity domain, the detection algorithms may fail.[3] The original range-velocity map fragment is demonstrated in Figure 3.48a.

As can be seen in Figure 3.48b, not all of the components can be efficiently concentrated using simple estimators (3.121) and (3.123), which stems from a certain frequency and spatial-frequency modulation of a moving target. The image after concentrating with more efficient estimators (3.148) and (3.153) can be seen in Figure 3.48c. The scatterers are precisely focused on the most prominent points.

While extensive, the presented analyses, results, and derivations do not cover the entire topic or exhaust all aspects related to the concentration of domain-selective W2D-FT. Selected radar applications of the discussed technique have been demonstrated, showcasing the performance, pros and cons of all of them. In the next section, we will focus on improving the quality of the W2D-FT distribution in the time-space domain while disregarding dependencies on frequency and spatial frequency.

3.8.3.2 Time and Space Reassignment

Similarly, as in the frequency and spatial-frequency reassignment, we can differentiate between only time or space reassignment, and we can incorporate these two into the joint time and space concentration of the W2D-FT. From the beginning, we can introduce time-reassigned W2D-FT as

$$T_x^h(t,r,\omega,\eta) = \int_{-\infty}^{\infty} \left|F_x^h(\tau,r,\omega,\eta)\right|^2 \delta(t-\hat{t}_x(\tau,r,\omega,\eta))\mathrm{d}\tau, \tag{3.159}$$

where $\hat{t}_x(t,r,\omega,\eta)$ is given by (3.126). Analogously, the space-reassigned W2D-FT reads as

$$P_x^h(t,r,\omega,\eta) = \int_{-\infty}^{\infty} \left|F_x^h(t,\rho,\omega,\eta)\right|^2 \delta(r-\hat{r}_x(t,\rho,\omega,\eta))\mathrm{d}\rho, \tag{3.160}$$

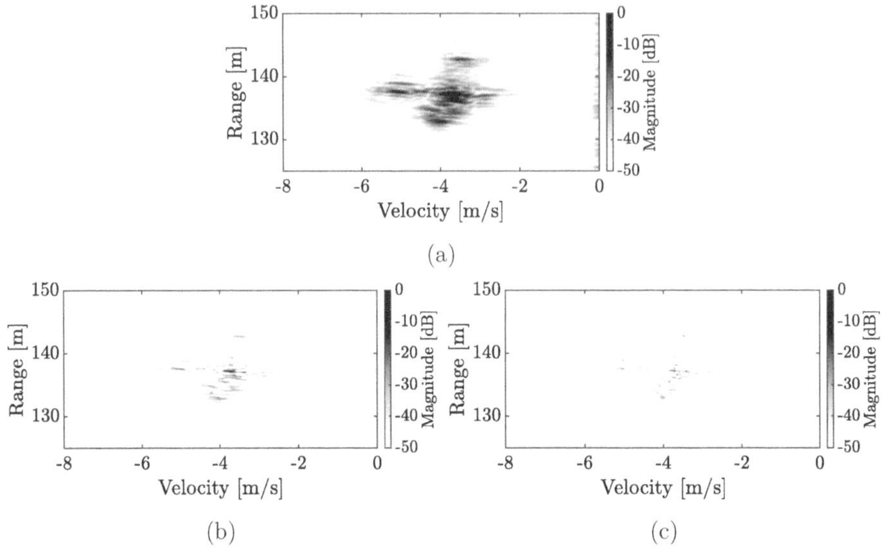

Figure 3.48 Close-up on the range-velocity map for the frequency-modulated continuous-wave radar and the KR-03 Puchatek glider as a target. (a) Original range-velocity map; (b) range-velocity map using simple frequency and spatial-frequency estimators; (c) range-velocity map using robust frequency and spatial-frequency estimators.

with $\hat{r}_x(t,r,\omega,\eta)$ defined by (3.128). Merging (3.159) and (3.160), we end up with the following formula for time and space concentration of the W2D-FT

$$D_x^h(t,r,\omega,\eta) = \int_{-\infty}^{\infty}\int_{-\infty}^{\infty} \left|F_x^h(\tau,\rho,\omega,\eta)\right|^2 \delta(t-\hat{t}_x(\tau,\rho,\omega,\eta), r-\hat{r}_x(\tau,\rho,\omega,\eta))\mathrm{d}\tau\mathrm{d}\rho. \tag{3.161}$$

Equation (3.161) is similar to (3.137), with the key difference that the former enhances W2D-FT readability in the time-space domain, while the latter focuses on frequency and spatial frequency. Disregarding single-domain concentration, the practical application of (3.161) can be illustrated using drone echoes recorded by the frequency-modulated continuous-wave radar already discussed in this chapter. Figure 3.49 showcases the concentration performance on a fragment of the range-velocity map.

The results clearly demonstrate that we can switch between full concentration across all four domains, concentration limited to the frequency and spatial-frequency domains, or concentration solely in the time and space domains. This flexibility enables a reduction in computational burden by allowing a focus on specific domains of interest. Naturally, neglecting concentration in certain domains can complicate the final interpretation, making analysis more challenging. However, some applications (such as those presented in this section) do not require a comprehensive examination of all W2D-FT dimensions. To quantitatively distinguish the differences between

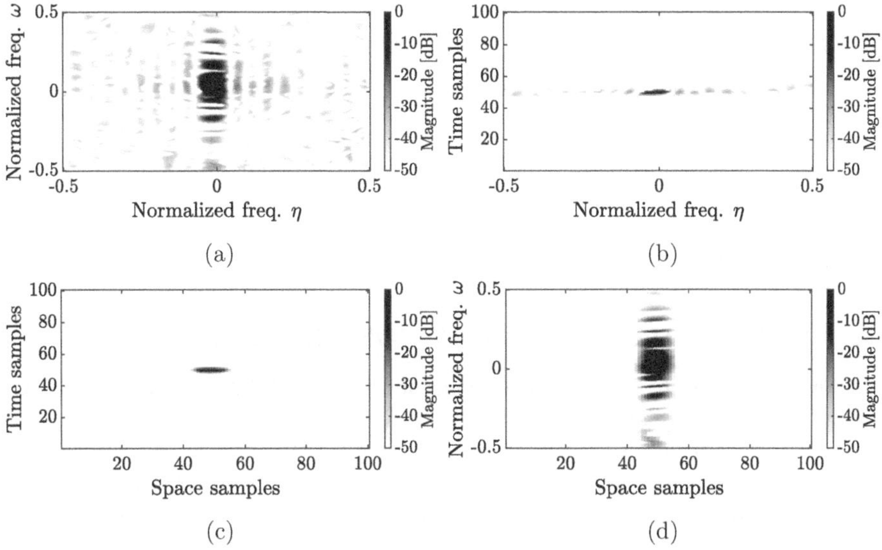

Figure 3.49 Real-life drone echo concentration using the domain-selective W2D-FT reassignment. (a) Spatial frequency-frequency distribution; (b) spatial frequency-time distribution; (c) space-time distribution; (d) space-frequency distribution.

Table 3.7
Third-Order Rényi Entropy for the Real-Life Drone Echo for Different Concentration Techniques

Concentration Variant	Rényi Entropy [bits]
No concentration	20.02
4D concentration	9.78
2D $t\&r$ concentration	10.46
2D $\omega\&\eta$ simple concentration	16.37
2D $\omega\&\eta$ robust concentration	7.78

the presented W2D-FT concentration variants, we compare their third-order Rényi entropies, as listed in Table 3.7.

From the outcomes, we can infer that, among the analyzed methods, the lowest entropy is achieved using the robust frequency and spatial-frequency estimators specifically, (3.151) and (3.155), as they account for target acceleration. A slightly higher entropy, by approximately two bits, is observed for the four-dimensional concentration. This difference arises from target acceleration not being considered in (3.129). A comparable entropy level is obtained for time and space reassignment. Beyond this, entropy increases for simple frequency and spatial-frequency concentration using (3.121) and (3.123), with the highest entropy occurring when no concentration is applied. Of course, other variations of domain-selective W2D-FT reassignment are possible, but they ultimately relate to time-frequency analysis, which was thoroughly examined in Chapter 2.

It has to be highlighted that the presented concentration methods focused only on energy concentration without discussion on the possibility of signal or a given component retrieval. The W2D-FT, as an extension of the STFT for higher dimensions, inherits its properties, such as invertibility. In the next section, we focus on taking advantage of this feature, incorporating the improved component separability by calculating the inverse W2D-FT and component extraction.

3.9 SIGNAL RETRIEVAL FROM THE CONCENTRATED W2D-FT

So far, our efforts have focused on concentrating the W2D-FT energy as effectively as possible by employing advanced estimators of instantaneous and spatial frequency. As a result, we have demonstrated that the W2D-FT energy can be concentrated either across all four dimensions (via reassignment) or within selected ones. However, stopping here would leave the reader wondering about the relationship between the W2D-FT and the STFT, since both transforms are invertible and allow for energy concentration in either all or selected dimensions.

Firstly, let us introduce a set of assumptions imposed on the two-dimensional signal $x(t,r)$. The window function $h(t,r)$ is assumed to be separable and Gaussian, with standard deviations σ_t and σ_r along the t and r dimensions, respectively. Each component of the signal is represented in the form $x_k(t,r) = A_k(t,r)e^{j\phi_k(t,r)}$ and is assumed to exhibit weak modulation in both variables. In particular, there exists a sufficiently small constant $\varepsilon > 0$ such that, for all (t,r), the second-order derivatives of the phase remain bounded according to

$$\sigma_t^2 \left| \frac{\partial^2 \phi_i(t,r)}{\partial t^2} \right| \leq \varepsilon, \qquad \sigma_r^2 \left| \frac{\partial^2 \phi_i(t,r)}{\partial r^2} \right| \leq \varepsilon, \tag{3.162}$$

while the amplitude variations are constrained by

$$\left| \frac{\partial A_i(t,r)}{\partial t} \right| \leq \varepsilon A_i(t,r), \qquad \left| \frac{\partial A_i(t,r)}{\partial r} \right| \leq \varepsilon A_i(t,r). \tag{3.163}$$

For simplicity, a single parameter ε is used in both dimensions, which implicitly assumes comparable dynamics in time and space. In a more general setting, separate parameters ε_t and ε_r could be introduced. In addition, the signal components are required to be sufficiently separated in the frequency ans spatial- frequency domains. Let the effective bandwidths of the Gaussian window be defined as

$$\Delta_t = \frac{\sqrt{2\log(2)}}{\sigma_t}, \qquad \Delta_r = \frac{\sqrt{2\log(2)}}{\sigma_r}. \tag{3.164}$$

Under this condition, the instantaneous frequency surfaces corresponding to different components must satisfy, for all (t,r) and for all $k \neq l$,

$$\left| \frac{\partial \phi_k(t,r)}{\partial t} - \frac{\partial \phi_l(t,r)}{\partial t} \right| \geq 2\Delta_t, \tag{3.165}$$

as well as

$$\left|\frac{\partial \phi_k(t,r)}{\partial r} - \frac{\partial \phi_l(t,r)}{\partial r}\right| \geq 2\Delta_r. \tag{3.166}$$

In what follows, we examine the invertible W2D-FT and emphasize its role in two-dimensional signal decomposition for the concentrated W2D-FT. This approach enables efficient extraction of nonstationary bivariate signal components, similar to invertible vertical synchrosqueezing combined with mode extraction. In particular, it allows the decomposition of any two-dimensional signal, including those employed in radar imaging, as illustrated later in this section.

As shown in (3.89), the W2D-FT enables signal reconstruction from its spectral representation to the bivariate time-range domain. This property follows directly from the fundamental principles of the Fourier transform, which were discussed in detail in Section 3.4. Let us recall the previously defined formula for frequency and spatial-frequency concentration in the W2D-FT (3.137)

$$S_x^h(t,r,\omega,\eta) = \int_{-\infty}^{\infty}\int_{-\infty}^{\infty} \left|F_x^h(t,r,\Omega,\zeta)\right|^2 \delta(\omega - \hat{\omega}_x(t,r,\Omega,\zeta), \eta - \hat{\eta}_x(t,r,\Omega,\zeta))\mathrm{d}\Omega\mathrm{d}\zeta. \tag{3.167}$$

From the energy concentration perspective, this formulation is valid and, as shown in examples, successfully improves component separability. Nevertheless, one of the key advantages of the Fourier transform lies in its invertibility. Since this property also holds for the W2D-FT, one can expect that frequency and spatial-frequency concentration, viewed as a four-dimensional counterpart of vertical synchrosqueezing, should likewise be invertible.

If we omit the energy integration in (3.167) and replace the squared magnitude of $F_x^h(t,r,\omega,\eta)$ with its complex-valued form, we obtain the following expression:

$$\dot{S}_x^h(t,r,\omega,\eta) = \int_{-\infty}^{\infty}\int_{-\infty}^{\infty} F_x^h(t,r,\Omega,\zeta)\delta(\omega - \hat{\omega}_x(t,r,\Omega,\zeta), \eta - \hat{\eta}_x(t,r,\Omega,\zeta))\mathrm{d}\Omega\mathrm{d}\zeta. \tag{3.168}$$

The invertible frequency and spatial-frequency concentrated W2D-FT enables full signal reconstruction, which can be obtained through the following inverse relation:

$$x(t,r) = \frac{1}{(2\pi)^2 h(0,0)} \int_{-\infty}^{\infty}\int_{-\infty}^{\infty} \dot{S}_x^h(t,r,\omega,\eta)\mathrm{d}\omega\mathrm{d}\eta. \tag{3.169}$$

As can be seen, the reconstruction formula (both for the W2D-FT and its concentrated form) ultimately reduces to the integration of the complex-valued distribution over the ω and η axes across t and r. In other words, the process consists of summing the two-dimensional spectrum at each time and space coordinate. By analogy with vertical synchrosqueezing discussed in Chapter 2, the signal can be reconstructed using (3.169) because its phase content remains unchanged throughout the procedure. The complex-valued transform coefficients do not migrate over t or r; instead, they are merely rearranged within the frequency-spatial-frequency domain. Their rearrangement does not affect the final summation; hence, the time-range domain signal preserves its full content, enabling the reconstruction of either the complete signal or a particular component.

Having addressed the reconstruction properties of the W2D-FT, it is natural to turn to another important topic, which is signal decomposition. As we have already established, both the W2D-FT and its concentrated variant can be inverted to yield a two-dimensional signal. However, the inverse formula (3.169) does not impose any restrictions on the integration limits, and in the form presented so far, it reconstructs the signal from the entire four-dimensional distribution. If, on the other hand, the integration domain is limited to the region occupied solely by a specific signal component, it becomes possible to decompose and denoise that component.

Let us therefore assume that the following formula enables the decomposition of a signal whose trajectory is associated with Ω. In the subsequent discussion, we will focus on the problem of determining this set Ω. The ith component of the signal can be reconstructed as [209]

$$\hat{x}_i(t,r) = \frac{1}{(2\pi)^2 h(0,0)} \int\limits_{\Omega_{i,t}(t,r)-\frac{\Delta\omega}{2}}^{\Omega_{i,t}(t,r)+\frac{\Delta\omega}{2}} \int\limits_{\Omega_{i,r}(t,r)-\frac{\Delta\eta}{2}}^{\Omega_{i,r}(t,r)+\frac{\Delta\eta}{2}} \dot{S}_x^h(t,r,\omega,\eta)\mathrm{d}\omega\mathrm{d}\eta, \tag{3.170}$$

where the parameters Δ_ω and Δ_η specify the neighborhood around the ith component that is included in the integration. In practice, the detected trajectory $\Omega_i(t,r)$ is used to build a four-dimensional binary mask. The quantities $\Omega_{i,t}(t,r)$ and $\Omega_{i,r}(t,r)$ correspond to the frequency and spatial-frequency coordinates of the ith component at the given time-space position (t,r). The mask takes the value 1 along the ridge of the component and in its vicinity, determined by Δ_ω and Δ_η, while being zero elsewhere.

Although the same concept could in principle be applied to the non-concentrated W2D-FT (3.86), in that case the uncertainty principle and the strong dependence on the analysis window make the choice of Δ_ω and Δ_η far from straightforward, requiring case-specific adjustments. By contrast, in the concentrated W2D-FT representation, the components exhibit sharp localization, which allows the mask to be restricted to the immediate vicinity of the detected ridge.

Having a solid background for signal decomposition from the W2D-FT, we can introduce the fundamental two-dimensional mode retrieval algorithm as an extension of the method known in the synchrosqueezing transform (2.157). To extract the ridge of a signal component, we define a two-dimensional curve that follows the trajectory of maximum energy. For this purpose, an algorithm designed to identify the dominant energy path $\Omega(t,r)$, which maximizes the energy while simultaneously enforcing smoothness through a total variation regularization term, expressed as

$$\hat{\Omega} = \underset{\Omega}{\operatorname{argmax}}\left(\int_{-\infty}^{\infty}\int_{-\infty}^{\infty} |\dot{S}_x^h(t,r,\Omega_t(t,r),\Omega_r(t,r))|^2 \mathrm{d}t\mathrm{d}r - \kappa\int_{-\infty}^{\infty}\int_{-\infty}^{\infty}\left(\left|\frac{\partial\Omega_t(t,r)}{\partial t}\right|^2 + \left|\frac{\partial\Omega_r(t,r)}{\partial r}\right|^2\right)\mathrm{d}t\mathrm{d}r\right), \tag{3.171}$$

where the parameter κ regulates the influence of the smoothness constraint on $\Omega(t,r)$. When $\kappa = 0$, the algorithm reduces to tracking the maxima of $\dot{S}_x^h(t,r,\omega,\eta)$, which is the setting adopted in the subsequent experiments of this work. Equation (3.171) enables the iterative extraction of multiple ridges and generalizes the time-frequency greedy algorithm originally proposed for the wavelet transform [127, 210].

To illustrate the basic principle of (3.171), we may first consider its two-dimensional projection. In that case, the problem reduces to simply following the strongest curve. Extending this idea to the full four-dimensional setting leads to the formulation of (3.171).

To begin, let us examine a straightforward yet insightful example of a two-dimensional nonstationary signal formed as a mixture of two chirps. In this simulation, discrete-time and space vectors of length 101 are defined over the interval $[-0.5,...,0.5]$. Based on these vectors, two quadratic-phase complex signals are generated. The first signal is defined as

$$x_1(t,r) = e^{j2\pi\left((20t^2-10t)+(20r^2-10r)\right)}. \tag{3.172}$$

In a similar manner, the second signal is given by

$$x_2(t,r) = e^{j2\pi\left((10t^2+10t)+(10r^2+10r)\right)}. \tag{3.173}$$

The resulting mixture takes the form

$$x(t,r) = x_1(t,r) + x_2(t,r). \tag{3.174}$$

The signal is additionally contaminated with weak noise, yielding an SNR of 50 dB. The corresponding two-dimensional spectrum, computed without windowing, is presented in Figure 3.50. In this figure, the two band-limited chirps clearly appear as distinct square-shaped regions in the frequency-spatial-frequency plane, consistent with the parameters defined in (3.172) and (3.173). This example highlights a typical nonstationary scenario, where two components, each characterized by time-varying frequency and spatial frequency, overlap within the signal.

Figure 3.51 presents the results of signal decomposition using the analyzed method. Initially, Figure 3.51a shows the cross-section of the W2D-FT energy distribution at $t = 0$ and $r = 0$. Next, Figure 3.51b displays the concentrated energy of the W2D-FT (using estimators (3.148) and (3.153)), clearly showing the highly focused

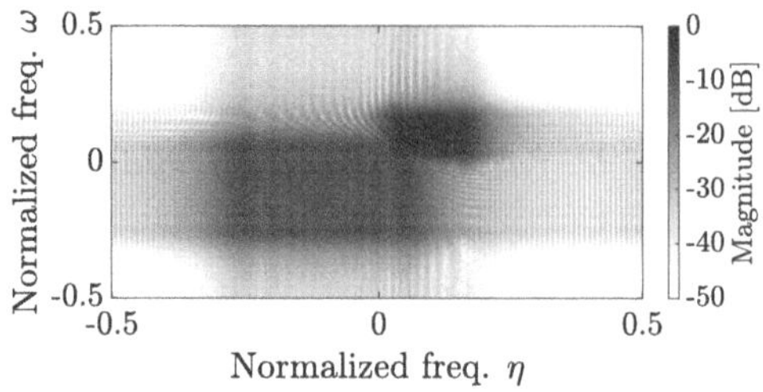

Figure 3.50 Two-dimensional Fourier spectrum of two chirps.

components. Because each component is strongly focused around its local instantaneous frequency and spatial frequency, its history can be easily retrieved as a curve. The input signal and the extracted components are presented in Figures 3.51c–h. As demonstrated, not only is the spectral magnitude conserved, but the signal's history in the time-range domain is conserved as well. The extracted components' spectra are presented in Figure 3.52. As shown, two signal modes were correctly decomposed.

Real-world radar signals are frequently nonstationary and multicomponent. We have demonstrated this characteristic repeatedly, such as during micro-Doppler analysis in Chapter 2. The situation remains the same in the two-dimensional context. If we recall the signal model for the frequency-modulated continuous-wave radar within the ISAR geometry, as detailed in Section 3.3, the expression is[4]

$$x_r(t) = A_x(t) \cdot \exp\left(j2\pi t\left(\frac{2r(t)B}{cT} - \frac{2v}{\lambda}\cos(\gamma(t))\right)\right). \tag{3.175}$$

By employing Taylor expansion, we reformulated the phase of (3.175) into the following polynomial:

$$\phi_{T,4} = \frac{4\pi BtR_0}{Tc} - \frac{4\pi t^2 v^2}{\lambda R_0} + \frac{2\pi Bt^3 v^2}{TcR_0} + \frac{2\pi t^4 v^4}{\lambda R_0^3}. \tag{3.176}$$

Two-dimensional radar signals, specifically those utilized for ISAR imaging, are organized into a two-dimensional distribution using a straightforward process called framing

$$x(t,r) = x_r(t + rT_f), \quad T_f > 0, \tag{3.177}$$

where T_f represents the frame duration. Combining (3.175) with (3.176) and (3.177) yields the expression

$$x(t,r) = A_x(t + rT_f)$$
$$\times \exp\left(j2\pi\left(\frac{2BR_0(t+rT_f)}{Tc} - \frac{2v^2(t+rT_f)^2}{\lambda R_0} + \frac{Bv^2(t+rT_f)^3}{TcR_0} + \frac{v^4(t+rT_f)^4}{\lambda R_0^3}\right)\right). \tag{3.178}$$

As r approaches 0, the expression (3.178) can be assumed to be independent on r. However, the signal becomes nonstationary in both t and r for higher target velocities, during maneuvers, or with increasing values of r. This effect is

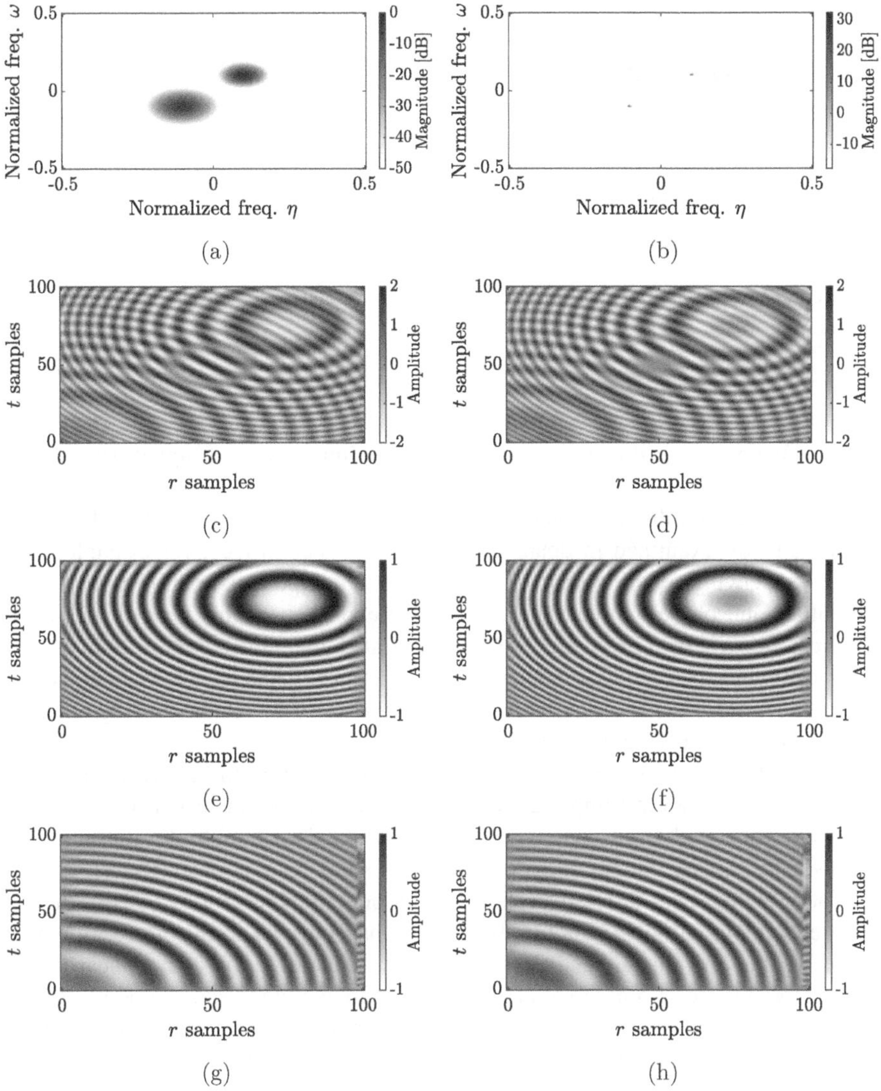

Figure 3.51 Results of the nonstationary simulated signal decomposition. (a) W2D-FT energy at $t = 0$ and $r = 0$; (b) concentrated W2D-FT energy at $t = 0$ and $r = 0$; (c) real and (d) imaginary part of the input signal; (g) real and (h) imaginary part of the first extracted signal component; (i) real and (j) imaginary part of the second extracted signal component.

particularly prominent when long integration times are used. While a longer integration time improves the SNR due to greater coherent gain, it also amplifies the influence of nonlinearities in the signal phase. In terms of target kinematic parameter estimation, this topic is discussed in Chapter 4. Now, let us focus on

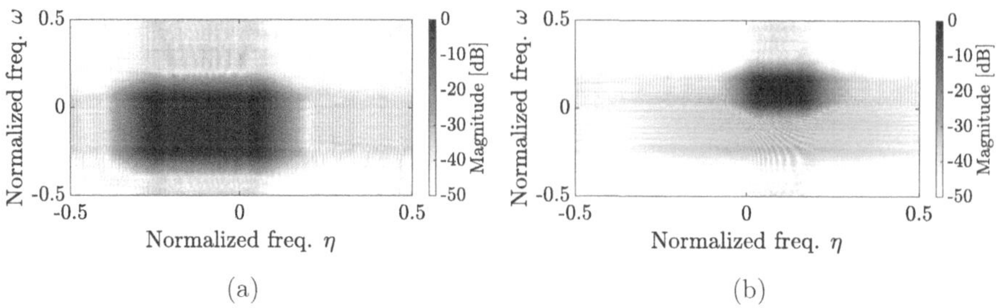

Figure 3.52 Spectra of the decomposed signal components. (a) First component; (b) second component.

ISAR imaging as an illustrative example of the importance of signal decomposition.

We now turn to the analysis of the range-velocity map for the KR-03 Puchatek glider, already examined in detail in the previous section (see, for example, Figure 3.48). Owing to the glider's motion, several components exhibit pronounced amplitude and phase nonlinearities, which are expected for this type of geometry. The results are shown in Figure 3.53. As illustrated, the almost complete image decomposition is achieved after 20 iterations (Figure 3.53g and h), while a sufficiently accurate reconstruction of the target image can already be obtained using only the first 10 components (Figures 3.53e and h).

The methodology, supported by both simulation studies and experimental data, allow us to draw several important conclusions. In real-world scenarios, ISAR images are generated from signals that typically consist of multiple nonstationary components. Under certain conditions, such as short integration intervals or when operating with radars of relatively low resolution, this nonstationarity can often be ignored without introducing noticeable distortions. However, as the observation time increases, the operational bandwidth widens, or the target follows a nonlinear trajectory, the influence of amplitude and frequency modulations becomes progressively more pronounced (see (3.178)). These effects directly impact the quality and interpretability of the reconstructed images. As a result, ISAR image decomposition cannot rely solely on purely harmonic representations, such as those employed in the commonly used CLEAN method. Instead, it must account for the presence of modulation in the received signals, which more accurately reflects the underlying physical phenomena and ensures a faithful reconstruction of both magnitude and phase characteristics.

In summary, we have presented a decomposition method for nonstationary components based on the four-dimensional W2D-FT and demonstrated its applicability to real radar data. The ISAR image was successfully reconstructed using only ten iterations. Nevertheless, extracting signal components from a four-dimensional representation is inherently challenging. The presented algorithm extends concepts known from the STFT and related transforms, such as synchrosqueezing, into a

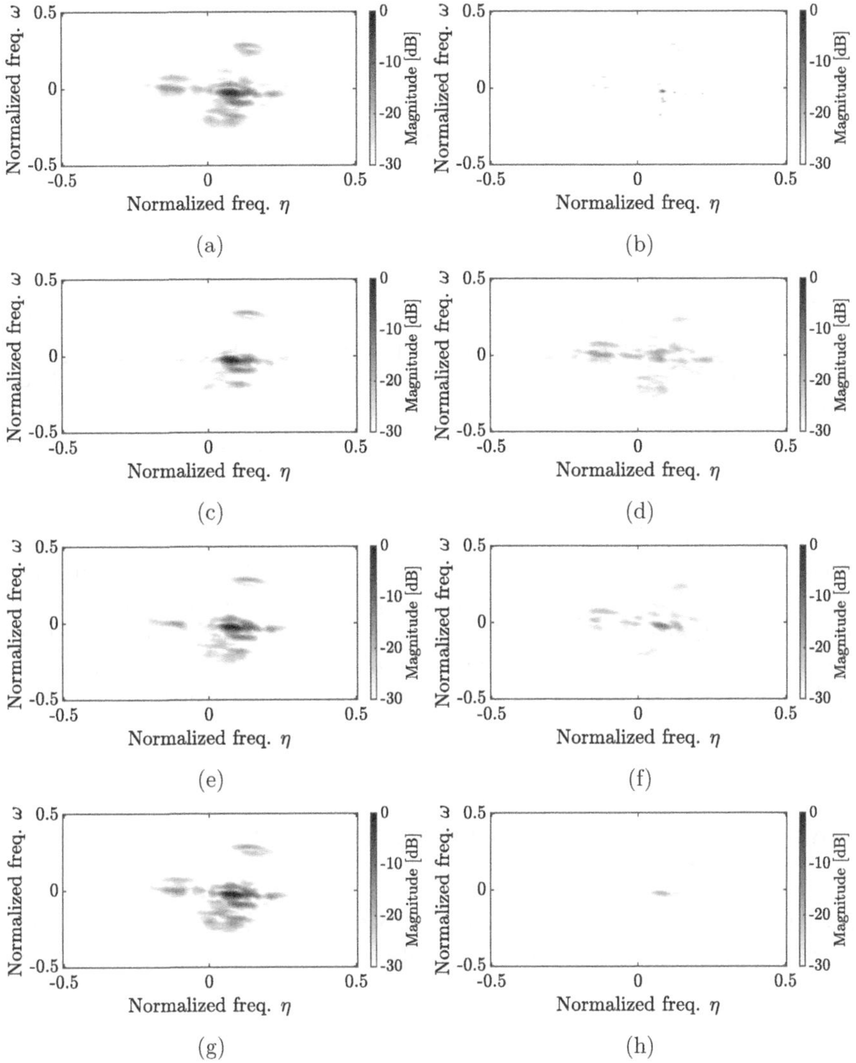

Figure 3.53 Results of the real-life ISAR image decomposition. (a) Original and (b) concentrated image processed using the W2D-FT; reconstructions from (c) 5, (e) 10, and (g) 20 extracted components with corresponding residuals shown in (d), (f), and (h). (Results reproduced from [209].)

higher-dimensional setting. While the four-dimensional domain offers greater opportunities for improving concentration and component separation, it also increases the complexity of the problem, which must be carefully considered in further developments. In the next section, we integrate W2D-FT concentration with multitaper spectrum estimation for noise suppression, aiming to enhance applications such as ISAR imaging.

3.10 MULTITAPER REASSIGNMENT IN W2D-FT

The multitaper method, introduced in Section 2.8, enhances spectral estimation by reducing variance by leveraging multiple orthonormal windows (tapers). This technique circumvents spectral leakage and improves robustness in noisy environments by averaging spectra computed for each taper. When combined with reassignment techniques, multitaper reassignment further sharpens time-frequency representations by redistributing spectral energy to its most probable location, significantly improving spectrum resolution and interpretability. This approach has proven effective in refining signal localization, particularly in scenarios where conventional spectral estimators struggle with noise suppression and energy dispersion.

Building on these advancements, this section extends the multitaper reassignment technique to higher dimensions, specifically applying it within the two- and four-dimensional W2D-FT concentration framework. By integrating energy concentration mechanisms into multitaper spectrum estimation, we strive to enhance resolution and mitigate noise effects in multidimensional signal processing. The practical utility of this method is demonstrated through its application to radar imaging, particularly in ISAR, where improved target representation and clutter reduction are crucial for high-precision analysis. Let us start with a bivariate case and focus solely on frequency and spatial-frequency representation, where radar images are typically demonstrated.

3.10.1 MULTITAPER DIMENSION-SELECTIVE W2D-FT REASSIGNMENT: ISAR IMAGE NOISE FLATTENING

The multitaper reassignment method can be extended to nonstationary and bivariate signals to make it suitable for ISAR imaging and radar applications. This problem of multitaper 2D spectrum estimation was addressed in [211], where bivariate spectra are averaged

$$\widehat{M}(\omega,\eta) = \frac{1}{K}\sum_{k=0}^{K-1}\left|M_x^{h_k}(\omega,\eta)\right|^2, \tag{3.179}$$

where

$$M_x^{h_k}(\omega,\eta) = \int_{-\infty}^{\infty}\int_{-\infty}^{\infty} x(t,r)h_k(t,r)e^{-j(\omega t+\eta r)}\mathrm{d}t\mathrm{d}r, \tag{3.180}$$

is a 2D Fourier transform of the signal $x(t,r)$ multiplied by the two-dimensional window $h_k(t,r)$. As an example, the orthonormality in time t and space r dimensions can be fulfilled by the previously introduced Hermite function. Bivariate orthonormal Hermite windows and their properties have been investigated in the literature [52, 212–214].

Let us recall the basics of the Hermite window and extend it to the two-dimensional case. The one-dimensional Hermite function along t is defined as [102]

$$h_k(t) = \frac{\left((t-\mathscr{D})^k g\right)(t)}{\sqrt{\pi 2^k k!}}, \tag{3.181}$$

where the Gaussian window is given by

$$g(t) = e^{-\frac{t^2}{2}}, \tag{3.182}$$

and $\mathscr{D}g(t)$ represents its derivative. A practical computation follows the recursive relation

$$h_k(t) = \frac{g(t)H_k(t)}{\sqrt{\pi 2^k k!}}, \tag{3.183}$$

where the Hermite polynomials satisfy

$$H_k(t) = 2tH_{k-1}(t) - 2(k-1)H_{k-2}(t), \quad k \geq 2, \tag{3.184}$$

with the initial values $H_0(t) = 1$ and $H_1(t) = 2t$.

Similarly, for the r domain, the Hermite function is defined as

$$h_l(r) = \frac{\left((r-\mathscr{D})^l g\right)(r)}{\sqrt{\pi 2^l l!}}, \tag{3.185}$$

where the Gaussian window is

$$g(r) = e^{-\frac{r^2}{2}}, \tag{3.186}$$

and the recursive computation as follows:

$$h_l(r) = \frac{g(r)H_l(r)}{\sqrt{\pi 2^l l!}}, \tag{3.187}$$

with the Hermite polynomials obeying

$$H_l(r) = 2rH_{l-1}(r) - 2(l-1)H_{l-2}(r), \quad l \geq 2, \tag{3.188}$$

with $H_0(r) = 1$ and $H_1(r) = 2r$.

Finally, the two-dimensional Hermite window is obtained as the product of the one-dimensional functions

$$h_{k,l}(t,r) = h_k(t)h_l(r). \tag{3.189}$$

When using the two-dimensional Hermite functions (3.189) in (3.179), we get the average of orthonormal spectra with flattened noise but spreads the signal components at the same time.

Remark 3.11. *Despite the possibility of mixing Hermite function orders in each dimension, in the evaluated examples, it was assumed that for $h_{k,l}(t,r)$, we always have $k = l$. Other cases are possible but out of our main priorities.*

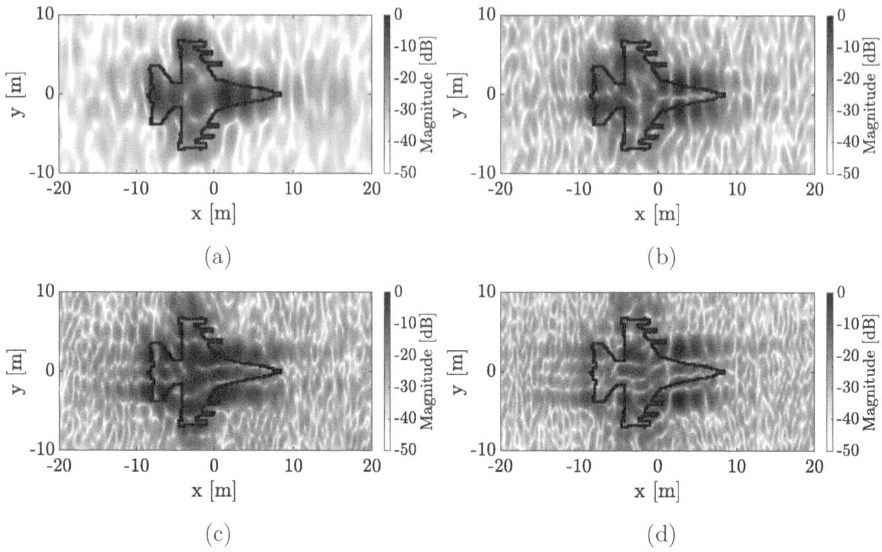

Figure 3.54 Bivariate spectra corresponding to ISAR images for different orders Hermite windows. (a) $k = 0$ and $l = 0$; (b) $k = 1$ and $l = 1$; (c) $k = 2$ and $l = 2$; (d) $k = 3$ and $l = 3$.

Let us investigate the bivariate multitaper spectrum estimation using the two-dimensional Hermite function. Figure 3.54 demonstrates the first four spectra obtained using Hermite windows of increasing order. The already discussed ISAR image for the signal bandwidth $B_2 = 100$ MHz and SNR = 0 dB.

The zeroth-order Hermite window ($k = 0, l = 0$) corresponds to an ordinary Gaussian window, providing a broad spectral response but with limited resolution. Such a situation corresponds to a classical windowed spectrum, e.g., the one we analyzed previously in Figure 3.44. As the order increases ($k, l > 0$), the window function introduces additional oscillations, leading to a more localized energy distribution on the plane.

Higher-order Hermite windows ($k = 3$, $l = 3$) improve the frequency and spatial-frequency resolution by concentrating energy more effectively. However, this enhancement comes at the cost of increased sidelobe levels, which may introduce additional artifacts in the reconstructed ISAR images. Intermediate orders ($k = 1$, $l = 1$ and $k = 2$, $l = 2$) provide a trade-off between resolution improvement and artifact suppression.

Averaging multiple ISAR images obtained using different Hermite window orders significantly impacts the final image quality by reducing noise and enhancing object representation. In Figure 3.55, we can analyze the averaged ISAR images for $K = 2, 4, 6$, and 8.

Averaging multiple ISAR images obtained using different Hermite window orders significantly impacts the final image quality by reducing noise and enhancing object representation. For lower values of K (e.g., $K = 2$), the image retains a high-noise

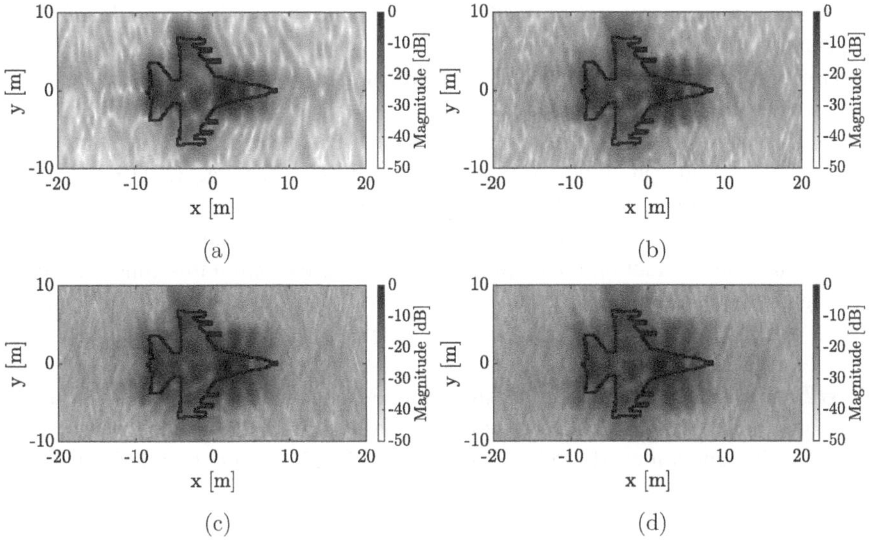

Figure 3.55 Averaged ISAR images for selected numbers K of averaged spectra. (a) $K = 2$; (b) $K = 4$; (c) $K = 6$; (d) $K = 8$.

level, resulting in speckle-like artifacts that obscure fine details. As K increases, noise is progressively suppressed due to the averaging effect, leading to a smoother background and improved target definition. This effect is particularly evident for $K = 8$, where noise fluctuations are considerably reduced, and the object's contours become more distinguishable.

However, excessive averaging introduces a trade-off. While noise alleviates, some finer structures of the target may become blurred due to the loss of high-frequency components. Therefore, selecting an optimal K is crucial to balance noise reduction with preserving structural details. In practical applications, a moderate K value (e.g., $K = 4$ or $K = 6$) often provides a good compromise between noise mitigation and maintaining sharp target features. These findings align with the conclusions inferred from the frequency-domain multitaper analysis evaluated in Chapter 2.

The method addresses the challenge of noise suppression; however, the impact of spectral smearing on the useful signal remains a concern, particularly in demanding applications such as high-resolution ISAR imaging. The primary goal of high-resolution imaging is to reveal details that lower-resolution systems fail to distinguish. Additionally, high-resolution radars are both costly and complex. Any reduction in image quality and scatterer separability undermines the effort invested in developing an advanced system.

Similar to the one-dimensional signal case, two-dimensional (or, more generally, multidimensional) multitapering flattens noise at the expense of frequency resolution. While this noise suppression can be significant, averaging multiple windows may degrade the clarity of the ISAR image. We combine the bivariate multitaper spectrum estimation method to overcome this limitation with W2D-FT concentration [215].

Noise in radar applications such as ISAR imaging is a tangible challenge, typically addressed using established methods designed for signals with a lower degree of freedom. ISAR images must be both sharp and noise-free to facilitate their perception and interpretation. As demonstrated in this chapter, W2D-FT concentration does not inherently differentiate between noise and useful signals; instead, it sharpens all fluctuations and harmonic terms. Thus, it cannot be considered a standalone noise-reduction technique. By leveraging the multitaper time-frequency reassignment method described in Section 2.8, we extend the orthonormal window-based spectrum averaging to achieve bivariate spectrum enhancement through noise suppression and component concentration.

W2D-FT concentration requires instantaneous frequency and spatial-frequency estimators to relocate the spectrum points according to (3.137). Aside from the general Hermite function window class, (3.121) and (3.123) require their derivatives. As discussed in Section 2.8, they can be computed as [111, 112]

$$\partial_t h(t) = \mathscr{T} h_k(t) - \sqrt{2(k+1)}h_{k+1}(t), \tag{3.190}$$

and

$$\partial_r h(r) = \mathscr{R} h_l(r) - \sqrt{2(l+1)}h_{l+1}(r), \tag{3.191}$$

with all symbols as stated so far. The relationship given by (3.189) remains valid regardless of whether the window function is differentiated in any dimension. Figure 3.56 presents an illustrative example of the first three orders of (3.189) along with their corresponding derivatives, as defined in (3.190) and (3.191), for the case where $k = l$. It is evident that as the function order k increases, oscillations become

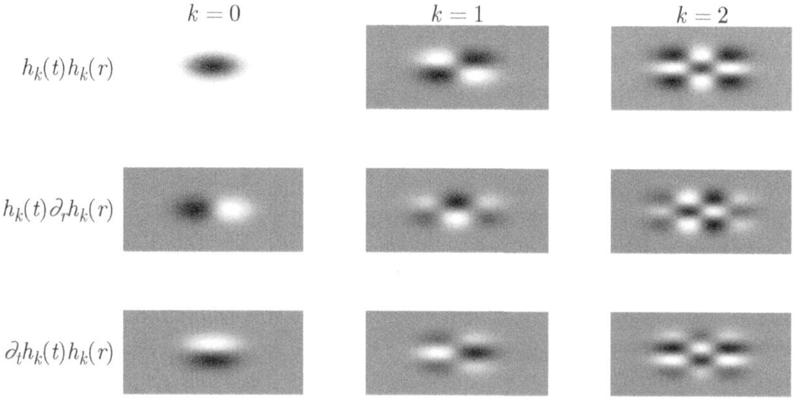

Figure 3.56 Example of bivariate Hermite functions for the first three orders $k = 0, 1, 2$ and their derivatives used for spectrum concentration. The example encompasses a situation where $k = l$, albeit it is not a must. The readers can validate other cases to figure out the influence of various situations when $k \neq l$.

more pronounced. This observation justifies the analytical computation of $\partial_t h(t)$ and $\partial_r h(r)$ instead of relying on numerical differentiation.

Having all required windows of the assumed order K, for each $k \in [0, K-1]$ the concentrated spectrum (3.137) can be computed and averaged as

$$\widehat{S}_x^h(t,r,\omega,\eta) = \frac{1}{K}\sum_{k=0}^{K-1} S_x^{h_k}(t,r,\omega,\eta), \tag{3.192}$$

where $S_x^{h_k}(t,r,\omega,\eta)$ is the dimension-selective W2D-FT (3.137) with kth-order Hermite function as a window.

The method processes a two-dimensional input signal. For each $k \in [0, K-1]$, three windows are initially generated: $\partial_r h_k(r)h_k(t)$, $h_k(r)\partial_t h_k(t)$, and $h_k(r)h_k(t)$. To simplify the explanation, consider the case of $k = 0$, as the procedure remains identical for higher k, differing only in the Hermite window order. Once these windows are prepared, they are systematically shifted across t and r over the entire signal $x(t,r)$. The windowed signal is used at each position to compute the bivariate spectrum, resulting in three spectral representations for every t and r. These spectra are then employed to estimate the spatial-frequency components according to (3.121) and (3.123), enabling energy redistribution to the most prominent local features. This process yields the focused image $S_x^{h_k}(t,r,\omega,\eta)$. The procedure is repeated for all $k = 1,\ldots,K-1$, and the final image is obtained by averaging the individual results as defined in (3.192).

We can analyze the method's performance using the same simulated ISAR image with $B = 100$ MHz as previously. Firstly, in Figure 3.57, we have a depiction of the concentrated ISAR images for $k = 0,1,2$, and 3. With a growing order of k, we see how local fluctuations resulting from noise take turns into more numerous local salient points. The process, akin to the one-dimensional case, stems from the increasing oscillations in the window distribution, leading to dispersing one peak into several adhering ones.

Ultimately, let us garner insight into the multitaper ISAR noise suppression for different orders of K. Figure 3.58 displays the comparison of the final ISAR images for different numbers of averaged spectra K.

The illustrated comparison shows the straightforward relation between the target's projection compactness and the noise dynamic. For example, for $K = 2$, we have a well-focused ISAR image, but it contains numerous local peaks resulting from noise. However, increasing K leads to the spectrum smearing, albeit the noise is flattened, reducing the probability of false alarms. For $K = 8$, the target completely fills the superimposed contour, and a noise blurs in the whole plane. However, the image resolution is worsened anytime when K increases.

The selection of the parameter K plays a crucial role in balancing noise suppression and image clarity in ISAR imaging. The observation for the bivariate spectrum is coherent with that for time-frequency analysis discussed in Section 2.8. As demonstrated in Figure 3.59, the contrast and entropy metrics deteriorate with increasing K, primarily due to the spectral blurring effect introduced by averaging multiple Hermite function windows. However, the rate of change in these metrics

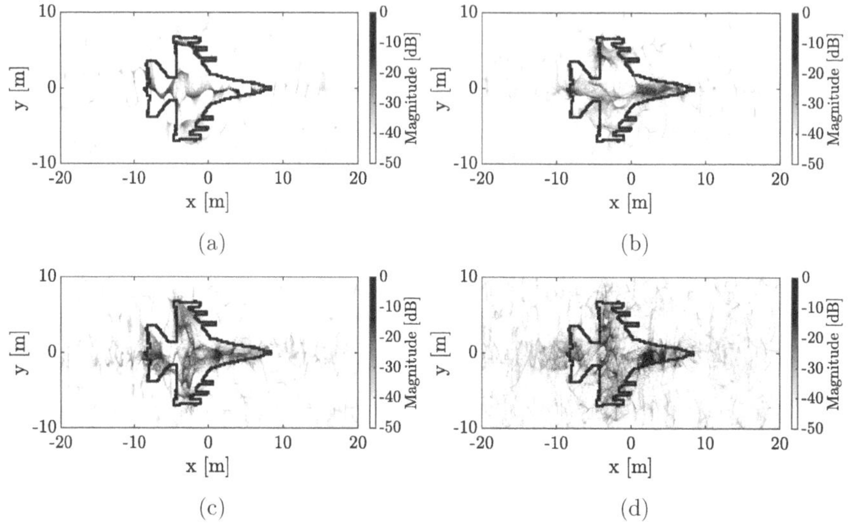

Figure 3.57 Concentrated ISAR images for the first four Hermite windows. (a) $k = l = 0$; (b) $k = l = 1$; (c) $k = l = 2$; (d) $k = l = 3$.

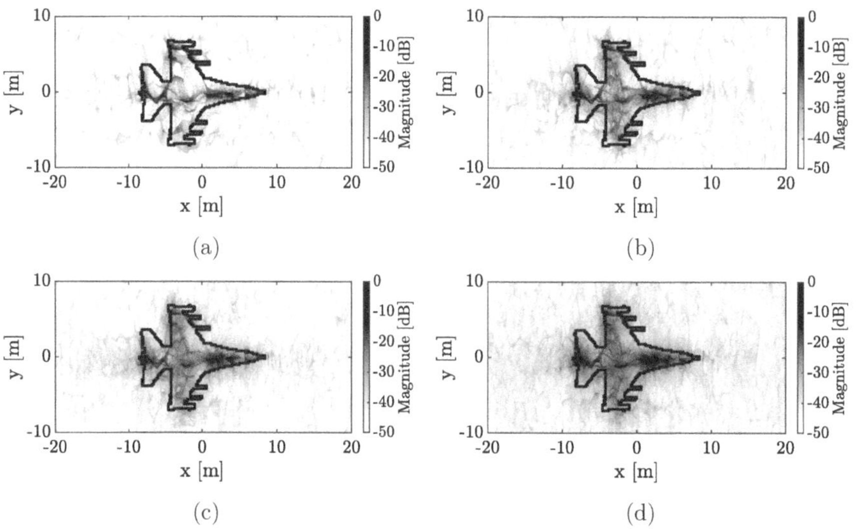

Figure 3.58 Simulated concentrated ISAR images of the with flattened noise – comparison of images for a different maximum order of the Hermite function K. (a) $K = 2$; (b) $K = 4$; (c) $K = 6$; (d) $K = 8$.

follows an asymptotic trend, suggesting that excessively high values of K result in increased computational cost without significant improvements in image quality.

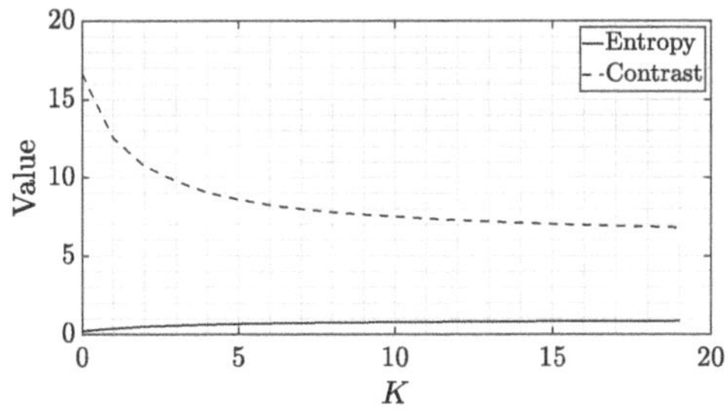

Figure 3.59 Contrast and entropy values for the analyzed ISAR image disturbed by noise giving SNR = 30 dB.

Consequently, it is essential to determine an optimal K that ensures effective noise flattening while minimizing computational overhead.

There is no universal rule for selecting K, as it depends on multiple factors, including image resolution and noise characteristics. Practical experiments indicate that in most cases, values in the range $K = 2 \sim 6$ are sufficient to achieve significant noise suppression. However, specific scenarios may necessitate an individualized approach. The primary considerations for choosing an appropriate K can be summarized as follows:

1. **Noise Level**: Higher noise levels require a larger K to adequately suppress fluctuations. A lower K may be insufficient for effective noise flattening in highly contaminated images.
2. **Noise Distribution**: The method assumes a Gaussian noise model. If the noise deviates from this assumption (e.g., due to multipath propagation or interference), increasing K may be necessary to mitigate such disturbances.
3. **Radar Resolution**: High-resolution radars distinguish closely spaced scatterers. Excessive averaging in high-resolution images can degrade detail separability, diminishing the radar's performance. Therefore, a lower K is often preferred for high-resolution radars to maintain fine details.
4. **Target Geometry**: The impact of averaging varies with target shape. Large target components may benefit from higher K due to enhanced noise suppression, whereas smaller features risk being smeared and becoming indistinguishable.
5. **Computational Complexity**: Higher values of K significantly increase processing time due to the need to compute multiple spectra. In real-time applications or computationally constrained scenarios, a lower K may be preferable.
6. **Desired Goal**: Image processing objectives influence K selection. If the goal is sharpness preservation, a lower K is suitable, whereas high-noise levels may necessitate a larger K despite the associated loss of fine details and increased computational load.

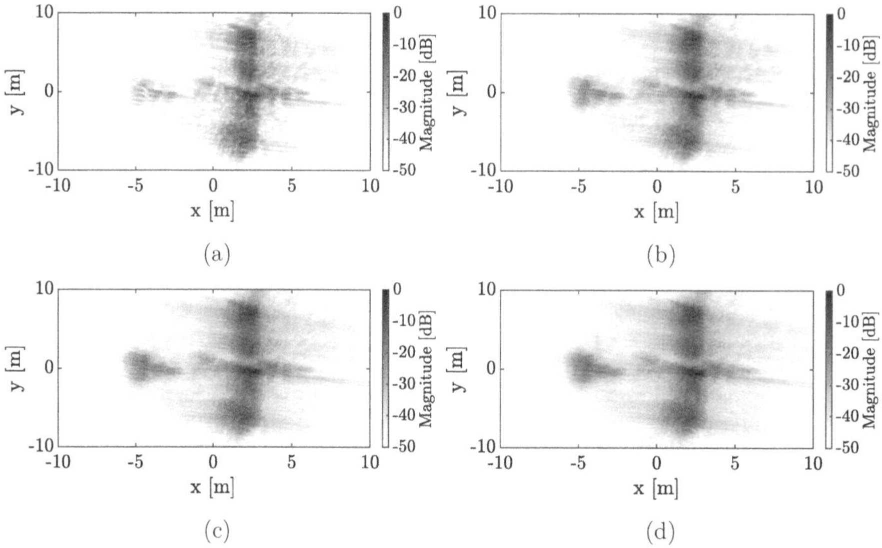

Figure 3.60 Real-life non-concentrated ISAR images of the KR-03 Puchatek glider with flattened noise – comparison of images for a different maximum order of the Hermite function K. (a) $K = 2$; (b) $K = 4$; (c) $K = 6$; (d) $K = 8$.

In summary, selecting an appropriate K requires balancing noise flattening, computational efficiency, and image clarity. While moderate values typically provide satisfactory results, specific applications may demand finetuning based on the abovementioned factors.

Let us close the analysis with a real-life example. The image of the KR-03 Puchatek glider was processed using the multitaper technique with and without concentration. The results without concentration are shown in Figure 3.60.

A comparative analysis of the ISAR images, obtained with varying maximum orders of the Hermite function K, reveals a clear trend of noise suppression as K increases. At $K = 2$, the image is characterized by a significant noise level, obscuring fine details and resulting in a grainy appearance. Progressing to $K = 4$, a noticeable reduction in noise dynamics is observed, resulting in a smoother representation and a more explicit definition of the glider's shape. Further increments in K to 6 and 8 continue to diminish noise, producing increasingly refined images. However, this noise reduction comes with a trade-off: higher K values may lead to over-smoothing, potentially erasing subtle details of the glider and obfuscation. After incorporating the multitaper technique with the W2D-FT concentration, we get the images displayed in Figure 3.61.

A comparative analysis of the concentrated ISAR images obtained with varying maximum orders of the Hermite function (K) reveals a similar trend to that observed in the previous set of images: a reduction in noise as K increases. However, the reinforced images exhibit generally higher sharpness and surpassing definition of details

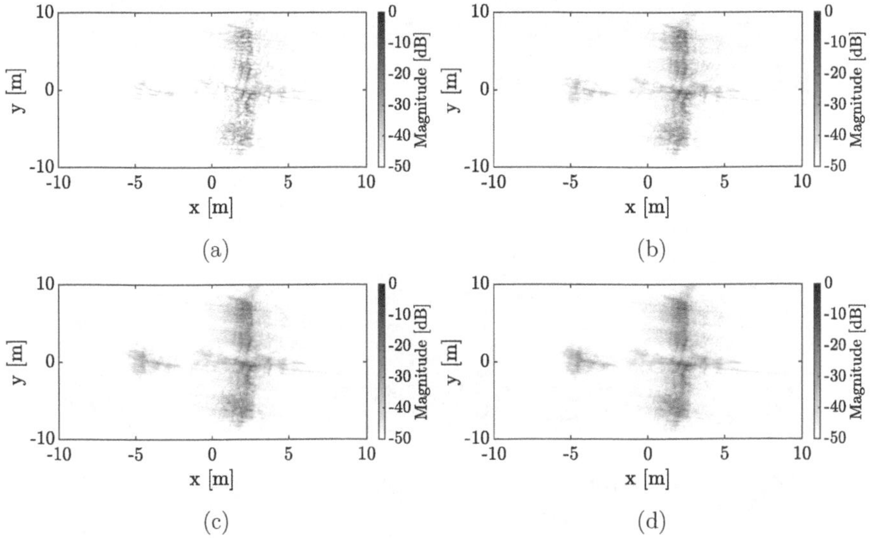

Figure 3.61 Real-life concentrated ISAR images of the KR-03 Puchatek glider with flattened noise – comparison of images for a different maximum order of the Hermite function K. (a) $K = 2$; (b) $K = 4$; (c) $K = 6$; (d) $K = 8$.

due to the applied concentration techniques. At $K = 2$, while noise is still noticeable, the glider's structure is more distinct than the corresponding image without concentration. As K gradually increases to $4, 6$, and 8, the noise is progressively smeared, and the image becomes increasingly smoother, similar to the previous images. However, due to concentration, even at higher K values, the loss of fine details is less pronounced, and the overall image quality remains high. The optimal K value, like in the case of non-concentrated images, represents a compromise between noise reduction and detail preservation. However, in this case, this compromise results in higher K values due to concentration, leading to superior image quality while minimizing the risk of losing essential details.

Remark 3.12. *SAR systems share a geometric configuration with ISAR, with the key distinction being that SAR systems image stationary targets, such as the Earth's or another planet's surface, while the radar platform itself moves. During flight, an air- or spaceborne radar illuminates the ground and processes the received signals to generate an image. Given the operational similarity between SAR and ISAR systems, their respective images can be processed using analogous techniques. Notably, the SAR image effectively represents a bivariate spectrum for frequency-modulated continuous-wave systems. Figure 3.62 illustrates an example of SAR images of a rural area processed using both conventional and multitaper techniques. Speckle noise reduction was achieved by averaging two spectra obtained with orthonormal Hermite windows. Speckle noise, a granular artifact inherent in SAR images, arises*

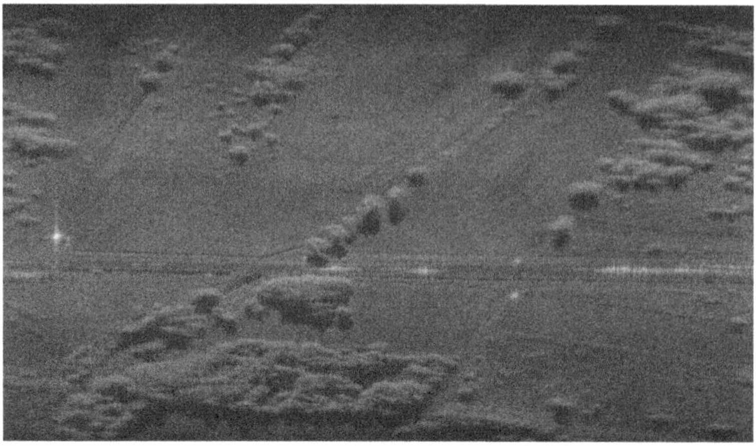

(a)

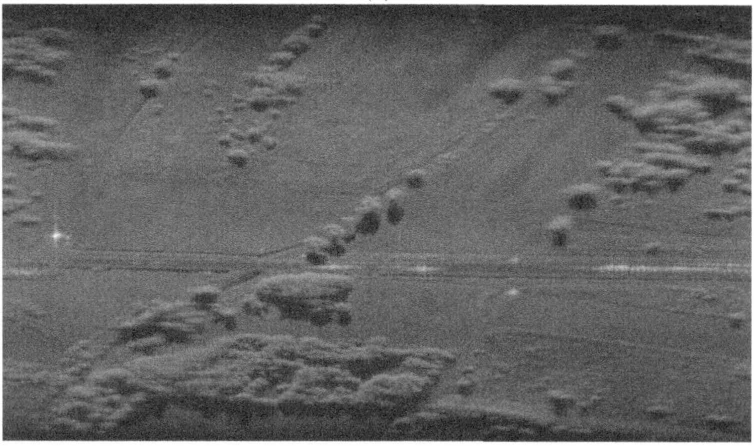

(b)

Figure 3.62 Real-life SAR image despeckling using the multitaper method. (a) Original image; (b) multitaper counterpart after averaging Hermite window-based spectra.

from the coherent processing of backscattered signals from multiple distributed scattering elements, thereby degrading image quality. Consequently, the filtered image exhibits improved target visibility and reduced granular texture. Although a non-Gaussian distribution characterizes speckle noise, this method successfully applies to SAR image despeckling and filtering.

In the case of SAR imaging, energy concentration does not make sense as we deform shapes (of buildings and roads, among others), making the image less resemble the actual surface. Thus, we can limit the analysis to averaging several spectra without concentration, but each obtained with a Hermite function of a different order.

Dimension-selective W2D-FT concentration combined with the multitaper spectrum estimation turned out to be a powerful and efficient method for noise flattening with simultaneous spectrum concentration. The presented analyses demonstrated how to practically make use of this approach when struggling with noisy radar ISAR data. Moreover, the promising prospect is speckle noise reduction in SAR data. Also, one can use it for range-velocity processing in frequency-modulated continuous-wave radars, as its processing comes down to bivariate spectrum analysis. The outcome can lead to improved target detection for high-noise environments. Now, let us generalize the multitaper noise flattening for the four-dimensional distribution of the W2D-FT.

3.10.2 MULTITAPER W2D-FT REASSIGNMENT IN FOUR DIMENSIONS: RANGE-VELOCITY MAP CASE STUDY

So far, we have analyzed bivariate spectrum estimation using the multitaper method with W2D-FT, as well as with simultaneous dimension-selective W2D-FT reassignment. Without loss of generality, the latter can be compared to multitaper spectrogram concentration in the frequency domain, as only frequency-related components are considered throughout the process. The time-frequency multitaper reassignment method described in Section 2.8 incorporates all dimensions into the concentration process. Therefore, in this section, we provide the foundation for the four-dimensional multitaper W2D-FT and its reassignment variant.

Knowing how to generate K bivariate Hermite functions according to (3.189), using each window $h_k(t,r), k \in [0,...,K-1]$, we have

$$\begin{aligned} F_x^{h_k}(t,r,\omega,\eta) &= \int_{-\infty}^{\infty}\int_{-\infty}^{\infty} x(\tau,\rho)h_k(\tau-t,\rho-r)e^{-j(\omega(\tau-t)+\eta(\rho-r))}\mathrm{d}\tau\mathrm{d}\rho \\ &= \int_{-\infty}^{\infty}\int_{-\infty}^{\infty} x(\tau+t,\rho+r)h_k(\tau,\rho)e^{-j(\omega\tau+\eta\rho)}\mathrm{d}\tau\mathrm{d}\rho, \end{aligned} \tag{3.193}$$

and its energy distribution

$$E_x^{h_k}(t,r,\omega,\eta) = \left|F_x^{h_k}(t,r,\omega,\eta)\right|^2. \tag{3.194}$$

The partially calculated results can be averaged using

$$\widehat{E}_x^h(t,r,\omega,\eta) = \frac{1}{K}\sum_{k=0}^{K-1} E_x^{h_k}(t,r,\omega,\eta). \tag{3.195}$$

Now, we can visualize the example under controlled conditions with the use of the simulated signal (3.90) contaminated by noise giving SNR $= 0$ dB. Assuming $K = 4$ as an example, the cross-sections' depiction is presented in Figures 3.63–3.65. The distributions were obtained using the same parameters, i.e., the Fourier transform size in two dimensions equaled 256 points, and the signal had 100×100 samples.

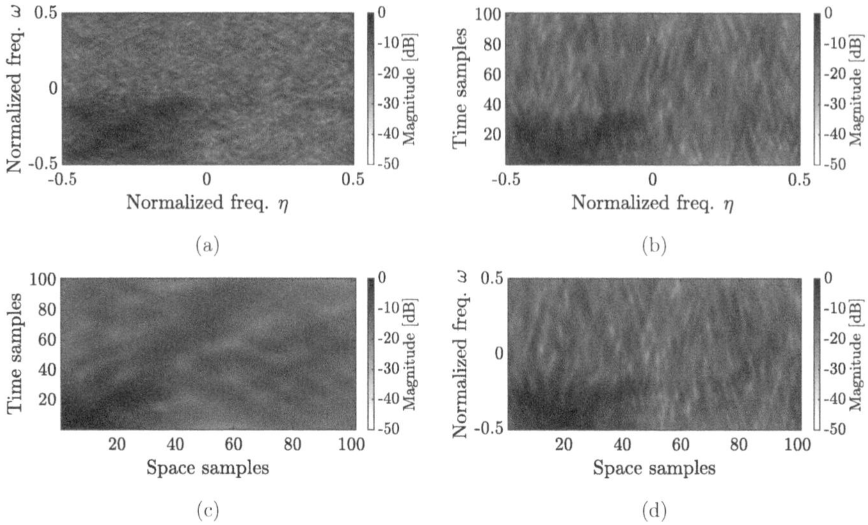

Figure 3.63 Cross-sections of the multitaper W2D-FT energy distribution for the first set of indices i.e. $t_c = 20$, $r_c = 20$, $\omega_c = 20$, and $\eta_c = 20$. (a) Spatial frequency-frequency distribution; (b) spatial frequency-time distribution; (c) space-time distribution; (d) space-frequency distribution.

In the consecutive cross-sections, we see the signal component contaminated by an intense noise. Also, averaging four distributions resulted in spectrum smearing, disqualifying unequivocal interpretation of the outcome. Although the multitaper spectrum estimation functionality allowed for noise variance suppression, we cannot easily estimate signal parameters for such a high level of disturbances.

A natural extension of (3.195) is its combination with the four-dimensional concentration, requiring a tiny extension of the hitherto introduced framework. The Hermite functions, their derivative, and other combinations were defined by (3.189), (3.190), and (3.191). To conduct the four-dimensional concentration of the W2D-FT for the Hermite window of a single order $k = l$, we need the following operators

$$\hat{\omega}_x^{h_k}(\omega,\eta,t,r) = \omega - \Im\left(\frac{F_x^{\partial_t h_k}(t,r,\omega,\eta)}{F_x^{h_k}(t,r,\omega,\eta)}\right). \tag{3.196}$$

$$\hat{\eta}_x^{h_k}(\omega,\eta,t,r) = \eta - \Im\left(\frac{F_x^{\partial_r h_k}(t,r,\omega,\eta)}{F_x^{h_k}(t,r,\omega,\eta)}\right). \tag{3.197}$$

$$\hat{t}_x^{h_k}(t,r,\omega,\eta) = t + \Re\left(\frac{F_x^{\mathscr{T} h_k}(t,r,\omega,\eta)}{F_x^{h_k}(t,r,\omega,\eta)}\right). \tag{3.198}$$

$$\hat{r}_x^{h_k}(t,r,\omega,\eta) = r + \Re\left(\frac{F_x^{\mathscr{R} h_k}(t,r,\omega,\eta)}{F_x^{h_k}(t,r,\omega,\eta)}\right). \tag{3.199}$$

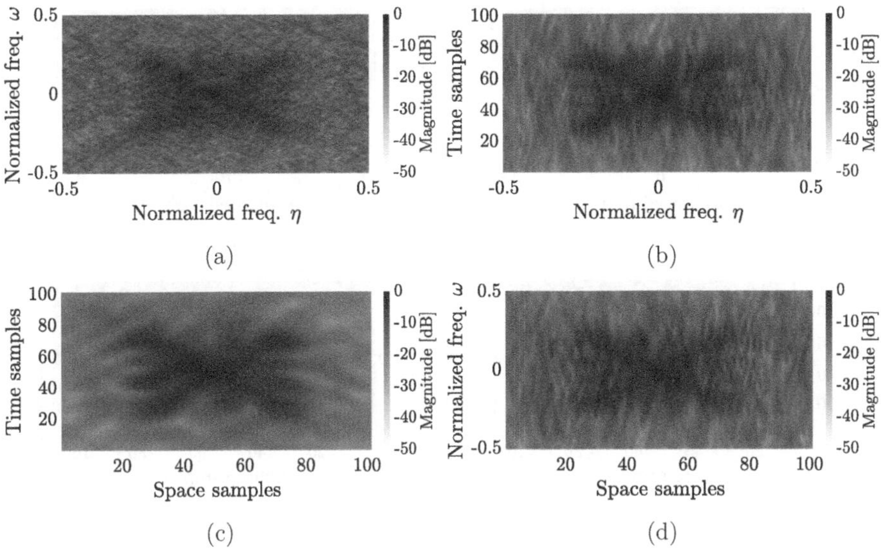

Figure 3.64 Cross-sections of the multitaper W2D-FT energy distribution for the second set of indices i.e. $t_c = 50$, $r_c = 50$, $\omega_c = 128$, and $\eta_c = 128$. (a) Spatial frequency-frequency distribution; (b) spatial frequency-time distribution; (c) space-time distribution; (d) space-frequency distribution.

Equations (3.196)–(3.199) substituted into the four-dimensional energy concentration formula give the following relationship

$$\begin{aligned} R_x^{h_k}(t,r,\omega,\eta) = &\int_{-\infty}^{\infty}\int_{-\infty}^{\infty}\int_{-\infty}^{\infty}\int_{-\infty}^{\infty} \left|F_x^h(\tau,\rho,\Omega,\zeta)\right|^2 \\ &\times \delta(t-\hat{t}_x^{h_k}(\tau,\rho,\Omega,\zeta), r-\hat{r}_x^{h_k}(\tau,\rho,\Omega,\zeta), \omega-\hat{\omega}_x^{h_k}(\tau,\rho,\Omega,\zeta), \\ &\eta-\hat{\eta}_x^{h_k}(\tau,\rho,\Omega,\zeta))\mathrm{d}\tau\mathrm{d}\rho\mathrm{d}\Omega\mathrm{d}\zeta, \end{aligned} \tag{3.200}$$

which can be averaged for all K windows

$$\widehat{R}_x^h(t,r,\omega,\eta) = \frac{1}{K}\sum_{k=0}^{K-1} R_x^{h_k}(t,r,\omega,\eta). \tag{3.201}$$

The signal illustrated in Figures 3.63–3.65 was processed according to (3.201), giving the intensely focused energy distribution with flattened noise, allowing better signal inspection. Let us look at Figures 3.66–3.68, where the cross-sections of the outcome are delineated.

The effect of concentration on the interpretation of results is evident when comparing the standard multitaper W2D-FT energy distributions (Figures 3.63–3.65) with the entirely concentrated distributions (Figures 3.66–3.68). In the non-concentrated cases, the signal components are less distinct due to the presence of

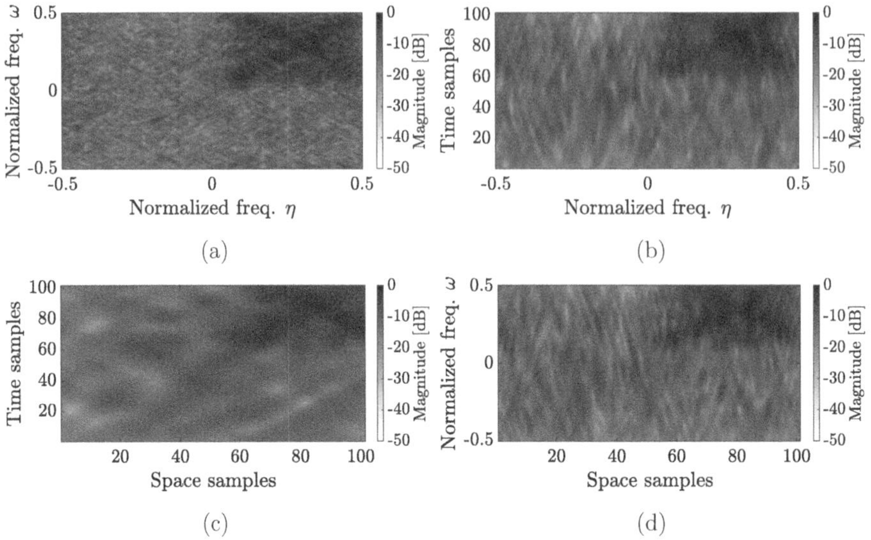

Figure 3.65 Cross-sections of the multitaper W2D-FT energy distribution for the third set of indices i.e. $t_c = 80$, $r_c = 80$, $\omega_c = 220$, and $\eta_c = 220$. (a) Spatial frequency-frequency distribution; (b) spatial frequency-time distribution; (c) space-time distribution; (d) space-frequency distribution.

residual noise and a more diffuse energy distribution. This can hinder the precise localization of time-frequency or space-frequency features.

After applying the concentration procedure, the noise level is significantly reduced, leading to a more refined and focused representation of the signal structure. The primary signal components become more pronounced in each cross-section, enhancing the interpretability of the underlying features. Specifically:

- In the spatial frequency-frequency and spatial frequency-time distributions, the signal components are more distinguishable, while the background noise appears flattened.
- The space-time distribution reveals sharper transitions, providing a more apparent distinction between different temporal and spatial variations.
- The space-frequency distribution exhibits a more localized and concentrated representation, allowing for a more accurate identification of dominant frequency components.

Table 3.8 garnered the third-order Rényi entropy for the original W2D-FT energy without concentration and multitapering, concentrated W2D-FT, and the same distributions with multitaper processing.

It can be observed that the concentration method significantly reduces the entropy values, indicating a higher energy concentration in the time-frequency domain. Non-concentrated distributions without multitaper exhibit higher entropy values (25.25 bits), suggesting a more dispersed energy distribution. After applying concentration,

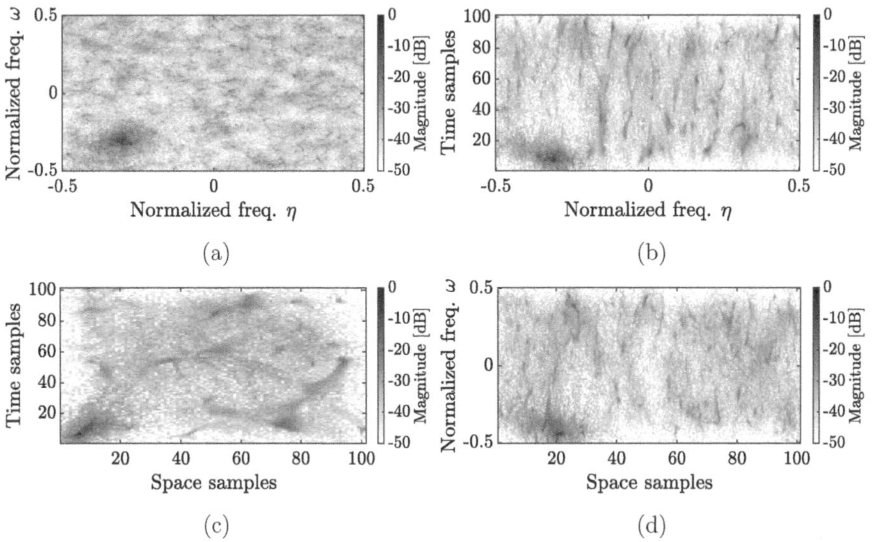

Figure 3.66 Cross-sections of the fully concentrated multitaper W2D-FT energy distribution for the first set of indices i.e. $t_c = 20$, $r_c = 20$, $\omega_c = 20$, and $\eta_c = 20$. (a) Spatial frequency-frequency distribution; (b) spatial frequency-time distribution; (c) space-time distribution; (d) space-frequency distribution.

Table 3.8
Third-Order Rényi Entropy of W2D-FT Distributions for Simulated Data

Distribution	Third-Order Rényi Entropy [bits]
Non-concentrated distribution without multitaper	25.25
Concentrated distribution without multitaper	15.15
Non-concentrated distribution with multitaper	25.45
Concentrated distribution with multitaper	16.26

the entropy decreases to 15.15 bits, confirming reinforced signal localization. The multitaper method slightly increases entropy in both cases, which follows from additional smoothing or the averaging effect of multiple spectral estimates. Nevertheless, overall readability significantly increases, allowing for better signal perception.

In summary, the multitaper W2D-FT four-dimensional concentration technique enhances the readability of the results by suppressing noise and improving the resolution of relevant signal structures. This facilitates a more accurate analysis of the time-frequency and space-frequency characteristics of the observed data.

Figure 3.69 ultimately presents the real-life radar signal processed using the multitaper technique with and without concentration (assuming $K = 4$). For the outcomes without concentration (Figure 3.69a and b), we see a meaningful energy dispersion resulting from averaging K spectra. When considering the concentrated cross-sections in Figure 3.69c and d, we can infer high energy concentration and less

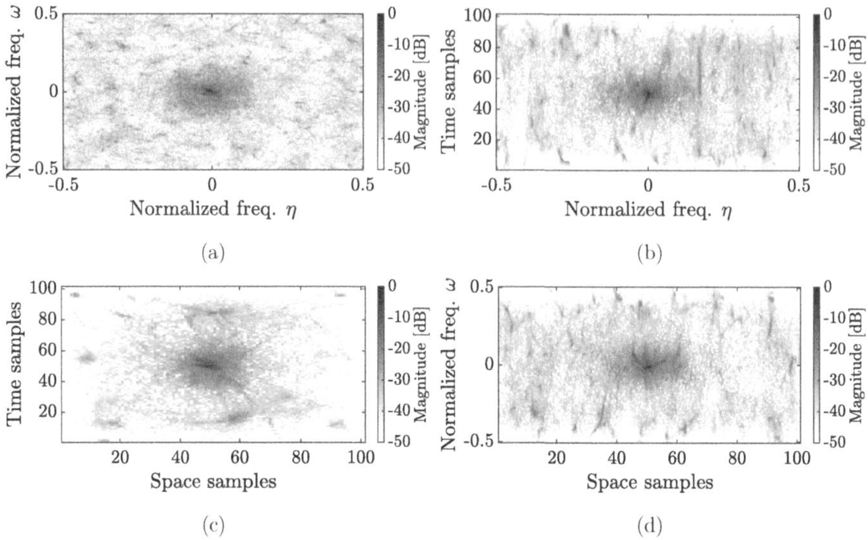

Figure 3.67 Cross-sections of the fully concentrated multitaper W2D-FT energy distribution for the second set of indices i.e. $t_c = 50$, $r_c = 50$, $\omega_c = 128$, and $\eta_c = 128$. (a) Spatial frequency-frequency distribution; (b) spatial frequency-time distribution; (c) space-time distribution; (d) space-frequency distribution.

Table 3.9

Third-Order Rényi Entropy of W2D-FT Distributions for Real-Life Frequency-Modulated Continuous-Wave Radar Signal

Distribution	Third-Order Rényi Entropy [bits]
Non-concentrated distribution without multitaper	19.45
Concentrated distribution without multitaper	9.80
Non-concentrated distribution with multitaper	18.98
Concentrated distribution with multitaper	10.72

noticeable energy smearing (especially in Figure 3.69c) allowing for straightforward signal parameter estimation. The Rényi entropies listed in Table 3.9 align with those for the simulated data.

In this section, we have explored the application of multitaper reassignment for the four-dimensional W2D-FT, validated using frequency-modulated continuous-wave radar data, and its dimension-selective variant demonstrated on ISAR data. The multitaper reassignment method enhances the concentration of the W2D-FT energy distribution, leading to improved interpretability and reduced spectral leakage.

For the fully four-dimensional case, the multitaper reassignment method significantly refines time-frequency representations by effectively reducing noise impact and enhancing signal component localization. The validation using frequency-modulated continuous-wave radar data confirms that this approach provides sharper

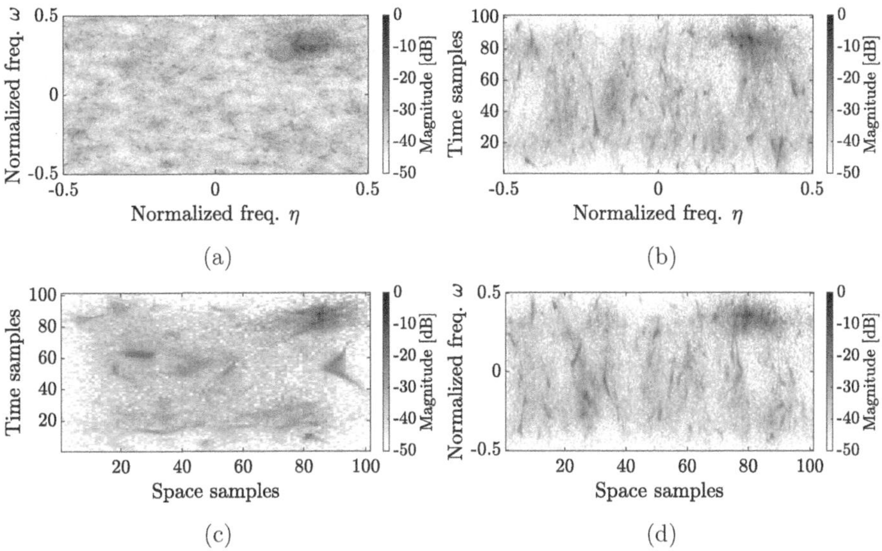

Figure 3.68 Cross-sections of the fully concentrated multitaper W2D-FT energy distribution for the third set of indices i.e. $t_c = 80$, $r_c = 80$, $\omega_c = 220$, and $\eta_c = 220$. (a) Spatial frequency-frequency distribution; (b) spatial frequency-time distribution; (c) space-time distribution; (d) space-frequency distribution.

spectral representations, which are particularly beneficial in radar signal analysis, such as target detection, motion estimation, and imaging.

In contrast, the dimension-selective W2D-FT reassignment was applied to ISAR data to improve spectral concentration in selected dimensions. This approach allows for flexible analysis, preserving the interpretability of signals while mitigating noise and enhancing contrast in specific domains. The ISAR results demonstrated that this technique is beneficial for feature extraction and object classification in high-resolution radar imaging.

Overall, the described methods contribute to enhancing W2D-FT by improving concentration and reducing noise interference. When combined with reassignment, the multitaper approach enables better readability of signal structures in the time-space-frequency-spatial-frequency domain. These improvements are essential for nonstationary radar signal analysis applications, where accurate and well-localized representations are crucial.

3.11 SUMMARY

We have explored the fundamental aspects of nonstationary two- and multidimensional signal analysis using Fourier transforms. This chapter covered various applications in radar signal processing, including SAR and ISAR imaging, frequency-modulated continuous-wave signal analysis, and ISAR image denoising

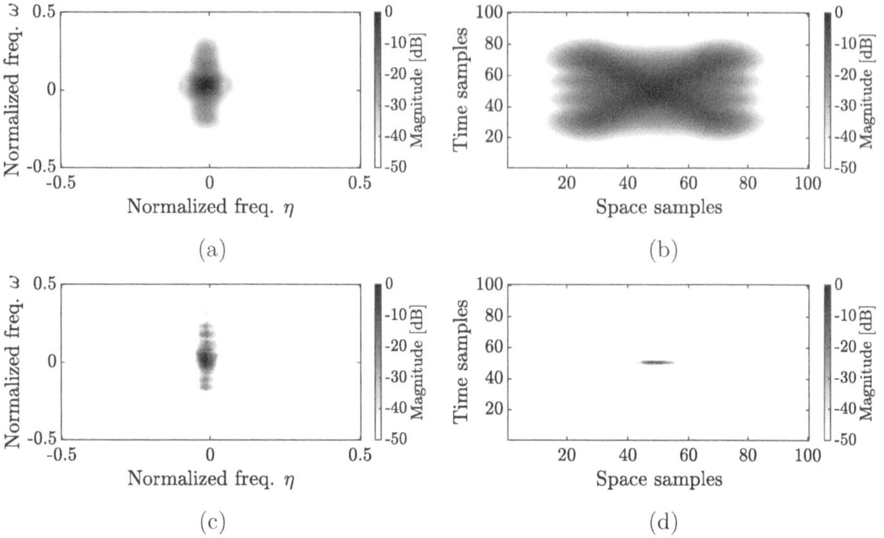

Figure 3.69 Real-life frequency-modulated continuous-wave radar signal processed with the multitaper and joint multitaper with four-dimensional concentration methods. (a) Spatial-frequency-frequency distribution for the multitaper W2D-FT spectrum estimation; (b) space-time distribution for the multitaper W2D-FT spectrum estimation; (c) spatial-frequency-frequency distribution for the concentrated multitaper W2D-FT; (d) space-time distribution for the concentrated multitaper W2D-FT.

and concentration. We examined how the bivariate spectrum can be derived from a time-domain signal and how high-resolution radar data can be visualized in three dimensions, encompassing time, frequency, and range.

In most cases, we conveyed the time-frequency framework by incorporating additional space and spatial-frequency dimensions. This led to developing one-, two-, and four-dimensional energy concentration techniques, enabling efficient and high-resolution radar imaging and target detection. In some scenarios, we integrated time-frequency analysis with an additional space axis for cubic data representation, while in others, we operated within a challenging four-dimensional space. We employed cross-sections and established connections with classical bivariate spectra and time-frequency representations to facilitate interpretation.

While this chapter provides a comprehensive overview of key concepts, it does not exhaust the subject. In particular, we have introduced a basic algorithm for mode extraction, which demonstrates the potential of component retrieval from the W2D-FT. However, this approach can be further refined and extended toward adaptive variants, better suited for diverse signal scenarios. Moreover, the four-dimensional nature of the W2D-FT opens the way to more sophisticated algorithms capable of exploiting its geometry and morphology. At the same time, such high-dimensional processing remains a nontrivial problem, leaving considerable room for further methodological advances. These directions may ultimately enable precise scatterer extraction from ISAR images, analogous to the component separation techniques based on the STFT discussed in Chapter 2.

A radar-related topic not yet analyzed, albeit closely linked to the methods discussed so far, is **target** parameter estimation in both active and passive radar systems. As we will explore in Chapter 4, the kinematic parameters of a moving target are embedded in the signal phase and can be extracted using techniques such as the Fourier transform. With this in mind, we now turn our focus to the estimation of target kinematic parameters.

3.12 CHAPTER IN A NUTSHELL

This chapter presented key concepts related to multidimensional radar signal analysis, emphasizing the role of Fourier-based techniques in improving signal representation and interpretation. The pivotal takeaways are:

- In radar processing, one-dimensional signals are often transformed into two-dimensional representations. This is achieved by segmenting the received signal into batches, effectively constructing a bivariate signal from a single-dimensional one.
- Bivariate signals play a crucial role in radar applications, including SAR and ISAR imaging, frequency-modulated continuous-wave radar, and passive radar (the latter being explored in Chapter 4).
- The analysis of two- and multidimensional signals requires specialized tools, with the Fourier transform being one of the most fundamental techniques.
- The two-dimensional Fourier transform alone is insufficient for nonstationary signals. The W2D-FT's counterpart utilizes a two-dimensional window moving across the bivariate signal, resulting in a four-dimensional complex-valued distribution.
- The W2D-FT can be optimized to achieve superior component separability, primarily through window function optimization.
- Energy concentration in the W2D-FT can be controlled separately in each dimension or jointly across multiple dimensions. Full four-dimensional concentration is also possible, significantly enhancing component separation.
- Nonstationary two-dimensional signals can be decomposed so that the dominant components are extracted and subsequently subtracted from the signal. This enables targeted component analysis and effective denoising.
- SAR and ISAR images can be refined to achieve high-resolution representations with improved readability.
- For noisy data, two-dimensional filtering techniques based on synchrosqueezing can enhance signal clarity, while four-dimensional multitaper spectrum estimation can further improve data interpretability by mitigating noise.

3.13 EXERCISES

1. Write a simple frequency-modulated continuous-wave radar simulator which, for defined parameters (carrier frequency, bandwidth, sweep repetition rate, integration time) and an object of given velocity and range, computes the range-velocity map using the two-dimensional Fourier transform.
2. Add a second object with identical parameters. Gradually vary their separation in range and observe the results, especially the ability to distinguish the two objects. Repeat the experiment for different signal bandwidths.
3. Repeat Exercise 2, but vary the velocity of one object while keeping the range constant. Repeat for different integration times.
4. Simulate a frequency-modulated continuous-wave radar signal in the ISAR geometry in Figure 3.6. Compute the range-velocity map. Vary the angle γ, the distance, and the velocity of the object and observe their influence.
5. Generate a real and complex two-dimensional harmonic signal. Compute the two-dimensional Fourier transform of each and display the magnitude. Compare the results.
6. Using a classical two-dimensional analysis window, observe its influence on the results of your frequency-modulated continuous-wave radar simulator. Focus on cases when two objects are close in range and/or in velocity. Compare the results obtained with and without the window.
7. Implement your own version of the W2D-FT.
8. Generate a two-dimensional signal with linear frequency modulation in both dimensions. Plot the two-dimensional spectrum and several selected cross-sections of the magnitude of the W2D-FT. Compare the results.
9. For the signal in Exercise 8, compute the Rényi entropy of the W2D-FT for different window widths in both dimensions. Determine the optimal window size and recompute the cross-sections of the magnitude.
10. Generate a two-dimensional window with additional frequency modulation (chirplet) in both dimensions. Given the modulation coefficients in the time and range dimensions, compute the W2D-CT and compare the results with those in Exercise 9.
11. Given an analytical formula for any two-dimensional window (e.g., the Gaussian or Hamming window), derive formulas for the first-order and second-order derivatives in both dimensions. Implement a procedure that generates the window, its derivatives, the window multiplied by first-order and second-order ramps in both dimensions, and mixed cases.
12. Compute the W2D-FT using the windows generated in Exercise 11.
13. Implement a four-dimensional reassignment procedure. Compare the results with those obtained using the script `run_W2DFT_4D.m` provided in the toolbox at `https://github.com/kabratkiewicz`.
14. Implement the dimension-selective reassignment of the W2D-FT. Analyze selected cross-sections and compare them with the four-dimensional reassignment.

15. Based on Chapters 2 and 3, derive estimators of instantaneous-phase acceleration (in both time and range) in the time-space-frequency-spatial-frequency domain. They may be used to design a new adaptive version of the AW2D-FT.
16. Using the script `run_INSPIRE.m`, extract two frequency-modulated two-dimensional components. Modify the input so that their spectra increasingly overlap. Observe when separation fails.
17. Generate a two-dimensional noise signal. Using the multitaper technique (`run_multitaper.m`), plot cross-sections of the magnitude of the W2D-FT for different numbers of tapers. Observe how the noise behaves as the number of tapers increases.
18. Generate a two-dimensional signal with linear frequency modulation of bandwidth at least half the sampling rate. Add a delayed copy with delays T and R in time and range. Vary T and R and observe cross-sections of the magnitude of the W2D-FT concentrated in four dimensions. Investigate the appearance of artifacts and the separability of components, and relate the separability to window parameters.
19. Integrate your self-implemented frequency-modulated continuous-wave radar simulator with the multitaper technique. Determine for which noise levels the simulated object is detectable on the original range-velocity map and after taper averaging. Test different numbers of tapers and multiple objects.
20. Add energy concentration techniques to the solution from Exercise 19. Evaluate their influence on detection capability.

NOTES

1 For instance: For a pulse radar, the received waveform occupies a bandwidth of $B = 30$ MHz. If sampled directly according to the Nyquist criterion, the required sampling rate is at least $f_s = 60$ MHz. In practice, however, down conversion to baseband can substantially reduce the sampling requirements. For an FMCW radar with $T = 10$ ms and the same waveform bandwidth, the beat frequency corresponding to a target at $R = 1$ km is only $\Delta f = 20$ kHz, which significantly relaxes the requirements on the analog-to-digital converter.

2 The demonstrated method is not a super-resolution technique, but it ensures better component separability.

3 Typical radar detection algorithms analyze the dynamics of surrounding points by adapting the detection threshold based on the statistical properties of neighboring cells, ensuring robust target detection while minimizing false alarms.

4 For detailed description, see Section 3.3.

4 Target Kinematic Parameter Estimation in Radars

4.1 INTRODUCTION

Chapters 2 and 3 focused on one- and two-dimensional signals, where we explored efficient methods for processing, visualizing, and analyzing radar signals. However, we have yet to examine the origins of signal nonstationarity and the factors that drive signal variations. Apart from nonstationarity resulting from the signal nature (e.g., chirps emitted by radars), amplitude and phase variations may occur when radar signals reflect from targets. Typically, the more complicated the target trajectory, the more variance it introduces in the radar signal. The dynamics of radar echoes depend not only on the target's velocity but also on its acceleration, rotation, and micro-motion effects, such as vibrations and structural deformations. These phenomena impact both the spectral content of the received signals and the radar systems' ability to estimate target parameters accurately.

In this chapter, we will put forward the factors that cause signal modulation and try to elucidate why these phenomena occur. In particular, we will look at the kinematic properties of objects and analyze how motion affects the radar's ability to detect them. Special attention will be given to the interaction between target motion and the Fourier-based representation of signals, as well as to how these effects manifest in the operational sense. It should be stressed that this chapter does not cover all radar-related topics. The available literature comprehensively invokes the subject of radars, radar architectures, and radar signal processing, to name only a few [7–10, 22, 23, 216]. Instead, we concentrate on the Fourier-related aspects dealing with signal non-stationarity and seek to estimate target parameters. The discussion is supplemented with examples illustrating the impact of nonstationarity on radar measurements and showcasing methods for mitigating its effects in signal processing.

The previous chapters paved the way for efficient Fourier transform-based methods for signal parameter estimation and spectrum readability improvement. The presented techniques facilitate the detection and characterization of signals by filtering, denoising, and improving signal concentration in the frequency domain. Given the complexity of radar signals and their variations over time, a solid understanding of these methods is crucial for interpreting the subsequent content of this chapter. Readers are expected to be familiar with the theoretical foundations and practical aspects of spectral analysis discussed earlier to fully grasp the challenges posed by nonstationary radar signals.

Nonstationarity in radar signals concerning maneuvering targets encompasses several effects, with the most prominent being:

DOI: 10.1201/9781003765783-4

- **Target Motion**: Movement of targets, including acceleration, high-order kinematic parameters, vibration, and micro-Doppler effects, induces time-dependent variations in the received signal. These variations affect both Doppler-based velocity estimation and time-frequency analysis.
- **Propagation Effects**: Atmospheric conditions, such as turbulence, ionospheric disturbances, or multipath propagation, modify the signal characteristics over time. Such effects introduce distortions that may obscure target detection and tracking.
- **Antenna Scanning and Beamforming**: Mechanically or electronically steered antennas introduce variations in signal properties due to changing illumination geometry. These variations can cause Doppler shifts, beam broadening, and sidelobe fluctuations.
- **Interference and Clutter**: Dynamic interference from other radio signal sources or fluctuating clutter, such as sea waves or foliage, alters the signal spectrum. The nonstationary nature of clutter can pose significant challenges in distinguishing targets from background reflections.
- **System-Induced Variations**: Internal radar system instabilities, including oscillator phase noise, frequency drift, and power fluctuations, contribute to nonstationarity. These effects may degrade coherence and introduce unwanted frequency components.

The selected examples, aided by real-life case studies, are discussed throughout this chapter. In particular, in Section 4.2, we first introduce (or revisit) the fundamental principles of range and velocity estimation in both active and passive radar systems within the context of Fourier analysis. Moving targets, especially those that maneuver, influence radar signal-processing results by spreading their echoes across multiple range and velocity cells. The conditions under which this effect occurs and potential solutions to mitigate its impact are explored in Section 4.3. Subsequently, we delve into kinematic signal parameter estimation in radar systems. We begin by examining target acceleration estimation in Section 4.4 and then generalize the discussion to higher-order motion components in Section 4.5, where we address the role of jerk, snap, and higher derivatives in radar measurements. Range-velocity map noise flattening with additional echo energy concentration is described in Section 4.6, where we incorporated the framework of multitaper spectrum estimation with radar signal processing.

By systematically analyzing these effects, this chapter aims to provide a deeper understanding of nonstationary radar signals and the signal-processing techniques used to handle them.

4.2 RANGE AND VELOCITY ESTIMATION IN RADARS

The basics of range and velocity estimation in active and passive radars were gathered briefly in Chapter 1 and explained to some extent in Chapter 3. Let us recall and discuss these relationships in more detail. Although the target range and velocity are easily interpretable, and we know and can imagine the meaning of these quantities, they are estimated both indirectly and using different strategies. From the beginning, in monostatic radars, we estimate the target range using the received signal delay with respect to the transmitted one. In passive/bistatic radars, in turn, the range is calculated from the delay of the echo signal with respect to the direct one (for more details, see Chapter 1 and the references therein). The cross-correlation function, the fundamental tool for range estimation, is widely employed by scientists and radar engineers

$$\chi(t) = \int_{-\infty}^{\infty} x(\tau)y^*(\tau - t)\mathrm{d}\tau = \int_{-\infty}^{\infty} x(\tau + t)y^*(\tau)\mathrm{d}\tau, \tag{4.1}$$

where $x(t)$ and $y(t)$ are the two analyzed signals and y^* is the complex conjugate of y. Assuming no Doppler frequency shift and a noiseless scenario, we can deem $x(t) = y(t - \tau_0)$, which suggests that $x(t)$ is a pure signal $y(t)$ with a delay τ_0. Substituting these into (4.1), we have

$$\chi(t) = \int_{-\infty}^{\infty} y(\tau + (t - \tau_0))\, y^*(\tau)\mathrm{d}\tau. \tag{4.2}$$

For typical radar waveforms (e.g., chirp, noise waveform, etc.) the cross-correlation function exhibits a distinct peak at $t = \tau_0$, i.e.,

$$\chi(t) = R_{yy}(t - \tau_0), \tag{4.3}$$

where $R_{yy}(t)$ denotes the autocorrelation function of $y(t)$. Equation (4.3), therefore, is the signal autocorrelation delayed with respect to τ_0 (see Figure 1.6). For efficient calculation, we can recall the convolution theorem proved in Chapter 2

$$\mathscr{F}\{\chi(t)\} = F_x(\omega)F_y^*(\omega), \tag{4.4}$$

which states that the Fourier transform of the cross-correlation is given by the product $F_x(\omega)F_y^*(\omega)$. To have the correlation result in the time domain, we need an additional inverse Fourier transform

$$\chi(t) = \mathscr{F}^{-1}\{F_x(\omega)F_y^*(\omega)\}. \tag{4.5}$$

For large signals typically processed in radars, (4.5) is more numerically efficient[1]. Although (4.5) works for any radar (mono- and bistatic), the frequency-modulated continuous-wave radar signals are processed in the way described by (3.9) in Chapter 3, as the signal mixing is usually done in the analog part of the system, allowing cheaper analog-to-digital converters. Nevertheless, (4.5) is also suitable for this kind of radar.

Remark 4.1. *Note that correlation and matched filtering are closely related operations and yield identical outputs under consistent definitions. Specifically, the matched filter is obtained by time-reversing and complex-conjugating the reference signal, i.e.,* $h(t) = y^*(-t)$*. Convolving the received signal* $x(t)$ *with* $h(t)$ *is therefore equivalent to computing the cross-correlation between* $x(t)$ *and* $y(t)$*.*

In the trivial (albeit impractical) case, the range is estimated indirectly as the delay τ corresponding to the peak of correlation (4.5)

$$\tau = \operatorname{argmax}\left(|\chi(t)|\right), \tag{4.6}$$

giving

$$R = \frac{c\tau}{M}, \tag{4.7}$$

where $M = 1$ for bistatic[2] radars and $M = 2$ for monostatic. Note that (4.6) holds for the noiseless case and a single target. When the signal is contaminated by noise, or several targets occur (both such situations are common in practice), one needs a more sophisticated detection algorithm like the constant false alarm rate methods [2].

The effectiveness of radar systems lies in their ability to estimate the range and velocity of multiple targets simultaneously. However, cross-correlation (or matched filtering) primarily determines the time shift required to align a received signal with a reference pattern. At this point, we can consider the dimensionality of cross-correlation. If we examine only the signal delay, the result is a one-dimensional function describing the signals' temporal relationship. However, we obtain a two-dimensional representation by also evaluating the Doppler shift. This operation is known as the cross-ambiguity function and is defined as [217–219]

$$\Psi(t,\omega) = \int_0^{T_i} x(\tau)\, y^*(\tau - t)\, e^{-j\omega\tau} \mathrm{d}\tau, \quad \Psi(t,\omega) \in \mathbb{C}, \tag{4.8}$$

and is usually represented as its squared absolute value

$$\Upsilon(t,\omega) = |\Psi(t,\omega)|^2. \tag{4.9}$$

$\Psi(t,\omega)$ (and of course $\Upsilon(t,\omega)$) represents a two-dimensional function of time delay t and Doppler angular frequency ω, where $x(t)$ is the received signal, $y(t)$ is the reference signal, and T_i denotes the integration time. After appropriate scaling, t corresponds to the range, while ω relates to velocity through the Doppler frequency. In monostatic radars, one has $R = \frac{c\tau}{2}$ and $V = -\frac{\lambda}{2} f_D$, whereas in bistatic configurations $R_b = c\tau$, $V_b = -\lambda f_D$, with $\omega = 2\pi f_D$. For these reasons, (4.8) is sometimes expressed as a straightforward function of range and velocity as

$$\Psi(R,V) = \int_0^{T_i} x(t)\, y^*\left(t - \frac{MR}{c}\right) e^{-j\frac{2\pi MVt}{\lambda}} \mathrm{d}t, \tag{4.10}$$

and these two formulations are used interchangeably in this book. We also see that $\omega = -M \cdot 2\pi \frac{V}{\lambda}$.

Equation (4.8) can be interpreted as the cross-correlation between two signals, where one signal is both time-shifted and frequency-shifted. Equivalently, it can be interpreted as matched filtering, i.e., convolution with a time-reversed and complex-conjugated version of the reference signal. Since, in practice, the time and frequency shifts of the signal reflected from a target are unknown, all possible shifts must be tested. The presence of the target is determined by identifying the peaks of (4.8), which correspond to the strongest echoes.

Remark 4.2. *Note that (4.8) can be written in four equivalent forms. These formulations arise due to the inherent symmetry in the cross-ambiguity function definition, where shifts in time and frequency can be applied to either of the signals involved. The key differences between these notations lie in the sign of the Doppler frequency shift and in which signal undergoes the time shift. These variations result from different conventions used in signal processing and radar literature. Each formulation is mathematically valid and can be chosen based on convenience in a given application.*

- ***Notation 1**: In this formulation, the reference signal $y(\tau)$ is shifted in time by t, and the exponential term retains the standard negative sign for the Doppler shift. This is the most commonly used definition in radar signal processing*

$$\Psi(t,\omega) = \int_0^{T_i} x(\tau)\, y^*(\tau - t)\, e^{-j\omega\tau}\mathrm{d}\tau. \tag{4.11}$$

- ***Notation 2**: Here, the Doppler shift is referenced to the delayed version of $y(\tau)$, meaning the exponential term includes $(\tau - t)$ instead of τ. This shift makes the frequency term dependent on the time lag*

$$\Psi(t,\omega) = \int_0^{T_i} x(\tau)\, y^*(\tau - t)\, e^{j\omega(\tau - t)}\mathrm{d}\tau. \tag{4.12}$$

- ***Notation 3**: In this variant, the signal $x(\tau)$ is shifted backward, leading to a change in the exponent structure. The negative sign in the Doppler term remains, ensuring consistency with standard Fourier conventions*

$$\Psi(t,\omega) = \int_0^{T_i} x(\tau + t)\, y^*(\tau)\, e^{-j\omega(\tau + t)}\mathrm{d}\tau. \tag{4.13}$$

- ***Notation 4**: Similar to the previous case, $x(\tau)$ is shifted, but now the exponential term adopts a positive-frequency sign. This notation is sometimes used in systems where phase shifts are interpreted in an alternative reference frame*

$$\Psi(t,\omega) = \int_0^{T_i} x(\tau + t)\, y^*(\tau)\, e^{j\omega\tau}\mathrm{d}\tau. \tag{4.14}$$

We have already encountered the cross-ambiguity function represented as bivariate spectra in the context of frequency-modulated continuous-wave radars (see

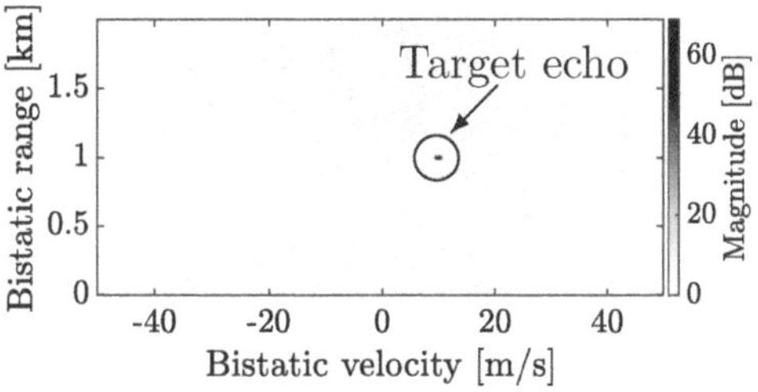

Figure 4.1 Range-velocity map obtained using the cross-ambiguity function using the simulated 5G signal.

Chapter 3). The same type of distribution applies to other radar systems. As an illustrative example, Figure 4.1 presents a simulated cross-ambiguity function for a passive bistatic radar. The target was illuminated by the simulated fifth-generation (5G) signal, and the bistatic range and velocity were set to 1 km and 10 m/s, respectively. As can be noticed, the outcome does not differ in interpretation from that for the frequency-modulated continuous-wave radar (see, e.g., Figure 3.17).

The cross-ambiguity function implicitly given by (4.8) is not used in practice due to its enormous computational load. We discuss the topic of practical utilization of the cross-ambiguity function in the further part of this chapter. Now, let us take a look at the effect of the target's movement and its impact on radar signal processing performance.

4.3 PROBLEM OF RANGE AND VELOCITY MIGRATION

The refresh rate in radars refers to how frequently the system updates target information. On one hand, a high refresh rate is crucial for reliable situational awareness. On the other hand, a longer observation time (known as integration time) enhances the detection of weaker targets and improves the overall SNR. However, excessively extending the integration time can cause range and velocity migration, ultimately degrading the accuracy of the results.

The rationale behind the need for a long integration time stems from the fundamental relation called the integration gain [9]

$$G_i = BT_i, \tag{4.15}$$

where B is the signal bandwidth and T_i is the integration time. Real considerations typically cause the signal bandwidth to be slightly smaller than the sampling rate $f_s = 1/t_s$, thus

$$G_i \approx f_s T_i = \frac{T_i}{t_s} = N, \tag{4.16}$$

where N is the number of samples used for the cross-ambiguity function calculation. Equation (4.15) tells us that the SNR in the cross-ambiguity function output is higher when we increase the integration time. Since the overall process is coherent, the

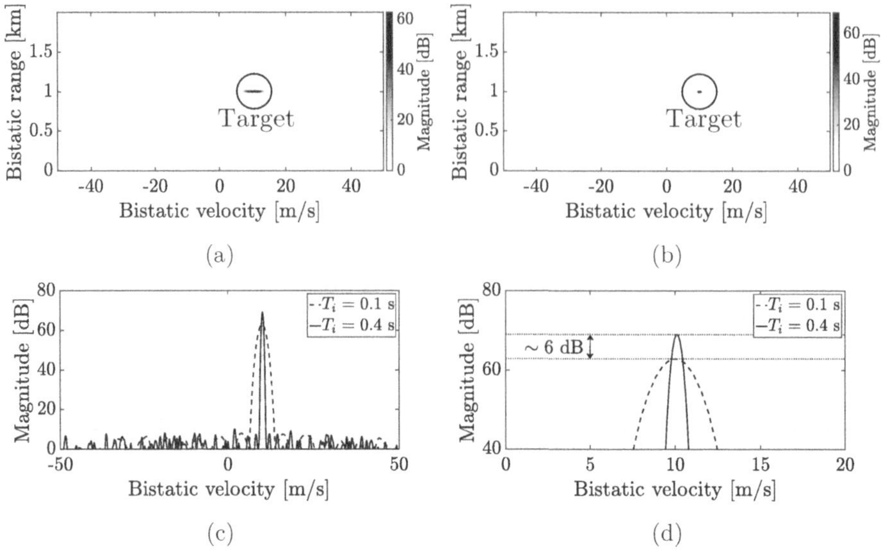

Figure 4.2 Integration gain visualization. (a) Range-velocity map for $T_i = 0.1$ s; (b) range-velocity map for $T_i = 0.4$ s; (c) cross-sections of (a) and (b) for the range cell corresponding to the target echo; (d) close-up on the peaks.

longer the integration time, the longer the signal part we analyze. When we observe the target whose range and velocity do not vary during the integration interval, we accumulate more energy from the echo when extending this observation time. To quantitatively compare this relationship, let us illustrate the same simulation result as in Figure 4.1 (with the target at $R = 1$ km and $V = 10$ m/s) but for two integration times as $T_i = 0.1$ s and $T_i = 0.4$ s. The outcomes delineated in Figure 4.2 lead to the twofold conclusion. Firstly, we see the difference in the velocity resolution, which for the bistatic configuration reads as (see Chapter 1)

$$\Delta V = \frac{\lambda}{T_i}. \tag{4.17}$$

Secondly, the effect of integration gain manifests itself as a higher echo peak corresponding to (4.16). In particular, we may write

$$\begin{aligned} \Delta G_i =& 10\log_{10}(f_s T_{i,2}) - 10\log_{10}(f_s T_{i,1}) \\ =& 10\log_{10}(50 \cdot 10^6 \cdot 0.4) - 10\log_{10}(50 \cdot 10^6 \cdot 0.1) \approx 6 \text{ dB}, \end{aligned} \tag{4.18}$$

which covers our simulations.

The outcomes, however, may suggest that we can increase the integration time freely, obtaining higher and higher integration gains. In practice, we identify two limitations that can possibly occur separately and jointly: range and velocity migration. Let us begin with the first one.

Figure 4.3 Simplified depiction of the range migration cause in (a) monostatic and (b) bistatic radars.

Range migration in radar systems occurs when a moving target changes its position within the coherent processing interval, causing its echo to shift across multiple range cells. This effect is particularly significant for high-speed or accelerating targets where traditional range-Doppler processing fails to maintain coherence and for excessively long integration times. As a result, energy from a single target disperses, reducing detection performance and blurring the radar result. A graphical interpretation of the result can be seen in Figure 4.3, where the target movement leads to a significant variation in the range (mono- and bistatic).

Let $y(t)$ be the reference signal in a passive bistatic radar. The received echo can be modeled as

$$x(t) = \mathscr{A} y\left(t - \frac{R+Vt}{c}\right) e^{j\omega_x t}, \tag{4.19}$$

where $\mathscr{A}$ is the (constant) amplitude and

$$R(t) = R + Vt, \tag{4.20}$$

denotes the target range assuming constant velocity V, and c is the propagation speed. The Doppler angular frequency is given by

$$\omega_x = -2\pi\frac{V}{\lambda}, \tag{4.21}$$

since we consider a bistatic configuration. Substituting (4.24) into (4.25) yields

$$\Psi(t,\omega) = \int_0^{T_i} \mathscr{A} y\left(\tau - \frac{R+V\tau}{c}\right) y^*(\tau - t) e^{j\omega_x\tau} e^{-j\omega\tau} \mathrm{d}\tau. \tag{4.22}$$

It follows that the CAF attains its maximum for $\omega = \omega_x$ and for a time delay matching the instantaneous propagation delay, i.e.,

$$t \approx \frac{R+V\tau}{c}. \tag{4.23}$$

The same derivations can be done for the monostatic radar system. Namely, the received echo can be modeled as

$$x(t) = \mathscr{A} y\left(t - \frac{2(R+Vt)}{c}\right) e^{j\omega_x t}, \tag{4.24}$$

where the factor 2 accounts for the two-way propagation path. The Doppler angular frequency is given by

$$\omega_x = -4\pi \frac{V}{\lambda}. \tag{4.25}$$

Substituting (4.24) and (4.25) into the CAF yields

$$\Psi(t,\omega) = \int_0^{T_i} \mathscr{A} y\left(\tau - \frac{2(R+V\tau)}{c}\right) y^*(\tau - t) e^{j\omega_x \tau} e^{-j\omega\tau} \mathrm{d}\tau. \tag{4.26}$$

The CAF attains its maximum for $\omega = \omega_x$ and for a delay matching the instantaneous round-trip propagation time, i.e.,

$$t \approx \frac{2(R+V\tau)}{c}. \tag{4.27}$$

As in the bistatic case, the time-varying delay causes the echo to migrate across range bins for longer integration times, leading to a spreading of the target energy in the range-velocity domain.

Since the propagation delay depends on time through $\tau(t) = \frac{M(R+Vt)}{c}$, where $M = 1$ for bistatic and $M = 2$ for monostatic configurations, the target response does not remain concentrated at a single delay value. For short integration times ($T_i \to 0$), the echo occupies a single range cell. However, as T_i increases, the time-varying delay causes the echo to migrate across range bins. This effect, known as range migration and shown in Figure 4.4, leads to a spreading of the target energy in the range-velocity domain, thereby reducing peak concentration and degrading detectability.

In summary, range migration occurs when the following criteria are met

- **Monostatic Radar**: Migration occurs when the target velocity V is sufficiently high such that the term VT_i exceeds the range resolution ΔR, i.e.,

$$VT_i > \Delta R. \tag{4.28}$$

- **Bistatic Radar**: Similar conditions apply, but with bistatic effects: the range migration depends on the bistatic velocity V_b, which is a function of both transmitter-target and receiver-target motion. The bistatic range migration condition follows

$$V_b T_i > \Delta R_b, \tag{4.29}$$

  where ΔR_b is the bistatic range resolution.

In the literature, we can find several algorithms that tackle this issue in manifold ways, e.g., [220,221].

Now, let us draw our attention to the second problem, namely, velocity migration. In order for the conditions for the velocity cell migration to occur, the object

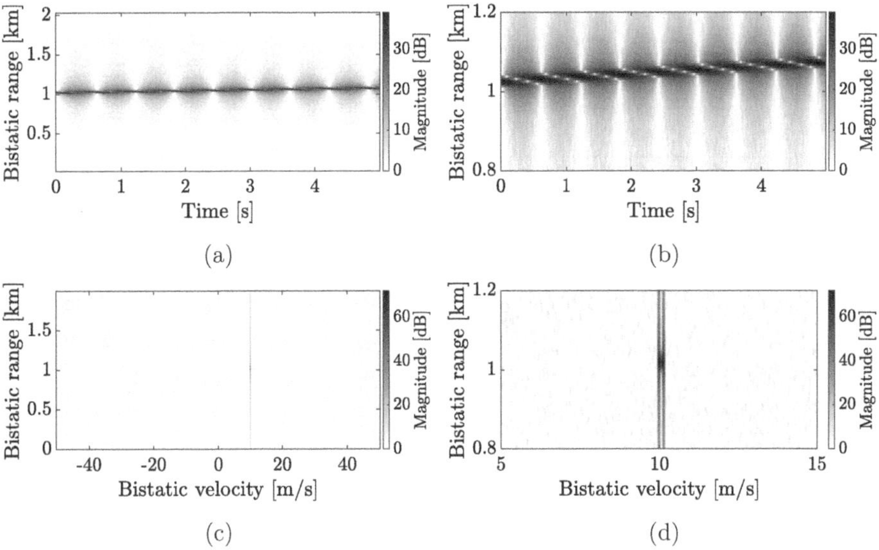

Figure 4.4 Range migration for high integration interval. (a) Whole range-time distribution; (b) close-up on the target echo; (c) whole range-velocity map; (d) close-up on the target echo.

must change its speed during observation. To investigate this problem deeply, let us introduce the general formula for the target range

$$R(t) = \sum_{n=0}^{\infty} \frac{a_n}{n!} t^n. \tag{4.30}$$

A typical assumption on a constant velocity yields

$$R(t) \approx a_0 + a_1 t, \tag{4.31}$$

where a_0 is the initial range (henceforth called R_0), and a_1 is the initial velocity so far denoted as V. However, in recent years, targets (especially those of special use) maneuver; thus, (4.31) is biased, and the range-velocity map is blurred with a target echo spread in velocity according to higher-order terms in (4.30). As an approximation for the maneuvering target, let us assume

$$R(t) \approx R_0 + a_1 t + \frac{a_2 t^2}{2}, \tag{4.32}$$

where a_2 corresponds to acceleration. Thus, we have an additional term in the echo signal phase

$$\alpha_x = -M \cdot 2\pi \frac{a_2}{\lambda}, \tag{4.33}$$

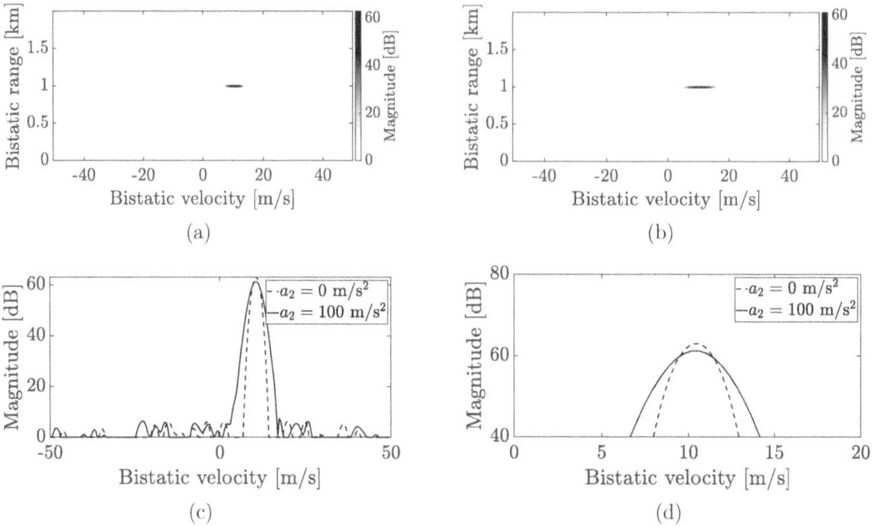

Figure 4.5 Velocity migration for accelerating target. (a) Range-velocity map for $a_2 = 0$ m/s^2; (b) range-velocity map for $a_2 = 100$ m/s^2; (c) cross-sections of the range-velocity maps across the range cell with the target echo; (d) close-up on the target echo.

where again $M = 1$ for the bistatic, and $M = 2$ for the monostatic configuration. The echo signal can be, therefore, written as

$$x(t) = \mathscr{A} y\left(t - \frac{M\left(R_0 + a_1 t + \frac{a_2 t^2}{2}\right)}{c}\right) e^{j\left(\omega_x t + \frac{\alpha_x t^2}{2}\right)}, \tag{4.34}$$

where

$$\omega_x = -M \cdot 2\pi \frac{a_1}{\lambda}, \quad \alpha_x = -M \cdot 2\pi \frac{a_2}{\lambda}. \tag{4.35}$$

Instead of a single peak, the CAF exhibits a spread along both delay and Doppler axes. The maximum follows a trajectory given approximately by

$$t \approx \frac{M\left(R_0 + a_1 \tau + \frac{a_2 \tau^2}{2}\right)}{c}, \quad \omega \approx \omega_x + \alpha_x \tau. \tag{4.36}$$

The effect of velocity migration is illustrated in Figure 4.5. The same simulation scenario was executed with an integration time of $T_i = 0.1$ s for two acceleration values: $a_2 = 0$ m/s^2 and $a_2 = 100$ m/s^2. Two key observations emerge from the results. First, in the range-velocity maps, the accelerating target exhibits a noticeable echo spread, causing the peak to be less distinct as its energy disperses across the velocity domain. Second, this spreading effect reduces the peak amplitude of the accelerating target, leading to an overall degradation in SNR.

Assuming that the target has acceleration a_2, the velocity change during the integration time T_i is given by

$$\Delta a_1 = T_i \cdot a_2. \tag{4.37}$$

Velocity cell migration occurs if the change Δa_1 is considerable relative to the velocity resolution cell Δa_1. The resolution cell for velocity is determined as

$$\Delta a_1 = \frac{\lambda}{M \cdot T_i}. \tag{4.38}$$

As a result, the constraint on the integration time for a given acceleration can be expressed as

$$T_i < \sqrt{\frac{\lambda}{M \cdot a_2}}. \tag{4.39}$$

Failure to satisfy this condition results in spreading the target echo along the velocity dimension. The longer the integration time, the more pronounced the echo dispersion in the velocity domain.

As we are already familiar with the problem of range and velocity migration in radars, we can focus on exploiting the Fourier transform to estimate the influence of these phenomena, and we will try to compensate for the negative factor on radar performance. The following sections seek to elucidate the problem in question in an intelligible way. Let us describe the issue of target kinematic parameter estimation.

4.4 ACCELERATION ESTIMATION

Range and velocity migration follow from the target kinematic parameters. We can observe a range walk solely when the object moves with a constant velocity. However, although the actual target velocity does not vary, we do not always observe it with the same estimated velocity within the observation interval. As we know from Chapter 1, the Doppler shift, as a quantity we can measure in radar, is proportional to the target velocity but is scaled by appropriate angles depending on whether we are discussing mono- or bistatic geometry. Nevertheless, even if the target moves with an ideally stable velocity, it may change due to the geometry. In such a case, we observe the echo as accelerating (or decelerating). Of course, maneuvers and intended acceleration certainly exist and can also impact the target echo observed on the range-velocity map. Since the ideal situation when the target velocity $\vec{V}$ is composed of the radial component V_r (that is actually measured by radar, see Figure 1.2), we can assume each target has *some* acceleration. The situation is common, especially in passive radars, where the geometry is more complicated. Thus, in practice, vast targets can be considered as accelerating in radars, although their actual velocity can be constant. This effect stems from the relationships

$$f_D = -\frac{2V}{\lambda}\cos(\varphi), \tag{4.40}$$

and

$$f_D = -\frac{2V}{\lambda}\cos(\theta)\cos(\beta/2), \tag{4.41}$$

for the mono- and bistatic radar, respectively. Here, φ denotes the angle between the object's motion vector and the propagation direction of the signal reflected from the target back to the radar, β is the bistatic angle, and θ is the angle between the velocity vector and the bistatic angle bisector. Figures 1.2 and 1.5 illustrate the angle definitions. While θ was previously used to denote the angle in all cases, here we distinguish between the two configurations: (4.40) employs φ for the monostatic case, whereas θ is retained for the bistatic geometry. When these angles vary, the Doppler shift also varies, thus being interpreted as a frequency rate and, therefore, acceleration.

The cross-ambiguity function (4.8) does not inherently support accelerating targets. As we discussed, we seek delays and constant Doppler shifts leading to the maximum correlation value. Although this approach is sufficient for relatively low integration times, its extension imposes the need to incorporate acceleration into the whole processing.

The problem of target kinematic parameter estimation was recognized in the literature manifold [222–231]. Both fast-moving and maneuvering targets and long integration times need to consider the target acceleration for reliable radar measurement and robust detection.

4.4.1 EXTENDED CROSS-AMBIGUITY FUNCTION

As we derived in the previous section, the echo is spread in velocity and sometimes in range when we observe tangible acceleration. To take this quantity into account, Malanowski proposed the so-called extended cross-ambiguity function [222], defined as

$$\Psi(t,\omega,\alpha) = \int_{0}^{T_i} x(\tau)\, y^*(\tau - t)\, e^{-j\left(\omega\tau + \alpha\frac{\tau^2}{2}\right)} \mathrm{d}\tau, \quad \Psi(t,\omega,\alpha) \in \mathbb{C}, \tag{4.42}$$

where α corresponds to the set of target accelerations and reads as

$$\alpha = -M \cdot 2\pi \frac{a_2}{\lambda}. \tag{4.43}$$

Since the method was initially proposed for passive radars [222], it can be easily applied to active monostatic systems, as proved in [223]. Thus, we can differentiate between these two configurations, and this is where M comes from in (4.43).

The extended cross-ambiguity function (4.42) can be understood as follows. For the received surveillance signal $x(t)$, we construct a two-dimensional filter bank that spans possible target velocities and accelerations. For each pair of these parameters, the reference signal is correlated with the received signal containing the target echo after demodulating it by the corresponding Doppler frequency and frequency rate. The result is a three-dimensional distribution with axes representing time delay t, Doppler angular frequency ω, and the frequency rate α associated with target acceleration. Typically, we observe the result using the squared magnitude, obtaining

$$\Upsilon(t,\omega,\alpha) = |\Psi(t,\omega,\alpha)|^2. \tag{4.44}$$

Remark 4.3. *The same as the basic cross-ambiguity function, the extended one can be written in four equivalent forms, as follows:*

$$\Psi(t,\omega,\alpha)=\int_0^{T_i} x(\tau)\,y^*(\tau-t)\,e^{-j\left(\omega\tau+\alpha\frac{\tau^2}{2}\right)}\mathrm{d}\tau. \quad (4.45)$$

$$\Psi(t,\omega,\alpha)=\int_0^{T_i} x(\tau)\,y^*(\tau-t)\,e^{j\left(\omega(\tau-t)+\alpha\frac{(\tau-t)^2}{2}\right)}\mathrm{d}\tau. \quad (4.46)$$

$$\Psi(t,\omega,\alpha)=\int_0^{T_i} x(\tau+t)\,y^*(\tau)\,e^{-j\left(\omega(\tau+t)+\alpha\frac{(\tau+t)^2}{2}\right)}\mathrm{d}\tau. \quad (4.47)$$

$$\Psi(t,\omega,\alpha)=\int_0^{T_i} x(\tau+t)\,y^*(\tau)\,e^{j\left(\omega\tau+\alpha\frac{\tau^2}{2}\right)}\mathrm{d}\tau. \quad (4.48)$$

When the value of α gets closer to the frequency rate caused by the target acceleration, we observe a higher and narrower peak associated with the target. The processing diagram is presented in Figure 4.6. The exponents are arbitrarily scaled so that all frequency-related components are scaled in m/s and m/s^2.

The reference signal is first modulated by an acceleration-dependent phase term for each value $a_2^{[1]}, a_2^{[2]}, \ldots, a_2^{[L]}$, where the superscript indicates a specific acceleration value drawn from an L-element set. Afterward, it passes through a Doppler filter

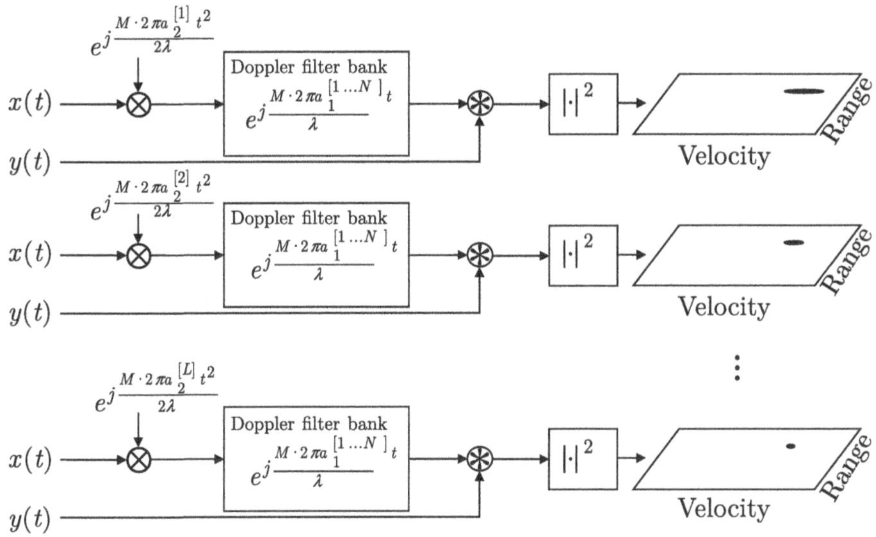

Figure 4.6 Extended cross-ambiguity function calculation scheme.

bank, undergoing additional modulation by N harmonic components corresponding to various Doppler frequency shifts. This doubly-modulated reference signal is then correlated with the received (surveillance) signal. The squared magnitude of the result forms a single slice of the extended cross-ambiguity function. This process is repeated L times, once for each assumed acceleration value.

Let us illustrate the performance of this method using the simulated passive radar scenario and the digital video broadcasting-terrestrial (DVB-T) signal as an illuminator. Let us deem the carrier frequency $f_c = 184.5$ MHz, the bandwidth $B = 7$ MHz, and the sampling rate $f_s = 8$ MSa/s. The target with an initial velocity of $a_1 = 40$ m/s was located at the bistatic range of $R = 1$ km. Its acceleration was set to $a_2 = 40$ m/s^2. Using the extended cross-ambiguity function (4.42) with the acceleration space embracing $a_2 \in [0, 100]$ m/s^2 with the step of 1 m/s^2, the target acceleration was estimated. This process produced a three-dimensional distribution, scaled accordingly to represent range, velocity, and acceleration, as shown in Figure 4.7a. In particular, both the full 3D distribution and its cross-sections are presented. The peak of the extended cross-ambiguity function occurs when the target acceleration in the received signal matches a value in the filter bank. At this point, the modulation term is effectively compensated, resulting in a narrower and more pronounced peak at the target echo's location. This behavior is clearly visible in Figure 4.7b–d. Additionally, a side effect of the demodulation process can be observed. For instance, weak clutter

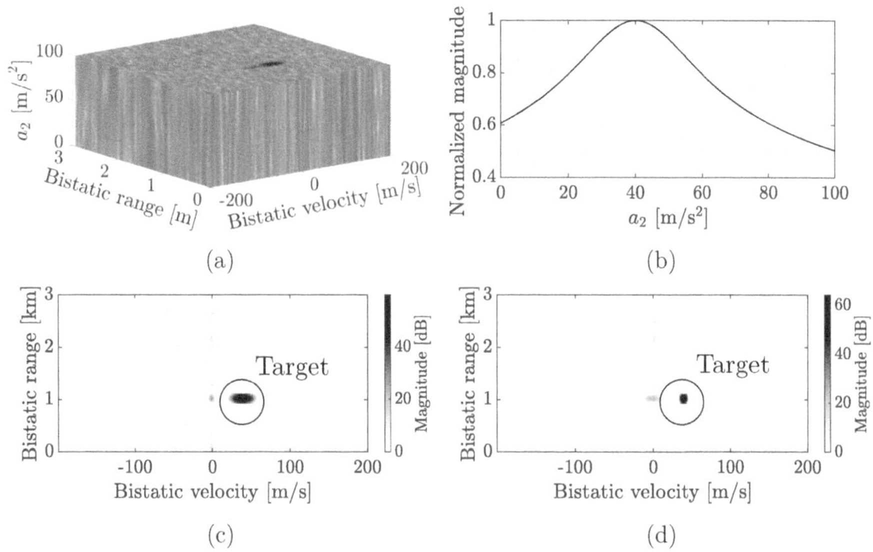

Figure 4.7 Extended cross-ambiguity function using the simulated signal example. (a) Three-dimensional range-velocity-acceleration distribution; (b) cross-section along the acceleration axis for the target range and velocity coordinates; (c) example of the range-velocity map for the mismatched target acceleration; (d) range-velocity map for target acceleration match.

around 0 m/s appears relatively narrow in Figure 4.7c, where the original signal is shown. However, as the modulation rate increases, components such as clutter (or potentially other targets) become increasingly spread when their true acceleration does not match the one used for demodulation. Thus, while the method improves the focus on a specific accelerating target, it may also cause deterioration in other signal components.

The concept is similar for both active and passive radars; therefore, let us omit excessive examples. Having the method validated in simulations, we can verify its performance in real-life scenarios. For this purpose, the DVB-T with the same parameters as in the simulations was used. The signal was acquired during the APART-GAS 2019 trials mentioned in Chapter 2. The measurement embraced observation of the amateur ~ 1.5 m long rocket (manufactured by the Polish company SpaceForest) by the passive radar. The missile was launched from the pad located at the airfield. During the flight, the reference signal (from the DVB-T illuminator) and the echo signal reflected from the rocket were recorded by the passive radar [232].

As this section mentions, we know from (4.15) that increasing the integration time leads to better target visibility. However, maneuvering targets like launching rockets have rapidly varying velocities, leading to the blur of their peak in the range-velocity map. This effect is observed in Figure 4.8a, where the target echo is spread in the velocity. The velocity migration stems from the relatively long integration time $T_i = 1$ s, which is a desirable value in passive radars, albeit simultaneously spurious to the output SNR. Target acceleration estimation yielded the outcome delineated in Figure 4.8b with the maximum value of $a_2 = 42$ m/s^2. After the acceleration compensation, we obtained the range-velocity map from Figure 4.8c. We see similar consequences to those in the simulated scenario. The clutter around 0 m/s is spread as the a_2 term modulates it. At the same time, the echo became narrower and higher, leading to unequivocal target parameter estimation. The numerically expressed improvement is shown in Figure 4.8d, where we gained over 4 dB in the peak magnitude and additionally reduced the peak width.

4.4.2 GENERALIZED RADON-FOURIER TRANSFORM FOR LONG-TIME COHERENT INTEGRATION

Although effective, the method described above, based on the extended cross-ambiguity function, remains constrained to relatively short coherent integration times. In scenarios where the observation duration is tangibly extended, the echo from a moving target migrates in both range and velocity. This phenomenon limits the performance of classical Doppler processing and necessitates more advanced methods. In such cases, the Radon-Fourier transform becomes highly beneficial [233–235].

The Radon-Fourier transform is a long-time coherent integration technique developed for radar target detection, particularly effective in conditions involving significant range migration and low SNRs. In contrast to traditional methods, such

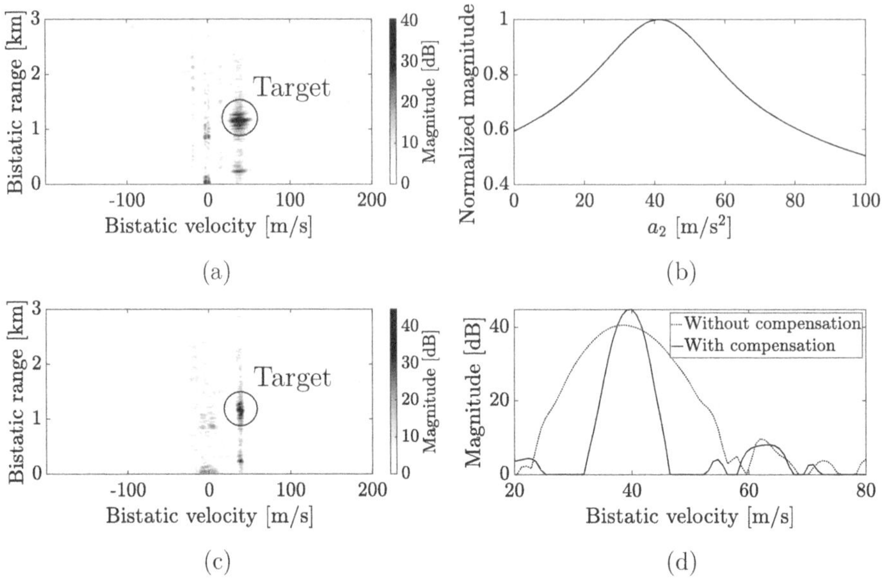

Figure 4.8 DVB-T-based passive radar signal-processing results with a rocket as an accelerating target. (a) Initial range-velocity map for $T_i = 1$ s; (b) acceleration estimation result using (4.42) with the maximum value indicating the target acceleration; (c) range-velocity map after the signal compensation using the estimated acceleration; (d) comparison of the target cross-section for the initial range-velocity map and after compensation.

as moving target detection or the extended cross-ambiguity function, which only compensate for Doppler shifts within fixed range bins, the Radon-Fourier transform compensates both Doppler-induced phase modulation and target range migration. This is achieved through marginalization of the range-time domain along the slow time dimension.

While the original formulation of the Radon-Fourier transform assumes a linear target motion model (i.e., constant velocity), the generalized Radon-Fourier transform [236] extends this framework to account for higher-order kinematic parameters, such as acceleration, jerk, and beyond. Therefore, the generalized Radon-Fourier transform is particularly well-suited for tracking highly maneuvering targets.

The generalized Radon-Fourier transform operates under the assumption that the target moves with a time-varying velocity model, which introduces nonstationary Doppler shifts that cannot be effectively handled using classical Fourier analysis. This limitation becomes especially apparent in the presence of accelerating targets.

In standard radar signal processing chains, operations are typically performed in the range-velocity domain. The generalized Radon-Fourier transform, however, can be applied directly to the range-compressed signal, regardless of how range compression is performed (e.g., via matched filtering, cross-correlation, or the Fourier transform, as in frequency-modulated continuous-wave systems).

The foundation of the generalized Radon-Fourier transform lies in the signal model after range compression $r(t,\tau)$, which is expressed as

$$r(t,\tau) = \text{sinc}\left(B\left(\tau - \frac{MR(t)}{c}\right)\right) \cdot e^{-j\frac{M2\pi R(t)}{\lambda}}, \tag{4.49}$$

where t and τ represent slow and fast time, respectively. The range trajectory is modeled as a polynomial

$$R(t) = R_0 + a_1 t + \frac{1}{2}a_2 t^2 + \cdots + \frac{1}{N}a_N t^N. \tag{4.50}$$

For clarity, we simplify the trajectory to include only the first two motion parameters

$$R(t) = R_0 + a_1 t + \frac{1}{2}a_2 t^2. \tag{4.51}$$

All other symbols retain their previously defined meanings.

It is important to note that the generalized Radon-Fourier transform assumes the range-compressed echo has the shape of a two-dimensional sinc function

$$\text{sinc}(t,\tau) = \frac{\sin(\pi t)}{\pi t} \cdot \frac{\sin(\pi\tau)}{\pi\tau}, \tag{4.52}$$

in contrast to earlier assumptions based on the Dirac delta function, which served as a simplification without affecting the generality of the approach.

The principle of the generalized Radon-Fourier transform is to estimate the range $\hat{R}_0$ and motion parameters $\hat{a}_1$ and $\hat{a}_2$ such that the energy in the range-time domain is maximized after marginalization along time. This process is formally expressed as

$$G(\dot{R}_0,\dot{a}_1,\dot{a}_2) = \int_{-\infty}^{\infty} r(t,\tau) \cdot \exp\left(j\frac{M2\pi R(t|\dot{R}_0,\dot{a}_1,\dot{a}_2)}{\lambda}\right) \mathrm{d}t, \quad G(\dot{R}_0,\dot{a}_1,\dot{a}_2) \in \mathbb{C}, \tag{4.53}$$

where $\dot{R}_0$, $\dot{a}_1$, and $\dot{a}_2$ denote the grid of hypothesized initial range, velocity, and acceleration parameters, t is a slow time, and τ is a fast time. The final parameter estimates (for the single target) are obtained via

$$(\hat{R}_0,\hat{a}_1,\hat{a}_2) = \underset{\dot{R}_0,\dot{a}_1,\dot{a}_2}{\text{argmax}} \left|G(\dot{R}_0,\dot{a}_1,\dot{a}_2)\right|. \tag{4.54}$$

Remark 4.4. *Both the extended cross-ambiguity function (4.42) and the generalized Radon-Fourier transform can be interpreted as maximum likelihood estimators under standard radar signal and noise assumptions.*

Let us consider a received signal modeled as

$$y(t) = \kappa x(t;\theta) + w(t), \tag{4.55}$$

where $x(t;\theta)$ denotes the reference signal modulated according to a parameter vector θ (e.g., delay, Doppler shift, acceleration), κ is a complex amplitude, and $w(t)$ is additive white Gaussian noise with zero mean and variance σ^2.

Under these assumptions, the likelihood function of the received signal given the parameters θ is

$$\mathscr{L}(\theta) \propto \exp\left(-\frac{1}{\sigma^2}\left|y(t)-\kappa x(t;\theta)\right|^2\right). \tag{4.56}$$

Maximizing the likelihood with respect to θ (and jointly with κ) leads to maximizing the squared magnitude of the inner product between the received and reference signals

$$\hat{\theta} = \underset{\theta}{\operatorname{argmax}}\left|\int y(t)x^*(t;\theta)\mathrm{d}t\right|^2, \tag{4.57}$$

which corresponds directly to the operation performed in the cross-ambiguity function or generalized Radon-Fourier transform domain.

Thus, both the extended cross-ambiguity function and the generalized Radon-Fourier transform can be viewed as implementations of the maximum likelihood estimator principle, where the matched filter is constructed for a bank of hypothesized parameters, and the one maximizing the output energy is selected.

This interpretation holds under the following assumptions:

- *The additive noise is Gaussian, white, and independent of the signal.*
- *The reference signal $x(t;\theta)$ is known and can be modulated for any θ.*
- *The observed echo $y(t)$ consists of a delayed and modulated version of the reference plus noise.*
- *Interference from other targets is negligible or regarded as noise.*

The concept behind the generalized Radon-Fourier transform is depicted in Figure 4.9. In the original domain, the target follows a trajectory typical of an accelerating object (with the velocity given by the Nth-order polynomial), resulting in nonlinear range migration over time. The transform aims to counteract this effect by compensating both for the range migration and associated phase shifts, effectively aligning the target's response to appear at a constant range throughout the observation interval. Once this compensation is applied, a coherent marginalization (i.e., accumulation) along the time axis of the range-time distribution produces a pronounced peak. This peak reaches its maximum precisely when the assumed grid parameters align with the actual kinematic parameters of the target.

In summary, the generalized Radon-Fourier transform estimates the kinematic parameters of a target by adaptively compensating for its motion in the range-time domain. The method, although initially proposed for active radars [233–235], also works for passive bistatic systems [189,237]. The iterative process introduces tweaks to both range and phase drifts across candidate parameters in the sets $\dot{R}_0, \dot{a}_1, \dot{a}_2$. When these candidate parameters match the actual target trajectory, the integrated signal energy reaches its peak, thus revealing the optimal estimates $\hat{R}_0, \hat{a}_1$, and $\hat{a}_2$.

Let us again analyze the DVB-T signal with the rocket echo. The previously shown results were obtained under the prerequisite of an integration time $T_i = 1$ s, which is relatively long for DVB-T-based passive radars, yet still quite short, particularly for low SNRs and maneuvering targets. If we extend the integration time

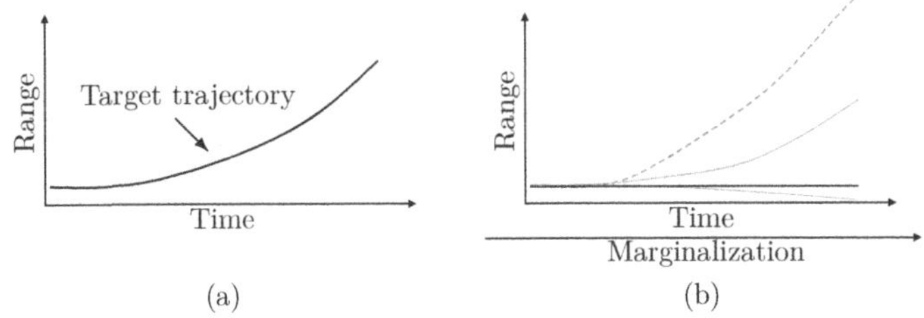

Figure 4.9 Generalized Radon-Fourier transform idea. (a) Target trajectory in the range-time domain; (b) Trajectory compensation for different kinematic parameters (in gray dashed lines), and the ideal situation, where the target parameters are precisely estimated, and the range curve becomes straight (black solid line). The latter case results in a strong concentrated peak after marginalization.

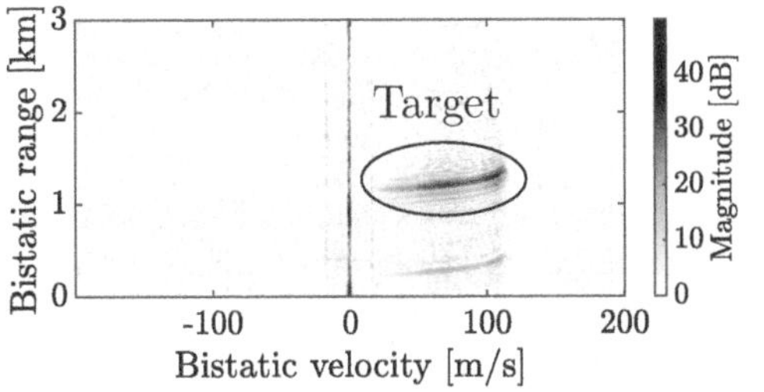

Figure 4.10 Range-velocity map for the DVB-T signal and the accelerating rocket with $T_i = 5$ s.

such that $T_i = 5$ s, we obtain an apparent range and velocity drift, as demonstrated in Figure 4.10.

Focusing on the initial flight phase, including the rocket launch, we can analyze its kinematic parameters for a wider time interval using the generalized Radon-Fourier transform. Figure 4.11 demonstrates the result.

In Figure 4.11a, the initial range-time distribution is presented, clearly showing the echo of a target migrating in range. Figure 4.11b and c illustrate the effects of applying the generalized Radon-Fourier transform with different randomly selected motion parameters. Specifically, the result in Figure 4.11b corresponds to the parameter set $\dot{a}_1 = 10$ m/s and $\dot{a}_2 = 10$ m/s^2, while Figure 4.11c reveals the outcome for $\dot{a}_1 = 90$ m/s and $\dot{a}_2 = 50$ m/s^2. In contrast, Figure 4.11d shows the range-time map after compensation using the optimal kinematic parameters. As a result, the target echo is straightened into a clear linear trajectory. The estimated initial range is $\hat{R}_0 = 1.04$ km, with the initial velocity and acceleration estimated at $\dot{a}_1 = 50$ m/s and $\dot{a}_2 = 10$ m/s^2, respectively. It is important to note that these estimates were obtained using long-time integration.

In comparison, the previous analysis based on the extended cross-ambiguity function considered only a limited 20% segment of the same recording. Consequently, the

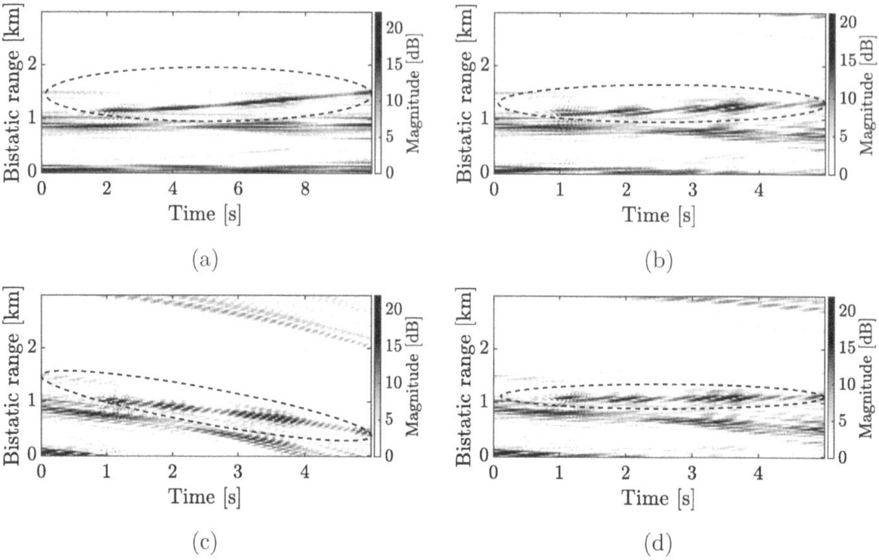

Figure 4.11 DVB-T-based passive radar signal-processing results with a rocket as an accelerating target. The target range history is marked using the black dashed oval. (a) Initial range-time distribution; (b) range-time distribution after compensation of $\dot{a}_1 = 10$ m/s and $\dot{a}_2 = 10$ m/s^2; (c) range-time distribution after compensation of $\dot{a}_1 = 90$ m/s and $\dot{a}_2 = 50$ m/s^2; (d) range-time distribution after compensation of $\dot{a}_1 = 50$ m/s and $\dot{a}_2 = 10$ m/s^2. The maximum value of (4.53) was obtained for the last example.

kinematic parameters estimated over a short-time window may differ from those obtained over a longer duration. This variation arises because the initial velocity can depend on the specific moment when the observation starts. Moreover, since the acceleration was not constant throughout the flight, its estimated value may also vary depending on the selected segment of the signal.

These examples demonstrate how the generalized Radon-Fourier transform compensates for range and velocity migration across the hypothesized parameter space. When the selected parameters from the grids $\dot{R}, \dot{a}_1, \dot{a}_2$ closely match the actual kinematic profile of the target, we observe a noticeable straightening of the target's trajectory in the range-time domain. Importantly, each transformation not only aligns the signal in delay but also compensates for phase shifts. As a result, marginalizing the compensated range-time distributions over time yields a clear peak, maximized when the applied parameters best match the target's proper motion, as seen in Figure 4.11d.

The generalized Radon-Fourier transform is precise, but its vulnerability is the tremendous computational cost. Similar to the extended cross-ambiguity function, both methods are iterative and, thus, highly inefficient. Moreover, the technique is inherently parametric, which means it benefits from tuning to the specifics of a given scenario. Ideally, the assumed kinematic parameter grid should include the actual target motion characteristics. While a more expansive parameter space increases the likelihood of successful detection, it may also lead to longer processing times. Conversely, a parameter set that is too narrow might not fully capture the target's dynamics, potentially resulting in suboptimal estimation. Therefore, a balanced choice of parameter space is recommended to ensure both accuracy and computational efficiency.

Despite the effectiveness of the previously described methods, one tangible limitation becomes apparent–their considerable computational complexity. This problem stems primarily from the need to perform exhaustive grid-based searches over multiple motion parameters, which can be computationally demanding, especially in real-time radar applications. In practical scenarios, modern radar systems must operate with high efficiency and responsiveness, often under constrained hardware and time budgets. Consequently, there is an apparent demand for algorithms that do not rely on discretized parameter grids and can estimate multiple target parameters simultaneously and adaptively. A promising alternative arises from the time-frequency signal-processing framework discussed extensively in Chapter 2. In that context, we highlighted the ability to estimate the local chirp rate, a phase acceleration at each time-frequency coordinate (t, ω), independently of the number or nature of signal components. This capability opens up the possibility of capturing complex kinematic behaviors without the need for prior parameter discretization.

Building on that idea, we now aim to extend this methodology into the range-velocity domain. By doing so, we seek to retain the flexibility and resolution of time-frequency techniques while addressing the challenges of multi-target parameter estimation in the radar context. Such an approach holds the potential to significantly reduce computational load while maintaining robustness in dynamic and cluttered environments.

4.4.3 ZAK TRANSFORM-BASED TARGET ACCELERATION ESTIMATION

In the previous discussion, we addressed the problem of detecting accelerating targets over long integration times. By compensating for Doppler shifts (via the extended cross-ambiguity function) or for both Doppler shifts and range migration (using the generalized Radon-Fourier transform), we could extend the coherent integration time T_i and thereby enhance detection performance.

However, the SNR is already sufficiently high in specific scenarios to ensure reliable detection. In such cases, our focus may shift from detection to extracting target kinematic parameters from a single range-velocity map obtained over a short integration period T_i.

Reducing T_i leads to two important consequences:

1. Range and velocity migration may become imperceptible, especially when using a low-resolution radar. Nonetheless, the target's kinematic information may still be encoded in the phase of the received echo, manifesting as a subtle phase gradient induced by acceleration or higher-order motion components.

2. In most radar systems, the integration time T_i coincides with the system's refresh rate. This results from typical architectural and processing constraints: the signal is acquired and processed over each interval T_i, and the outcome is delivered before processing the next incoming data segment. Consequently, shortening T_i imposes stricter processing performance and efficiency requirements since each result must be computed in real time before the next segment becomes available.

Given these observations, we assume for the remainder of this section that the echo delay remains approximately constant during the integration interval, i.e.,

$$\tau = \frac{R(t)}{c} \approx \text{const.}, \tag{4.58}$$

which allows us to denote $R(t)$ simply as R. This assumption is justified for several reasons. First, even a rapidly maneuvering target cannot significantly change its position over short integration times. Second, we consider scenarios with relatively high SNR, where direct parameter extraction from the echo is feasible. In low SNR conditions, the previously discussed long-time integration methods remain more robust and effective than those introduced in the following part of this section.

The cross-ambiguity function is hardly ever implemented directly. Its main downside – enormous computational burden – disqualifies the straightforward utilization of (4.8) in practice, especially in real-time systems [189, 217]. Thus, approximation methods that deem a linear time-invariant system are used. The specific case is the Woodward approximation of the cross-ambiguity function. Let us derive this by employing the Zak transform.

Consider the Zak transform (which we already defined in Chapter 3) of the signal $x(t)$. For clarity and consistency, we recall its original definition [185, 186]

$$\mathscr{Z}_x(t,\omega) = \sqrt{T}\sum_{k=-\infty}^{\infty} x(t+kT)e^{-j\omega kT}, \quad k \in \mathbb{Z}, \quad T \in \mathbb{R}_+. \tag{4.59}$$

We neglect $\sqrt{T}$ in line with our previous assumptions.

Previous studies have explored the relationship between the ambiguity function and the Zak transform, particularly in the context of waveform design [238, 239]. These results suggest that the Zak framework can also be applied effectively to range-velocity processing. In practical implementations, the cross-ambiguity function is typically computed via cross-correlation of the received signal $x(t)$ with a reference one $y(t)$, given by

$$\chi(t) = \int_{-\infty}^{\infty} x(\tau)y^*(\tau - t)\,\mathrm{d}\tau = \int_{-\infty}^{\infty} x(\tau + t)y^*(\tau)\,\mathrm{d}\tau. \tag{4.60}$$

However, to derive the range-velocity map practically, the correlation in (4.60) is not applied to the entire signals but instead performed on small time segments. Both $x(t)$ and $y(t)$ are divided into overlapping or non-overlapping blocks, and cross-correlations are computed block-wise. These partial correlations are then processed using a windowing function, and then the Fourier transform is applied across the blocks. Windowing helps suppress sidelobes and enhances the visibility of weak target echoes located near stronger scatterers.

To illustrate this process mathematically, consider the initial block-wise cross-correlation

$$\hat{\chi}(t,0) = \int_0^{T_b} x(\tau)y^*(\tau - t)\,\mathrm{d}\tau = \int_0^{T_b} x(\tau + t)y^*(\tau)\,\mathrm{d}\tau, \tag{4.61}$$

where T_b is the block duration. Shifting the blocks by a constant step $\zeta \in \mathbb{R}$ yields:

$$\hat{\chi}(t,\zeta) = \int_{\zeta}^{T_b+\zeta} x(\tau)y^*(\tau - t)\,\mathrm{d}\tau = \int_{\zeta}^{T_b+\zeta} x(\tau + t)y^*(\tau)\,\mathrm{d}\tau, \tag{4.62}$$

which can be rewritten as

$$\hat{\chi}(t,\zeta) = \int_0^{T_b} x(\tau+\zeta)y^*(\tau+\zeta-t)\mathrm{d}\tau = \int_0^{T_b} x(\tau+\zeta+t)y^*(\tau+\zeta)\mathrm{d}\tau. \tag{4.63}$$

Introducing a window function $h(\zeta)$, we define the weighted correlation as

$$\hat{\chi}_h(t,\zeta) = \hat{\chi}(t,\zeta)h(\zeta). \tag{4.64}$$

Assuming discretization of the window positions as $\zeta = kT$ for integer $k \in [-K/2, K/2)$, we obtain the final range-velocity map using the Fourier transform

$$\Psi(t,\omega) = \sum_{k=-\frac{K}{2}}^{\frac{K}{2}-1} \hat{\chi}(t,kT)h(kT)e^{-j\omega kT}. \tag{4.65}$$

Substituting from (4.63), we arrive at

$$\begin{aligned}\Psi(t,\omega) &= \sum_{k=-\frac{K}{2}}^{\frac{K}{2}-1} \int_0^{T_b} x(\tau+kT)y^*(\tau+kT-t)\,h(kT)e^{-j\omega kT}\mathrm{d}\tau \\ &= \sum_{k=-\frac{K}{2}}^{\frac{K}{2}-1} \int_0^{T_b} x(\tau+kT+t)y^*(\tau+kT)h(kT)e^{-j\omega kT}\mathrm{d}\tau.\end{aligned} \tag{4.66}$$

To simplify notation, let $y_r^*(\tau+kT) := y^*(\tau+kT)h(kT)e^{-j\omega kT}$. Then

$$\begin{aligned}\Psi(t,\omega) &= \sum_{k=-\frac{K}{2}}^{\frac{K}{2}-1} \int_0^{T_b} x(\tau+kT+t)y^*\,(\tau+kT)\,h(kT)e^{-j\omega kT}\mathrm{d}\tau \\ &= \sum_{k=-\frac{K}{2}}^{\frac{K}{2}-1} \int_0^{T_b} x(\tau+kT+t)y_r^*(\tau+kT)\mathrm{d}\tau.\end{aligned} \tag{4.67}$$

This expression can be reformulated by adjusting the integration variable as $T' = kT - t$, then

$$\begin{aligned}
&\int_0^{T_b} x(\tau+kT+t)y_r^*(\tau+kT)\mathrm{d}\tau = \int_0^{T_b} x(\tau+kT)y_r^*(\tau+kT-t)\mathrm{d}\tau \\
&= \int_0^{T_b} x(\tau+kT)y_r^*(\tau+T')\mathrm{d}\tau = \int_0^{T_b} x(\tau+kT)y^*(\tau+T')h(T')e^{-j\omega T'}\mathrm{d}\tau \\
&= \int_0^{T_b} x(\tau+kT)y^*(\tau+kT-t)h(kT-t)e^{-j\omega(kT-t)}\mathrm{d}\tau,
\end{aligned} \tag{4.68}$$

and, finally

$$\begin{aligned}
\Psi(t,\omega) &= \sum_{k=-\frac{K}{2}}^{\frac{K}{2}-1} \int_0^{T_b} x(\tau+kT)y^*(\tau+kT-t)h(kT-t)e^{-j\omega(kT-t)}\mathrm{d}\tau \\
&= \sum_{k=-\frac{K}{2}}^{\frac{K}{2}-1} \int_0^{T_b} x(\tau+kT+t)y^*(\tau+kT)h(kT)e^{-j\omega kT}\mathrm{d}\tau.
\end{aligned} \tag{4.69}$$

Equation (4.69) offers a practical formulation that closely resembles the Zak transform as a form of cross-correlation processing. If the window function is set to unity, then for each index k, the integral represents a repeated segment-wise correlation. Due to the assumption of constant delay $\tau \approx R/c$ during the observation interval, these correlations are effectively aligned, and their summation modulated by the exponential term $e^{-j\omega kT}$ constitutes a discrete Zak transform. Hence, the notation $\mathscr{Z}^h(t,\omega)$ is introduced in place of $\Psi(t,\omega)$ to highlight this connection.

This process is schematically illustrated in Figure 4.12, where the target echo appears as a localized peak in each range profile. These profiles collectively form a two-dimensional representation. The dimension indexed by t is called fast time, while the block index k corresponds to slow time. The application of the window function precedes the Fourier transform, ultimately producing the range-velocity map.

The structure of (4.69), particularly its left-hand side, is not new in radar literature. This formulation is sometimes referred to as Woodward's approximation of the cross-ambiguity function [217, 219], and its implementation corresponds to the so-called "batches algorithm" [240]. Although this approach offers excellent computational efficiency for real-time processing, it does introduce a trade-off: detection performance may degrade in the presence of maneuvering targets. This limitation arises from the underlying assumption of signal stationarity. As such, targets undergoing non-uniform motion are not optimally mapped within the range-velocity domain. This also justifies the earlier assumption of constant delay.

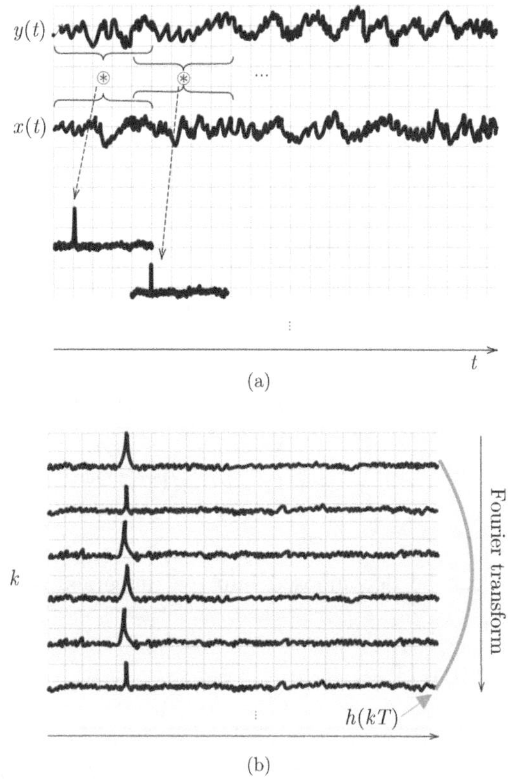

Figure 4.12 Schematic of range-velocity map generation in radar systems. (a) Formation of range profiles via block-wise correlation using any signal type (the displayed example uses a noise-like waveform); (b) range-velocity map computation using Zak transform. (Reproduced from [189], licensed under CC BY 4.0.)

The signal reflected from a moving target and impinging on the radar receiver can be modeled as

$$x(t) = A_x(t)e^{j\phi_x(t)}y\left(t-\frac{R}{c}\right) = x_e(t)y\left(t-\frac{R}{c}\right), \tag{4.70}$$

where $y\left(t-\frac{R}{c}\right)$ is the reference signal delayed according to the target range R, and $x_e(t)$ represents a modulation term related to target dynamics.

Differentiating the expression in (4.70) with respect to time yields

$$\frac{\mathrm{d}x(t)}{\mathrm{d}t} = \frac{\mathrm{d}y\left(t-\frac{R}{c}\right)}{\mathrm{d}t}x_e(t) + y\left(t-\frac{R}{c}\right)\frac{\mathrm{d}x_e(t)}{\mathrm{d}t}. \tag{4.71}$$

The partial derivative of the Zak-based transform $\mathscr{Z}^h(t,\omega)$ with respect to t is then given by

$$\frac{\partial \mathscr{Z}^h(t,\omega)}{\partial t} = \sum_{k=-\frac{K}{2}}^{\frac{K}{2}-1} \int_0^{T_b} \frac{\partial x(\tau+kT+t)}{\partial t} y^*(\tau+kT)h(kT)e^{-j\omega kT}\,\mathrm{d}\tau. \tag{4.72}$$

Substituting (4.71) into (4.72) yields

$$\begin{aligned}
-\mathscr{Z}^{\mathscr{D}h}(t,\omega) + j\omega\mathscr{Z}^h(t,\omega) = &\sum_{k=-\frac{K}{2}}^{\frac{K}{2}-1} \int_0^{T_b} \left[\frac{\partial y\left(\tau+kT+t-\frac{R}{c}\right)}{\partial t} x_e(\tau+kT+t) \right. \\
&\left. + y\left(\tau+kT+t-\frac{R}{c}\right) \frac{\partial x_e(\tau+kT+t)}{\partial t} \right] \\
&\times y^*(\tau+kT)h(kT)e^{-j\omega kT}\,\mathrm{d}\tau,
\end{aligned} \tag{4.73}$$

where $\mathscr{D}^n h = \frac{\mathrm{d}^n h(t)}{\mathrm{d}t^n}$.

Equation (4.73) can be decomposed into two integrals, denoted A and B, as follows:

$$\begin{aligned}
&j\omega\mathscr{Z}^h(t,\omega) - \mathscr{Z}^{\mathscr{D}h}(t,\omega) + \\
&\overbrace{\sum_{k=-\frac{K}{2}}^{\frac{K}{2}-1} \int_0^{T_b} y\left(\tau+kT-\frac{R}{c}\right) x_e(\tau+kT) \frac{\partial y^*(\tau+kT-t)}{\partial t} h(kT-t)e^{-j\omega(kT-t)}\,\mathrm{d}\tau}^{A} \\
&\underbrace{= \sum_{k=-\frac{K}{2}}^{\frac{K}{2}-1} \int_0^{T_b} \frac{\partial y\left(\tau+kT+t-\frac{R}{c}\right)}{\partial t} x_e(\tau+kT+t) y^*(\tau+kT)h(kT)e^{-j\omega kT}\,\mathrm{d}\tau}_{B} \\
&+ \sum_{k=-\frac{K}{2}}^{\frac{K}{2}-1} \int_0^{T_b} y\left(\tau+kT+t-\frac{R}{c}\right) \frac{\partial x(\tau+kT+t)}{\partial t} y^*(\tau+kT)h(kT)e^{-j\omega kT}\,\mathrm{d}\tau.
\end{aligned} \tag{4.74}$$

Since $A = B$, this leads to the following tractable relation

$$\begin{aligned}
\mathscr{Z}^{\mathscr{D}h}(t,\omega) = &j\omega\mathscr{Z}^h(t,\omega) - \sum_{k=-\frac{K}{2}}^{\frac{K}{2}-1} \int_0^{T_b} y\left(\tau+kT+t-\frac{R}{c}\right) \frac{\partial x_e(\tau+kT+t)}{\partial t} \\
&\times y^*(\tau+kT)h(kT)e^{-j\omega kT}\,\mathrm{d}\tau.
\end{aligned} \tag{4.75}$$

To describe the time-varying echo characteristics, let us express the echo term as

$$x_e(t) = A_x(t)e^{j\phi_x(t)}, \tag{4.76}$$

where $A_x(t)$ and $\phi_x(t)$ represent the instantaneous amplitude and phase, respectively. These functions can be approximated via Taylor series expansion around the time variable τ

$$\log(A_x(t)) = \sum_{n=0}^{N} \frac{(\log A_x)^{(n)}(\tau)}{n!}(t-\tau)^n, \tag{4.77}$$

$$\phi_x(t) = \sum_{n=0}^{N} \frac{\phi_x^{(n)}(\tau)}{n!}(t-\tau)^n. \tag{4.78}$$

Substituting (4.77) and (4.78) into (4.75) results in the expanded expression

$$\begin{aligned}\mathscr{X}^{\mathscr{D}h}(t,\omega) =& j\omega\mathscr{X}^h(t,\omega) - \sum_{k=-\frac{K}{2}}^{\frac{K}{2}-1}\int_0^{T_b} y\left(\tau+kT+t-\frac{R}{c}\right)\\ &\times \frac{\partial}{\partial t}\exp\left(\sum_{n=0}^{N}\frac{(\log A_x)^{(n)}(\tau)}{n!}(t+kT)^n + j\sum_{n=0}^{N}\frac{\phi_x^{(n)}(\tau)}{n!}(t+kT)^n\right)\\ &\times y^*(\tau+kT)h(kT)e^{-j\omega kT}\mathrm{d}\tau.\end{aligned} \tag{4.79}$$

Let us consider the case where the amplitude and phase of the echo signal, as defined in (4.76), follow second-order polynomial models, corresponding to $N = 2$. Under this assumption, the amplitude is modeled using a Gaussian function:

$$A_x(t) = \mathscr{A}_x e^{-\frac{(t-t_x)^2}{2T_x^2}}, \tag{4.80}$$

where $\mathscr{A}_x$ denotes the constant amplitude scale, t_x represents the time offset, and T_x is the temporal width of the echo.

The phase term, in turn, takes the form

$$\phi_x(t) = \varphi_x + \omega_x t + \frac{1}{2}\alpha_x t^2, \tag{4.81}$$

where $\varphi_x = 2\pi a_0$ is the initial phase offset, and the terms $\omega_x = -M2\pi a_1/\lambda$ and $\alpha_x = -M2\pi a_2/\lambda$ correspond to the linear and quadratic phase modulation components related to the target's velocity and acceleration, respectively.

Incorporating (4.80) and (4.81) into the definition of the echo signal, the time derivative of $x(t)$ reads as

$$\frac{\mathrm{d}x_e(t)}{\mathrm{d}t} = (q_x t + p_x)x_e(t), \tag{4.82}$$

where the coefficients are given by:

$$q_x = -\frac{1}{T_x^2} + j\alpha_x, \tag{4.83}$$

and

$$p_x = \frac{t_x}{T_x^2} + j\omega_x. \tag{4.84}$$

Substituting (4.82) into (4.79) and grouping the formula, we end up with the following identity

$$\mathscr{Z}^{\mathscr{D}h}(t,\omega) = -q_x\mathscr{Z}^{\mathscr{T}h}(t,\omega) - (q_xt + p_x - j\omega)\mathscr{Z}^h(t,\omega). \tag{4.85}$$

Here, we see a distinctive analogy to our previous time-frequency relations, particularly (2.151). If we contemplate (4.85) and (2.151), we can notice a direct analogy as the short-time Fourier transform and the Zak transform-based approximation of the cross-ambiguity function result in time-frequency signal representations. Therefore, we can take advantage of the previously formulated framework for signal parameter estimation and apply it to the analyzed case. As such, we can derive the derivative of (4.85) with respect to time and angular frequency and write them as follows:

$$\mathscr{Z}^{\mathscr{D}^2h}(t,\omega)(t,\omega) = -q_x\mathscr{Z}^{\mathscr{D}\mathscr{T}h}(t,\omega) - (q_xt + p_x - j\omega)\,\mathscr{Z}^{\mathscr{D}h}(t,\omega), \tag{4.86}$$

and

$$\mathscr{Z}^{\mathscr{D}\mathscr{T}h}(t,\omega) + \mathscr{Z}^h(t,\omega) = (q_xt + p_x - j\omega)\mathscr{Z}^{\mathscr{T}h}(t,\omega), \tag{4.87}$$

where $\mathscr{Z}^{\mathscr{T}h}(t,\omega)$ and $\mathscr{Z}^{\mathscr{D}\mathscr{T}h}(t,\omega)$ are the Zak transforms with the time-weighted window and the window derivative multiplied by the time ramp, respectively.

Incorporating (4.85) with (4.86) into one linear system yields

$$\hat{\omega}_x^{[t]}(t,\omega) = \omega - \Im\left(\frac{\mathscr{Z}^{\mathscr{D}^2h}(t,\omega)\mathscr{Z}^{\mathscr{T}h}(t,\omega) - \mathscr{Z}^{\mathscr{D}\mathscr{T}h}(t,\omega)\mathscr{Z}^{\mathscr{D}h}(t,\omega)}{\mathscr{Z}^{\mathscr{D}h}(t,\omega)\mathscr{Z}^{\mathscr{T}h}(t,\omega) - \mathscr{Z}^{\mathscr{D}\mathscr{T}h}(t,\omega)\mathscr{Z}^h(t,\omega)}\right), \tag{4.88}$$

and

$$\hat{\alpha}_x^{[t]}(t,\omega) = \Im\left(\frac{\mathscr{F}_x^{\mathscr{D}^2h}(t,\omega)\mathscr{F}_x^h(t,\omega) - \left(\mathscr{F}_x^{\mathscr{D}h}(t,\omega)\right)^2}{\mathscr{F}_x^{\mathscr{D}h}(t,\omega)\mathscr{F}_x^{\mathscr{T}h}(t,\omega) - \mathscr{F}_x^{\mathscr{D}\mathscr{T}h}(t,\omega)\mathscr{F}_x^h(t,\omega)}\right). \tag{4.89}$$

Similarly, combining (4.85) with (4.87) into another system, leads to

$$\hat{\omega}_x^{[\omega]}(t,\omega) = \omega + \Im\left(\frac{\mathscr{Z}^{\mathscr{D}\mathscr{T}h}(t,\omega)\mathscr{Z}^{\mathscr{T}h}(t,\omega) - \mathscr{Z}^{\mathscr{D}h}(t,\omega)\mathscr{Z}^{\mathscr{T}^2h}(t,\omega) + \mathscr{Z}^{\mathscr{T}h}(t,\omega)\mathscr{Z}^h(t,\omega)}{\left(\mathscr{Z}^{\mathscr{T}h}(t,\omega)\right)^2 - \mathscr{Z}^{\mathscr{T}^2h}(t,\omega)\mathscr{Z}^h(t,\omega)}\right), \tag{4.90}$$

and

$$\hat{\alpha}_x^{[\omega]}(t,\omega) = \Im\left(\frac{\mathscr{Z}^{\mathscr{D}\mathscr{T}h}(t,\omega)\mathscr{Z}^h(t,\omega) + \left(\mathscr{Z}^h(t,\omega)\right)^2 - \mathscr{Z}^{\mathscr{T}h}(t,\omega)\mathscr{Z}^{\mathscr{D}h}(t,\omega)}{\left(\mathscr{Z}^{\mathscr{T}h}(t,\omega)\right)^2 - \mathscr{Z}^{\mathscr{T}^2h}(t,\omega)\mathscr{Z}^h(t,\omega)}\right). \tag{4.91}$$

Estimated factors can be transformed on m/s^n through scaling by $\lambda/2\pi$ as follows:

$$\hat{a}_2(t,\omega) = -\hat{\alpha}_x(t,\omega)\cdot\lambda/(M\cdot 2\pi) = -\Im(\hat{r}_x)\cdot\lambda/(M\cdot 2\pi)\ \mathrm{m/s^2}, \tag{4.92}$$

$$\hat{a}_1(t,\omega) = -\hat{\omega}_x(t,\omega)\cdot\lambda/(M\cdot 2\pi) = -\Im(\hat{s}_x)\cdot\lambda/(M\cdot 2\pi)\ \mathrm{m/s}. \tag{4.93}$$

It is evident that the resulting estimators take the same form as those derived from the short-time Fourier transform. Both techniques operate in the time-frequency domain. Although their underlying principles differ slightly—the short-time Fourier transform applies a sliding window and converts signal segments into the frequency domain, while the Zak transform-based approximation of the cross-ambiguity function processes Fourier transforms of correlated signal blocks—their results are formally consistent. Estimators (4.88) and (4.90) are skipped now, but the problem of the instantaneous frequency estimation is discussed in Section 4.6 in this chapter.

The overall processing scheme is illustrated in Figure 4.13. In practical implementation, the radar signal first undergoes range compression, which may be performed via cross-correlation or the Fourier transform, depending on the specific radar type. Concurrently, all necessary window functions are precomputed and stored in memory to reduce processing time. Following range compression, the signal is windowed in the slow-time domain using each of the predefined windows. Subsequently, a Fourier transform is applied to estimate the target velocity. The resulting range-velocity maps (each corresponding to a different window) are then used collectively to estimate the target acceleration, yielding the final output.

Akin to the extended cross-ambiguity function and the generalized Radon-Fourier transform, the Zak transform-based method applies to various radar modalities. While the previous examples focused on passive radar systems, we now briefly turn to a frequency-modulated continuous-wave radar scenario to demonstrate the versatility of the approach. Figure 4.14 presents the processing outcome for a system configured with a carrier frequency of $f_c = 24$ GHz, bandwidth $B = 1$ GHz, pulse duration $T_p = 1$ ms, and sweep repetition rate $f_{\mathrm{SRR}} = 1000$ pulses per second. The simulated target is positioned at the range of $R_0 = 50$ m, moving with an initial velocity of $a_1 = 1$ m/s and a constant acceleration of $a_2 = 0.25$ m/s^2.

Firstly, we see a target echo at the classical range-velocity map in Figure 4.14a. Next, in Figure 4.14b, we see the acceleration estimate using (4.89) as an example. As we can observe, we have the outcome for the whole distribution simultaneously. It admits numerous targets being apparent in the radar beam, and each acceleration will be estimated at once. However, its readability is limited since the estimation result covers the entire range-velocity distribution. Therefore, we can govern the distribution saturation so that the strongest targets are shown while the noise of a low magnitude is neglected. The saturated (by the range-velocity map magnitude) acceleration estimate is shown in Figure 4.14c. The close-up of the target echo with the estimate corresponding to that set in the simulation is shown in Figure 4.14d.

In Chapter 3, and specifically in Figure 3.19, we analyzed a real-life echo generated by a drone. Figure 4.15a presents the corresponding acceleration estimation across the entire range-velocity domain. As shown, the method successfully recovers the target acceleration and distribution throughout the whole space. A zoomed-in view around the target region reveals a well-localized peak corresponding to the actual acceleration value. By identifying the peak location, we obtain an estimated acceleration of $\hat{a}_2 = 0.33$ m/s^2. Extensive experiments reported in [223] confirm a

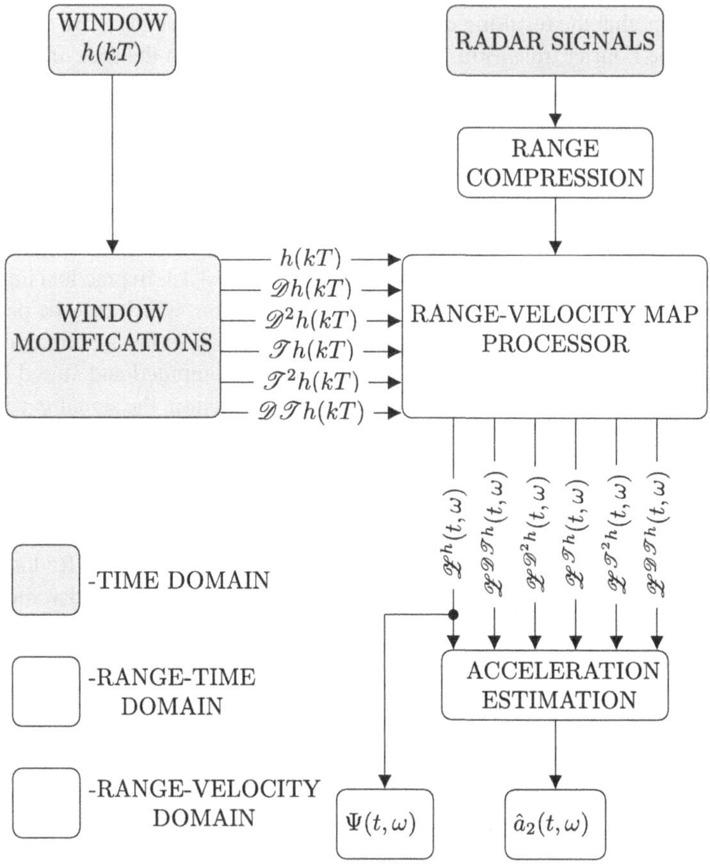

Figure 4.13 Target acceleration estimation using the Zak transform-based method. The diagram incorporates all six window modifications, even though each estimator utilizes only five. This intentional generalization ensures the schematic remains applicable to all estimator configurations. (Reproduced from [223], licensed under CC BY 4.0.)

strong consistency between this method and the extended cross-ambiguity function while achieving over 100-fold reduction in computational time.

Although this approach is more susceptible to noise than the two previously discussed techniques, its advantages can be summarized as follows:

- The method is non-iterative and gridless. We do not operate under a prerequisite of any space in which we search the target parameters. Therefore, there is no risk that the target acceleration will be outside the considered grid.
- While the extended cross-ambiguity function and the generalized Radon-Fourier transform were iteratively optimized to a given target, the Zak

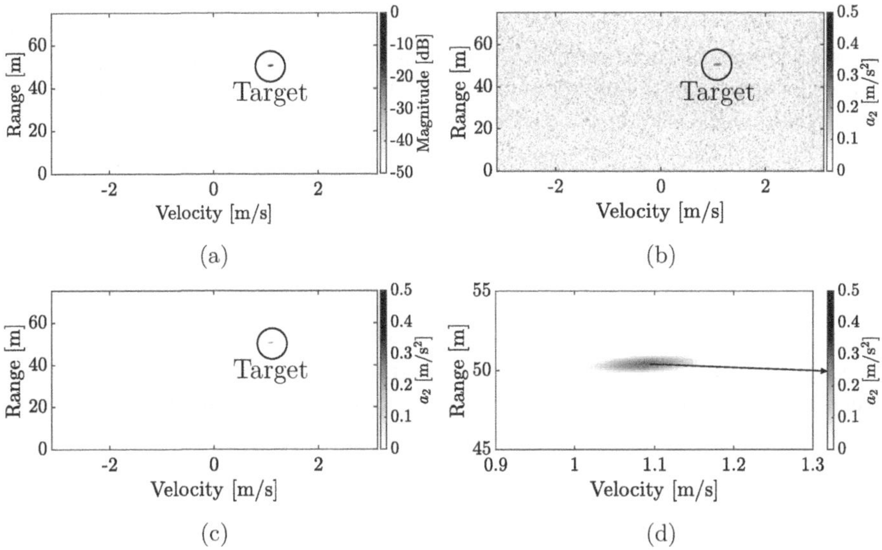

Figure 4.14 Target acceleration estimation using the Zak transform-based cross-ambiguity function approximation in the frequency-modulated continuous-wave radar. (a) Range-velocity map; (b) acceleration estimate for the whole distribution; (c) saturated acceleration estimate; (d) close-up on the target of the saturated acceleration distribution.

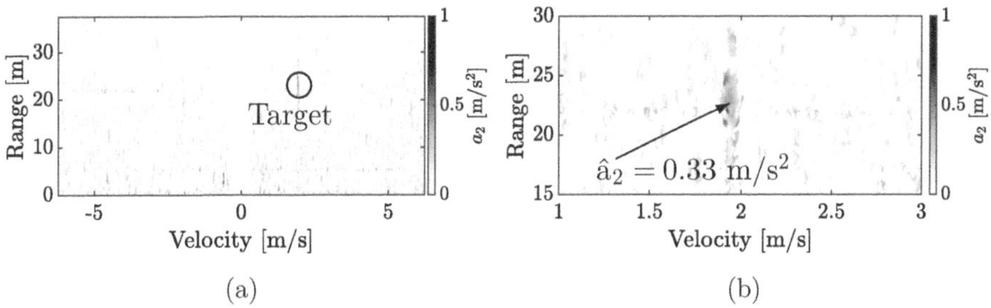

Figure 4.15 Saturated acceleration estimate using the real-life signal example with a drone echo. The range-velocity map of the same signal can be seen in Figure 3.19. (a) Saturated acceleration estimate for the whole range-velocity distribution; (b) close-up on the target.

transform-based method estimates the acceleration for any number of targets.

- From the above factors, the Zak transform-based approach has a constant computational complexity and always requires five range-velocity maps to give the result.

At this point, one may question why the analysis has been restricted to the second-order phase term, which corresponds to acceleration. In fact, all the methods discussed so far–including the extended cross-ambiguity function, the generalized

Radon-Fourier transform, and the Zak transform-based approach–exhibit estimation bias when higher-order kinematic components (such as jerk or snap) are present in the echo signal. These limitations arise from the intrinsic assumptions embedded in the signal models used by each technique. The following section delves deeper into this issue and examines how such higher-order dynamics affect performance and estimation accuracy.

4.5 HIGH-ORDER TARGET KINEMATIC PARAMETER ESTIMATION

An echo signal model truncated to the second-order phase term associated with constant acceleration is often sufficient to characterize the motion of a target, especially when the integration time T_i is relatively short. Under more challenging conditions, such as limited observation intervals, higher-order phase components (associated with jerk, snap, among others) usually make only a negligible contribution to the overall phase variation. This simplification follows directly from the kinematic expansion in (4.30), where each successive term is weighted by a higher power of time. As T_i decreases, the influence of higher-order terms diminishes rapidly, making acceleration the dominant dynamic contributor. Consequently, limiting the motion model to the second-order term offers a practical balance between computational efficiency and estimation precision.

Notwithstanding, this prerequisite may not hold in all scenarios. The higher-order motion components can no longer be ignored when dealing with highly maneuvering targets such as agile drones, guided munitions, or fast-accelerating vehicles. In such cases, the target's signal phase deviates from the parabolic model implied by constant acceleration, and the estimator misinterprets the residual energy from unmodeled phase curvature. This mismatch introduces bias and degrades the accuracy of acceleration estimation, as the estimator attributes nonlinear phase behavior to second-order dynamics.

From a detection perspective, ignoring higher-order kinematic contributions may also reduce the signal's coherence during integration. The uncompensated phase evolution causes the reflected signal to defocus, leading to energy spread on the range-velocity map. As a result, the peak amplitude in matched filters or correlation-based detectors is reduced, impeding the target detection, particularly under low SNR conditions. To mitigate this issue, extending the motion model to include higher-order derivatives may be necessary, especially in long integration regimes or when observing targets known to exhibit rapid maneuvers. Incorporating third- or fourth-order phase terms can improve the signal's alignment during coherent processing and the accuracy of motion parameter estimation. However, this comes at the cost of increased computational expenditure and the need for more sophisticated estimation algorithms.

To enable the estimation of higher-order kinematic parameters using the previously discussed methods, the target echo model must be extended to incorporate additional phase terms corresponding to those higher-order dynamics. In this context, we consider third-order (jerk) and fourth-order (snap) motion components, though the presented framework is general and can accommodate any polynomial order.

For the extended cross-ambiguity function, the modified formulation becomes

$$\Psi(t,\omega,\alpha,\beta,\gamma)=\int_0^{T_i} x(\tau)\, y^*(\tau-t)\, e^{-j\left(\omega\tau+\alpha\frac{\tau^2}{2}+\beta\frac{\tau^3}{3}+\gamma\frac{\tau^4}{4}\right)}\mathrm{d}\tau,\ \Psi(t,\omega,\alpha,\beta,\gamma)\in\mathbb{C}, \tag{4.94}$$

where the sets of modulation parameters are defined as

$$\alpha=-M\cdot 2\pi\frac{a_2}{\lambda}, \tag{4.95}$$

$$\beta=-M\cdot 2\pi\frac{a_3}{\lambda}, \tag{4.96}$$

$$\gamma=-M\cdot 2\pi\frac{a_4}{\lambda}, \tag{4.97}$$

with a_2, a_3, and a_4 denoting the target's acceleration, jerk, and snap, respectively.

All the advantages and drawbacks of the acceleration-only extended cross-ambiguity function (4.42) remain applicable to this higher-order formulation. Specifically, the algorithm requires predefined search ranges for the parameters α, β, and γ, corresponding to the putative motion characteristics of the target. Without prior information, one must assume sufficiently broad and finely sampled parameter spaces to ensure coverage of all possible target trajectories. However, expanding the grid meaningfully increases the computational cost and renders real-time implementation impractical.

This trade-off is similarly observed in the generalized Radon-Fourier transform framework. As an iterative method that searches over multidimensional parameter grids, the inclusion of additional degrees of freedom, namely jerk and snap, leads to an exponential increase in processing complexity. The extended formulation of the generalized Radon-Fourier transform is expressed as

$$G(\dot{R}_0,\dot{a}_1,\dot{a}_2,\dot{a}_3,\dot{a}_4)=\int_{-\infty}^{\infty} r(t,\tau)\cdot\exp\left(j\frac{M2\pi R(t|\dot{R}_0,\dot{a}_1,\dot{a}_2,\dot{a}_3,\dot{a}_4)}{\lambda}\right)\mathrm{d}t, \tag{4.98}$$

where $\dot{R}_0$, $\dot{a}_1$, $\dot{a}_2$, $\dot{a}_3$, and $\dot{a}_4$ define the hypothesized values of range, velocity, acceleration, jerk, and snap, respectively, and the extended range model is given by

$$R(t)=R_0+a_1t+\frac{1}{2}a_2t^2+\frac{1}{3}a_3t^3+\frac{1}{4}a_4t^4. \tag{4.99}$$

The optimal parameter set, corresponding to the best match between the signal and the hypothesized trajectory, is then estimated as

$$(\hat{R}_0,\hat{a}_1,\hat{a}_2,\hat{a}_3,\hat{a}_4)=\underset{\dot{R}_0,\dot{a}_1,\dot{a}_2,\dot{a}_3,\dot{a}_4}{\operatorname{argmax}}\left|G(\dot{R}_0,\dot{a}_1,\dot{a}_2,\dot{a}_3,\dot{a}_4)\right|. \tag{4.100}$$

Although the extension to higher-order kinematics improves modeling accuracy for complex target motion, especially under long integration times or rapidly maneuvering targets, it also introduces practical limitations. These include increased dimensionality in the search space, which affects both memory usage and computation time. Moreover, estimation bias in lower-order parameters (e.g., acceleration) may arise when higher-order components are present in the signal but not included in the estimation model. Therefore, a careful trade-off must be made between model fidelity and algorithmic feasibility, particularly in real-time or resource-constrained radar systems.

Special consideration must be given to the Zak transform-based methodology, as the derivations presented in the previous section are valid solely with the constraint of a precisely defined signal model (in the same way as we derived the estimators in high-order vertical synchrosqueezing in Section 2.9.1 in Chapter 2) [189], and the assumptions are violated when higher-order kinematic parameters occur. To accommodate higher-order motion components, let us generalize (4.76) given by

$$x_e(t) = A_x(t)e^{j\phi_x(t)}, \tag{4.101}$$

through expressing it as a fourth-order polynomial model of the signal envelope

$$A_x(t) = \mathscr{A}_x \cdot \exp\left(-\frac{(t-t_x)^2}{2T_x^2} - \Delta_x\frac{(t-t_x)^3}{3T_x^2} - \Sigma_x\frac{(t-t_x)^4}{4T_x^2}\right), \tag{4.102}$$

where $\mathscr{A}_x$ denotes the constant signal magnitude, t_x is the time shift, and T_x defines the temporal width of the echo. The three exponential terms: $-\frac{(t-t_x)^2}{2T_x^2}$, $-\Delta_x\frac{(t-t_x)^3}{3T_x^2}$, and $-\Sigma_x\frac{(t-t_x)^4}{4T_x^2}$ represent the second-, third-, and fourth-order amplitude modulation components, respectively. The corresponding phase function is formulated as

$$\phi_x(t) = \varphi_x + \omega_x t + \alpha_x t^2/2 + \beta_x t^3/3 + \gamma_x t^4/4, \tag{4.103}$$

where $\varphi_x = 2\pi a_0$ denotes the initial phase, and the subsequent parameters $\omega_x = -M2\pi a_1/\lambda$, $\alpha_x = -M2\pi a_2/\lambda$, $\beta_x = -M2\pi a_3/\lambda$, and $\gamma_x = -M2\pi a_4/\lambda$ correspond to the first through fourth-order phase modulation terms, accounting respectively for the target's velocity, acceleration, jerk, and snap. Having the signal model, we can formulate its derivative as

$$\frac{\mathrm{d}x_e(t)}{\mathrm{d}t} = \left(p_x t^3 + q_x t^2 + r_x t + s_x\right)x_e(t). \tag{4.104}$$

where

$$p_x = \frac{-\Sigma_x}{T_x^2} + j\gamma_x, \tag{4.105}$$

$$q_x = \frac{-\Delta_x + 3\Sigma_x t_x}{T_x^2} + j\beta_x, \tag{4.106}$$

$$r_x = -\frac{1 - 2\Delta_x t_x + 3\Sigma_x t_x^2}{T_x^2} + j\alpha_x, \tag{4.107}$$

$$s_x = \frac{t_x - \Delta_x t_x^2 + \Sigma_x t_x^3}{T_x^2} + j\omega_x. \tag{4.108}$$

In the previous section, we substituted the signal model with a term responsible for acceleration, obtaining (4.85). Putting the extended signal model into (4.79) yields

$$\begin{aligned}\mathscr{Z}^{\mathscr{D}h}(t,\omega) = &- p_x \mathscr{Z}^{\mathscr{T}^3 h}(t,\omega)\\ &-(3p_x t + q_x)\mathscr{Z}^{\mathscr{T}^2 h}(t,\omega)\\ &-(3p_x t^2 + 2q_x t + r_x)\mathscr{Z}^{\mathscr{T} h}(t,\omega)\\ &-(p_x t^3 + q_x t^2 + r_x t + s_x - j\omega)\mathscr{Z}^{h}(t,\omega).\end{aligned} \tag{4.109}$$

In general, the $(n-3)$th-order derivative (for $n \geq 4$) of (4.109) with respect to ω is

$$\begin{aligned}\mathscr{Z}^{\mathscr{T}^{n-3}\mathscr{D}}(t,\omega) + (n-4)\mathscr{Z}^{\mathscr{T}^{n-3}}(t,\omega) = &-p_x \mathscr{Z}^{\mathscr{T}^{n}}(t,\omega)\\ &-(3p_x t + q_x)\mathscr{Z}^{\mathscr{T}^{n-1}}(t,\omega)\\ &-(3p_x t^2 + 2q_x t + r_x)\mathscr{Z}^{\mathscr{T}^{n-2}}(t,\omega)\\ &-(p_x t^3 + q_x t^2 + r_x t + s_x - j\omega)\mathscr{Z}^{\mathscr{T}^{n-3}}(t,\omega),\end{aligned} \tag{4.110}$$

We end up with the linear system known from Section 2.9.1 from Chapter 2. For the sake of clarity, let us define it again for the Zak transform

$$\begin{bmatrix} \mathscr{Z}^{h}(t,\omega) & \mathscr{Z}^{\mathscr{T}h}(t,\omega) & \mathscr{Z}^{\mathscr{T}^2h}(t,\omega) & \mathscr{Z}^{\mathscr{T}^3h}(t,\omega)\\ \mathscr{Z}^{\mathscr{T}h}(t,\omega) & \mathscr{Z}^{\mathscr{T}^2h}(t,\omega) & \mathscr{Z}^{\mathscr{T}^3h}(t,\omega) & \mathscr{Z}^{\mathscr{T}^4h}(t,\omega)\\ \mathscr{Z}^{\mathscr{T}^2h}(t,\omega) & \mathscr{Z}^{\mathscr{T}^3h}(t,\omega) & \mathscr{Z}^{\mathscr{T}^4h}(t,\omega) & \mathscr{Z}^{\mathscr{T}^5h}(t,\omega)\\ \mathscr{Z}^{\mathscr{T}^3h}(t,\omega) & \mathscr{Z}^{\mathscr{T}^4h}(t,\omega) & \mathscr{Z}^{\mathscr{T}^5h}(t,\omega) & \mathscr{Z}^{\mathscr{T}^6h}(t,\omega) \end{bmatrix}$$
$$\times \begin{bmatrix} -p_x t^3 - q_x t^2 - r_x t - s_x + j\omega\\ -3p_x t^2 - 2q_x t - r_x\\ -3p_x t - q_x\\ -p_x \end{bmatrix} = \begin{bmatrix} \mathscr{Z}^{\mathscr{D}h}(t,\omega)\\ \mathscr{Z}^{\mathscr{D}\mathscr{T}h}(t,\omega) + \mathscr{Z}^{h}(t,\omega)\\ \mathscr{Z}^{\mathscr{D}\mathscr{T}^2h}(t,\omega) + 2\mathscr{Z}^{\mathscr{T}h}(t,\omega)\\ \mathscr{Z}^{\mathscr{D}\mathscr{T}^3h}(t,\omega) + 3\mathscr{Z}^{\mathscr{T}^2h}(t,\omega) \end{bmatrix}, \tag{4.111}$$

which, assuming the matrix to be invertible, leads to

$$\begin{aligned}\begin{bmatrix} v_1\\ v_2\\ v_3\\ v_4 \end{bmatrix} &= \begin{bmatrix} -p_x t^3 - q_x t^2 - r_x t - s_x + j\omega\\ -3p_x t^2 - 2q_x t - r_x\\ -3p_x t - q_x\\ -p_x \end{bmatrix}\\ &= \begin{bmatrix} \mathscr{Z}^{h}(t,\omega) & \mathscr{Z}^{\mathscr{T}h}(t,\omega) & \mathscr{Z}^{\mathscr{T}^2h}(t,\omega) & \mathscr{Z}^{\mathscr{T}^3h}(t,\omega)\\ \mathscr{Z}^{\mathscr{T}h}(t,\omega) & \mathscr{Z}^{\mathscr{T}^2h}(t,\omega) & \mathscr{Z}^{\mathscr{T}^3h}(t,\omega) & \mathscr{Z}^{\mathscr{T}^4h}(t,\omega)\\ \mathscr{Z}^{\mathscr{T}^2h}(t,\omega) & \mathscr{Z}^{\mathscr{T}^3h}(t,\omega) & \mathscr{Z}^{\mathscr{T}^4h}(t,\omega) & \mathscr{Z}^{\mathscr{T}^5h}(t,\omega)\\ \mathscr{Z}^{\mathscr{T}^3h}(t,\omega) & \mathscr{Z}^{\mathscr{T}^4h}(t,\omega) & \mathscr{Z}^{\mathscr{T}^5h}(t,\omega) & \mathscr{Z}^{\mathscr{T}^6h}(t,\omega) \end{bmatrix}^{-1}\\ &\times \begin{bmatrix} \mathscr{Z}^{\mathscr{D}h}(t,\omega)\\ \mathscr{Z}^{\mathscr{D}\mathscr{T}h}(t,\omega) + \mathscr{Z}^{h}(t,\omega)\\ \mathscr{Z}^{\mathscr{D}\mathscr{T}^2h}(t,\omega) + 2\mathscr{Z}^{\mathscr{T}h}(t,\omega)\\ \mathscr{Z}^{\mathscr{D}\mathscr{T}^3h}(t,\omega) + 3\mathscr{Z}^{\mathscr{T}^2h}(t,\omega) \end{bmatrix}.\end{aligned} \tag{4.112}$$

Therefore, the estimators can be obtained recursively:

$$\hat{p}_x = -v_4, \tag{4.113}$$

$$\hat{q}_x = -3\hat{p}_x\hat{t}(t,\omega) - v_3, \tag{4.114}$$

$$\hat{r}_x = -3\hat{p}_x[\hat{t}(t,\omega)]^2 - 2\hat{q}_x\hat{t}(t,\omega) - v_2, \tag{4.115}$$

$$\hat{s}_x = -\hat{p}_x[\hat{t}(t,\omega)]^3 - \hat{q}_x[\hat{t}(t,\omega)]^2 - \hat{r}_x\hat{t}(t,\omega) - v_1 + j\omega, \tag{4.116}$$

where

$$\hat{t}(t,\omega) = \frac{\mathscr{Z}^{\mathscr{T}h}(t,\omega)}{\mathscr{Z}^h(t,\omega)}. \tag{4.117}$$

To derive (4.117), we begin by recalling the expression for the Zak transform-based radar processing defined in (4.69):

$$\mathscr{Z}^h(t,\omega) = \sum_{k=-\frac{K}{2}}^{\frac{K}{2}-1} \int_0^{T_b} x(\tau+kT)y^*(\tau+kT-t)h(kT-t)e^{-j\omega(kT-t)}\mathrm{d}\tau. \tag{4.118}$$

To estimate the time delay t, we differentiate (4.118) with respect to the frequency variable ω, yielding

$$\frac{\partial \mathscr{Z}^h(t,\omega)}{\partial \omega} = \frac{\partial}{\partial \omega} \sum_{k=-\frac{K}{2}}^{\frac{K}{2}-1} \int_0^{T_b} x(\tau+kT)y^*(\tau+kT-t)h(kT-t)e^{-j\omega(kT-t)}\mathrm{d}\tau. \tag{4.119}$$

To identify the stationary point of $\mathscr{Z}^h(t,\omega)$ with respect to ω, we set the derivative in (4.119) equal to zero

$$\frac{\partial}{\partial \omega} \sum_{k=-\frac{K}{2}}^{\frac{K}{2}-1} \int_0^{T_b} x(\tau+kT)y^*(\tau+kT-t)h(kT-t)e^{-j\omega(kT-t)}\mathrm{d}\tau = 0. \tag{4.120}$$

Evaluating the derivative of the exponential term leads to

$$-j(kT-t) \sum_{k=-\frac{K}{2}}^{\frac{K}{2}-1} \int_0^{T_b} x(\tau+kT)y^*(\tau+kT-t)h(kT-t)e^{-j\omega(kT-t)}\mathrm{d}\tau = 0. \tag{4.121}$$

Recognizing that multiplication by kT in the sum corresponds to a weighted version of $\mathscr{Z}^h(t,\omega)$, we define this new quantity as $\mathscr{Z}^{\mathscr{T}h}(t,\omega)$, yielding the relation

$$\mathscr{Z}^{\mathscr{T}h}(t,\omega) = t\mathscr{Z}^h(t,\omega). \tag{4.122}$$

Solving for the time delay t, we obtain the estimator in the form

$$\hat{t}(t,\omega) = \frac{\mathscr{Z}^{\mathscr{T}h}(t,\omega)}{\mathscr{Z}^h(t,\omega)}. \tag{4.123}$$

Remembering that M euqlas 1 and 2 for the bistatic and monostatic configurations, respectively, estimated factors $\hat{\gamma}_x(t,\omega), \hat{\beta}_x(t,\omega), \hat{\alpha}_x(t,\omega)$ and $\hat{\omega}_x(t,\omega)$ can be transformed on m/s^n by taking the imaginary part $\Im$ and scaling by $\lambda/2\pi$ as follows:

$$\hat{a}_4(t,\omega) = -\hat{\gamma}_x(t,\omega)\cdot\lambda/(M\cdot 2\pi) = -\Im(\hat{p}_x)\cdot\lambda/(M\cdot 2\pi)\ \mathrm{m/s}^4, \tag{4.124}$$

$$\hat{a}_3(t,\omega) = -\hat{\beta}_x(t,\omega)\cdot\lambda/(M\cdot 2\pi) = -\Im(\hat{q}_x)\cdot\lambda/(M\cdot 2\pi)\ \mathrm{m/s}^3, \tag{4.125}$$

$$\hat{a}_2(t,\omega) = -\hat{\alpha}_x(t,\omega)\cdot\lambda/(M\cdot 2\pi) = -\Im(\hat{r}_x)\cdot\lambda/(M\cdot 2\pi)\ \mathrm{m/s}^2, \tag{4.126}$$

$$\hat{a}_1(t,\omega) = -\hat{\omega}_x(t,\omega)\cdot\lambda/(M\cdot 2\pi) = -\Im(\hat{s}_x)\cdot\lambda/(M\cdot 2\pi)\ \mathrm{m/s}. \tag{4.127}$$

In contrast to the extended cross-ambiguity function and the Radon-Fourier transform, we obtain the estimate for the entire (t,ω) (or equivalently, range-velocity) distribution at the same time, regardless of the number of targets in the observed area.

Analogous reasoning can be applied by considering the derivative with respect to time instead of frequency. The reader is encouraged to conduct a similar analysis in the time domain to explore complementary insights (see [132] for a similar analysis).

Remark 4.5. *It is important to emphasize that the presence of higher-order components in the echo signal inherently introduces bias into the estimation of lower-order motion parameters. As a result, accurate parameter estimation requires a gradual decomposition approach, starting from the highest-order term. Specifically, one should first estimate the highest-order kinematic parameter, compensate for its influence by appropriately demodulating the received signal, and then proceed recursively with the estimation and compensation of lower-order terms. Only such a sequential decomposition ensures that the estimates of lower-order parameters remain unbiased and accurately reflect the true dynamics of the target. In practice, the compensation for a given target can be performed as follows:*

1. *Estimate* $\hat{a}_N(t,\omega)$.
2. *Read the value for the target under interest from a given* (t,ω) *location.*
3. *Demodulate the echo signal by the exponent with a conjugated value* $\hat{a}_N(t,\omega)$.
4. *Go to point* 1. *with* $N-1$.

Let us experimentally verify how the higher-order terms affect the lower ones. Consider the DVB-T signal with the carrier frequency of $f_c = 184.5$ MHz and the bandwidth of $B = 8$ MHz. The signal illuminates the target at the range $R_0 = 1$ km with the following kinematic parameters: $a_1 = 50$ m/s, $a_2 = 10$ m/s^2, $a_3 = 100$ m/s^3, and $a_4 = 500$ m/s^4. Figure 4.16 demonstrates the saturated target acceleration estimate in the range-velocity domain. In particular, Figure 4.16a shows the outcome when higher-order terms (jerk and snap) were not compensated. In Figure 4.16b, we see the same distribution, but for the signal demodulated according to $\hat{a}_4$ and then $\hat{a}_3$.

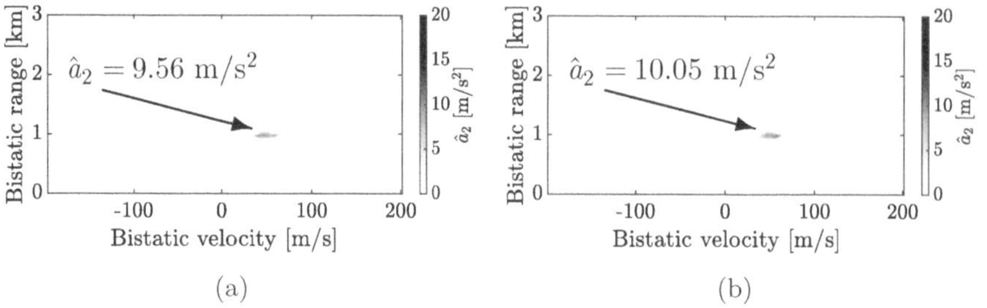

Figure 4.16 Saturated acceleration estimates without (a) and with (b) higher-order kinematic parameter compensation. When the target has non-negligible high-order kinematic parameters, their sequential compensation is crucial for precise target parameterization. The estimated values were read from the maximum for a given peak.

The presented experiment clearly demonstrates that higher-order kinematic parameters–such as jerk and snap–should not be disregarded, particularly when their values are non-negligible. In practical scenarios, the extent of target maneuverability is typically unknown in advance. As a result, incorporating third- and fourth-order (or higher) motion descriptors into the signal-processing chain can substantially reinforce target detection performance. Accounting for these higher-order dynamics enables more accurate modeling of complex motion patterns, enhancing both detection sensitivity and estimation accuracy, especially in the presence of maneuvering targets.

Although the Zak transform-based approach seems more complicated than the extended cross-ambiguity function and the generalized Radon-Fourier transform, it is characterized by the lowest computational cost.

Let us now examine the computational complexity of the analyzed methods for high-order target kinematic parameter estimation. As shown in (4.69), the first step involves cross-correlation of the received signals, which requires $\mathcal{O}(KL)$ operations, where $L = T_b \cdot f_s$ denotes the number of range bins, f_s is the sampling rate, and K is the number of slow-time samples, also equal to the size of the Fourier transform. While this correlation could technically be performed using FFT-based techniques, this possibility is not taken into account in the present analysis. The second step consists of computing L discrete Fourier transforms, each with complexity $\mathcal{O}(K \log_2 K)$, resulting in a total computational load of

$$\mathscr{C} = \mathcal{O}(KL) + \mathcal{O}(LK \log_2 K). \tag{4.128}$$

To estimate high-order kinematic parameters, the Zak transform-based method requires the generation of eleven range-velocity maps with different windowing functions, followed by solving a linear system (4.112). Importantly, these additional operations do not alter the overall asymptotic complexity given in (4.128), as they are performed on precomputed data. A key advantage of the approach lies in the fact that range compression, responsible for the bulk of the initial processing effort, is performed only once. Since this step is an integral part of range estimation in radar systems, it is considered a common operation and is thus excluded from the complexity assessment. This follows the reasoning presented in [180], where range compression is similarly omitted to better reflect the practical computational burden of parameter estimation. In this context, the Zak transform-based technique operates entirely on already range-compressed signals, which justifies this assumption.

The asymptotic complexities of all considered methods are summarized in Table 4.1, expressed in Landau notation.

The Zak transform-based method owes its efficiency to the fact that parameter estimation is performed in a gridless domain rather than over a discrete grid, which, in the case of the generalized Radon-Fourier transform, must span all possible combinations of motion parameters. In contrast, although the extended cross-ambiguity function also involves significant computational effort, it avoids the need for evaluating multiple range and velocity hypotheses but requires an FFT for each candidate combination of acceleration, jerk, and snap.

Let us verify the last analyzed method using the practical example of the passive radar system and the accelerating rocket. Figure 4.17 delineates the range-velocity map for the selected 0.5 s-long signal. We can see a target at the bistatic range of 1.19 km with a bistatic velocity of 72.13 m/s. The kinematic parameters were decomposed using the Zak transform-based approach, as shown in Figure 4.18. After frequency modulation compensation, the range-velocity map takes on the form as in Figure 4.18d.

Table 4.1
Computational Complexity Comparison [189]

Method	Computational Complexity
Zak Transform-Based Method	$\mathcal{O}(LK\log_2 K)$
Generalized Radon-Fourier Transform	$\mathcal{O}(R_x V_x A_x J_x S_x K)$
Extended Cross-Ambiguity Function	$\mathcal{O}(A_x J_x S_x KL) + \mathcal{O}(LK\log_2 K)$

K – number of slow-time points/FFT size, T_b – signal block length, $L = T_b \cdot f_s$ – number of range bins, R_x, V_x, A_x, J_x, S_x – cardinalities of discrete parameter sets for range, velocity, acceleration, jerk, and snap, respectively.

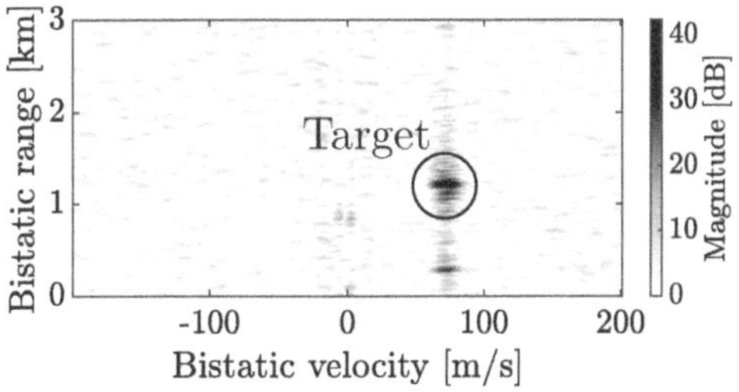

Figure 4.17 Selected range-velocity map for $T_i = 0.5$ s for the accelerating rocket.

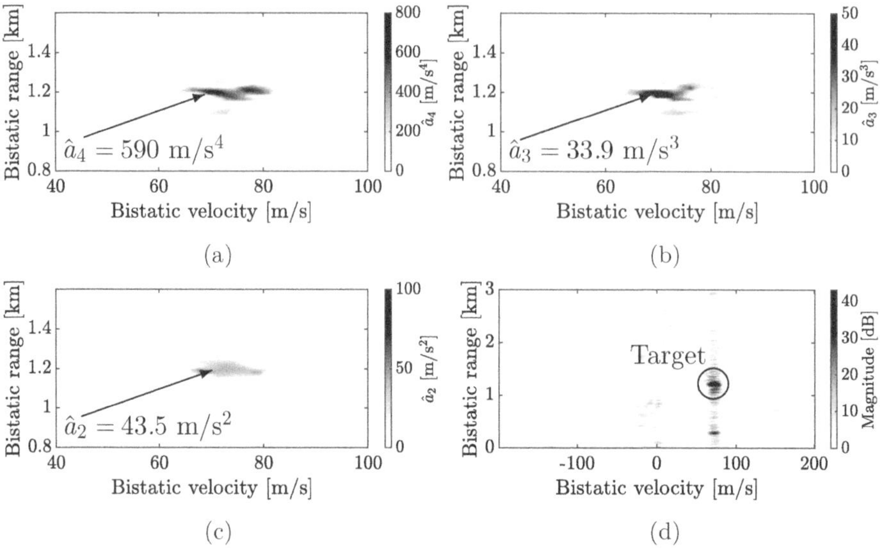

Figure 4.18 High-order target kinematic parameter estimation using the real-life passive radar signal. (a) Snap estimate; (b) jerk estimate; (c) acceleration estimate; (d) range-velocity map after high-order kinematic parameter compensation.

The enhancement achieved through frequency compensation is illustrated in Figure 4.19, which presents cross-sections of the target echo along the velocity axis. A clear outcome is narrowing the peak, resulting in a more compact and focused echo centered around the actual radial velocity due to the high-order kinematic parameter compensation. Furthermore, Figure 4.19b shows the close-up of the peak of the original signal with those obtained after demodulating the signal using the estimated acceleration only and then using acceleration, jerk, and snap. Notably, the impact of compensating for higher-order terms proves to be marginal. In this particular case, the gain amounted to only 0.02 dB, with the peak widths nearly identical. This example supports the rationale that estimating higher-order motion parameters is not always essential, as the computational effort required may not be justified by the relatively minor improvement in echo clarity or detectability. Nonetheless, the illustrated result is only a short part of the whole signal. A thorough examination of the complete dataset (which, for obvious reasons, cannot be fully presented in this book due to space limitations) would reveal that the inclusion of higher-order motion derivatives can indeed provide local benefits.

Accurate estimation of target kinematic parameters is a fundamental element of contemporary radar systems. The techniques discussed enable the characterization of a target's motion, thereby enhancing both detection performance and the quality of subsequent tracking. Although compensating the signal based on estimated motion parameters results in echo compression (particularly along the velocity axis), the

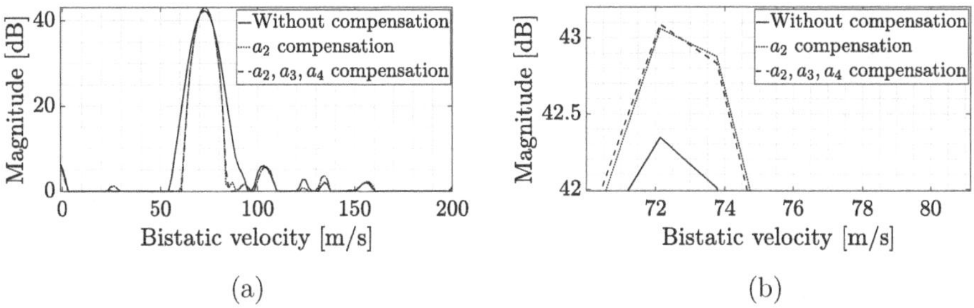

Figure 4.19 Cross-section of the range-velocity map for the original signal and after high-order kinematic parameter compensation. (a) General overview of the cross-sections for the original range-velocity map and after compensation; (b) close-up on the target peak.

ability to resolve closely spaced targets remains limited. To address this issue, the following section introduces a method inspired by time-frequency reassignment, which aims to simultaneously narrow target peaks and alleviate the influence of background noise, thereby improving overall target separability.

4.6 MULTITAPER SPECTRUM ESTIMATION IN RADARS – EFFICIENT WAY TO NOISE FLATTENING AND SIMULTANEOUS ECHO SHARPENING

In the simple scenario (excluding range and velocity migration), a target observed by a radar occupies a single cell in the range-velocity map according to its position relative to the radar and its radial or bistatic velocity. The target echo is spread on the plane and usually takes more than a single range-velocity cell. Typically, the velocity resolution is described by

$$\Delta V = \frac{\lambda}{MT_i}. \tag{4.129}$$

In practice, we usually apply a bell-shape window prior to the Fourier transform computation, which leads to the peak widening. Moreover, the assumption of constant velocity is rarely satisfied in practice, as targets generally undergo small velocity variations.

A natural way to increase the velocity resolution is to increase the integration time T_i in (4.129). From the previous section, we know that this operation entails some limitations as the target can migrate in range and velocity for extensive T_i. Thus, the integration time cannot be overextended to avoid echo migration. On the other hand, short integration time may affect the detection capability since the integration gain (4.15) decreases. The echo is weaker and can be covered by ambient noise, disabling its detection. To solve these two problems, we can incorporate the time-frequency framework into range-velocity processing to alleviate echo peak concentration and the simultaneous noise flattening using orthonormal windows.

Orthonormal windows in the form of the Hermite functions were discussed in Chapters 2 and 3. Without going into a profound discussion (the readers can see a bit wider description in the previous chapters and [103–105]), the Hermite window as a generalization of the Gaussian one is defined as

$$h_k(t) = \frac{\left((t-\mathscr{D})^k g\right)(t)}{\sqrt{\pi^{1/2}2^k k!}}, \tag{4.130}$$

where the Gaussian function is defined as $g(t) = \exp\left(-t^2/2\right)$, and $\mathscr{D}g(t) = \frac{\mathrm{d}g(t)}{\mathrm{d}t}$ represents its derivative. In practical applications, this can be efficiently computed using a recursive approach

$$h_k(t) = \frac{g(t)H_k(t)}{\sqrt{\pi^{1/2}2^k k!}}, \tag{4.131}$$

where $H_k(t)$ are Hermite polynomials, which obey the recurrence relation

$$H_k(t) = 2tH_{k-1}(t) - 2(k-1)H_{k-2}(t), \quad k \geq 2, \tag{4.132}$$

with initial conditions $H_0(t) = 1$ and $H_1(t) = 2t$. Two Hermite windows of orders k and l, where $k \neq l$, are orthonormal. Averaging spectra of the signal windowed by kth and lth order Hermite functions leads to [98]

- Widening the main peak,
- White Gaussian noise flattening.

Let us make use of these features in target kinematic parameter estimation in radars.

As we introduced in Section 4.4.3, the range-velocity map for radar can be formulated as

$$\begin{aligned}\mathscr{X}^h(t,\omega) &= \sum_{k=-\frac{K}{2}}^{\frac{K}{2}-1} \int_0^{T_b} x(\tau+kT)y^*(\tau+kT-t)\,h(kT-t)e^{-j\omega(kT-t)}\mathrm{d}\tau \\ &= \sum_{k=-\frac{K}{2}}^{\frac{K}{2}-1} \int_0^{T_b} x(\tau+kT+t)y^*(\tau+kT)\,h(kT)e^{-j\omega kT}\mathrm{d}\tau,\end{aligned} \tag{4.133}$$

with all symbols as stated so far. Again, the echo signal can be written as

$$x(t) = A_x(t)e^{j\phi_x(t)}y\left(t-\frac{R}{c}\right) = x_e(t)y\left(t-\frac{R}{c}\right), \tag{4.134}$$

where $A_x(t)$ and $\phi_x(t)$ denote the time-varying amplitude and phase, and $x_e(t)$ represents the target-induced modulation.

Differentiating (4.134) with respect to time yields

$$\frac{\mathrm{d}x(t)}{\mathrm{d}t} = \frac{\mathrm{d}y\left(t-\frac{R}{c}\right)}{\mathrm{d}t}x_e(t) + y\left(t-\frac{R}{c}\right)\frac{\mathrm{d}x_e(t)}{\mathrm{d}t}. \tag{4.135}$$

Assuming constant amplitude and velocity of the echo, the modulation term becomes

$$x_e(t) = e^{j\omega_x t}, \tag{4.136}$$

and its derivative is

$$\frac{\mathrm{d}x_e(t)}{\mathrm{d}t} = j\omega_x x(t), \tag{4.137}$$

where $\omega_x = -M2\pi a_1/\lambda$ denotes the angular Doppler frequency for a target moving with velocity a_1.

By differentiating (4.133) with respect to time and substituting (4.137), we obtain [189]

$$\mathscr{X}^{\mathscr{D}h}(t,\omega) = (j\omega - \omega_x)\mathscr{X}^h(t,\omega), \tag{4.138}$$

with $\mathscr{X}^{\mathscr{D}h}(t,\omega)$ denoting the range-velocity map computed using the derivative of the window. This leads to the Doppler frequency estimate

$$\hat{\omega}_x(t,\omega) = \omega - \Im\left(\frac{\mathscr{X}^{\mathscr{D}h}(t,\omega)}{\mathscr{X}^h(t,\omega)}\right). \tag{4.139}$$

Conveying the conceptual framework from time-frequency reassignment and synchrosqueezing techniques, the range-velocity distribution can be effectively sharpened by reallocating its spectral energy toward local ridges, using (4.139) as the reallocation rule. Therefore, we define the concentrated range-velocity distribution as

$$C^h(t,\omega) = \int_{-\infty}^{\infty} \left|\mathscr{X}^h(t,\Omega)\right| \delta\left(\omega - \hat{\omega}_x(t,\Omega)\right) \mathrm{d}\Omega, \tag{4.140}$$

where $\delta(\cdot)$ is the Dirac delta distribution. Unlike the classical time-frequency reassignment method [111, 112], the current approach applies reassignment only along the frequency (velocity) axis, preserving the range dimension (analogy to synchrosqueezing). While this enhances Doppler resolution and target separability, it may also increase noise variance and lead to false detections.

To mitigate this effect, a hybrid method that combines reassignment with multitaper averaging can be used. Since the range-velocity map is constructed via Fourier analysis in the slow-time domain, multitaper spectral estimation and spectral sharpening techniques can be employed. The multitaper range-velocity map is defined using Thomson's method as

$$\widehat{Z}(t,\omega) = \frac{1}{K}\sum_{k=0}^{K-1}\left|\mathscr{X}^{h_k}(t,\omega)\right|^2, \tag{4.141}$$

where $\mathscr{X}^{h_k}(t,\omega)$ represents the range-velocity map computed with K orthonormal windows $h_k(t)$, such as Hermite functions.

While this approach alleviates spectral fluctuations, it may degrade frequency resolution. To address this, we define the multitaper-concentrated range-velocity map as

$$\widehat{C}(t,\omega) = \frac{1}{K}\sum_{k=0}^{K-1}\left|C^{h_k}(t,\omega)\right|^2, \tag{4.142}$$

where $C^{h_k}(t,\omega)$ is the concentrated range-veolcity distribution given by (4.140) with the kth-order Hermite function as a window.

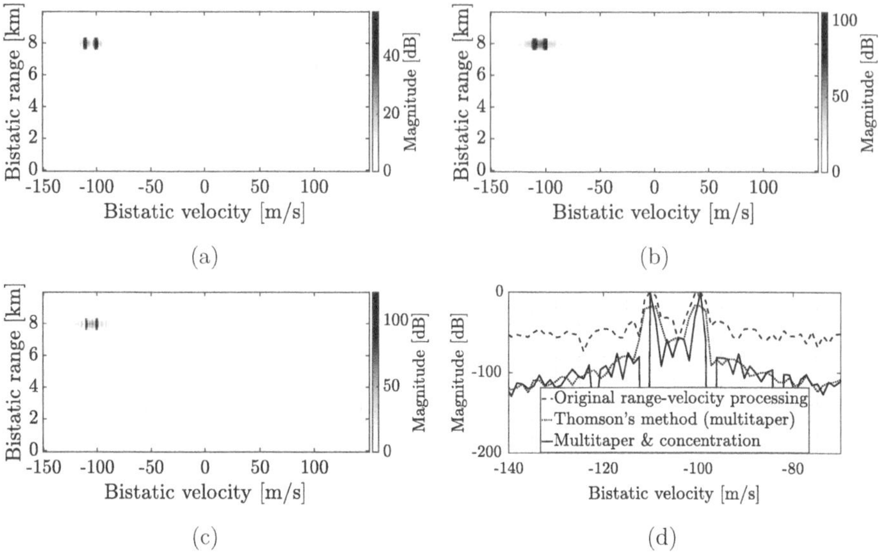

Figure 4.20 Passive radar simulation with two closely located targets and multitaper processing assuming $K = 2$. (a) Original range-velocity map; (b) range-velocity map for the multitaper (Thomson's) method; (c) range-velocity map for the multitaper technique with concentration; (d) cross-sections for the target echo.

Next, let us examine the simulation results for a scenario involving two closely spaced targets in the range-velocity plane. The transmitted signal is assumed to originate from a digital audio broadcasting (DAB) system, operating at a carrier frequency of $f_c = 210$ MHz with a bandwidth of $B = 1.536$ MHz and the sampling rate $f_s = 2$ MSa/s. The simulations were performed under a passive bistatic radar configuration. Both targets were located at the same bistatic range, $R = 8$ km, while their bistatic velocities were set to $V_1 = -100$ m/s and $V_2 = -110$ m/s, respectively. Figures 4.20–4.22 present the effect of varying K, the number of range-velocity maps averaged using Hermite windows of increasing order. The results clearly illustrate how this parameter influences both the target echoes' separability and the noise floor's suppression.

The presented results lead to several important observations. As the number of averaged range-velocity maps K increases, the noise surrounding the target echoes becomes progressively more flattened. In the case of Thomson's method (i.e., the multitaper approach without spectrum concentration), this flattening is achieved; however, it comes at the cost of increased echo smearing. Specifically, the target's response becomes more dispersed due to the use of higher-order Hermite windows, which compromises the accuracy of target localization. In contrast, when the multitaper technique is combined with spectral concentration, both noise suppression and peak sharpening are simultaneously achieved. This synergy allows for more precise

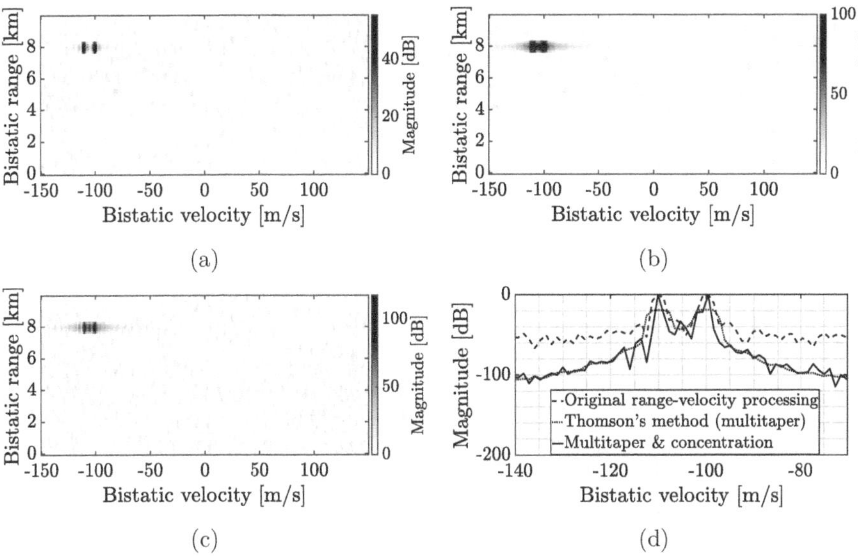

Figure 4.21 Passive radar simulation with two closely located targets and multitaper processing assuming $K = 4$. (a) Original range-velocity map; (b) range-velocity map for the multitaper (Thomson's) method; (c) range-velocity map for the multitaper technique with concentration; (d) cross-sections for the target echo.

separation of target echoes without the degradation of resolution. Notably, the difference in the noise floor between conventional range-velocity processing and the multitaper-concentration method reaches up to approximately 50 dB, underscoring the significant performance gain enabled by this approach.

Nevertheless, the parameter K serves as a tuning factor and must be carefully settled based on the specific scenario. Choosing a value that is too large may only offer a marginal echo magnitude reinforcement while significantly increasing computational complexity and causing a broadening of the target echo peak. Conversely, a value of K that is too small may result in insufficient noise flattening due to high variance and limited smoothing.

Setting $K \in [2, 4]$ typically provides a reasonable trade-off between Doppler resolution and noise variance in practical applications. This range ensures noticeable flattening of the noise floor without overly degrading the sharpness of the target response. Importantly, in scenarios involving multiple closely spaced targets, the presented method contributes to their improved separability, enhancing target detection and parameter estimation performance.

Next, let us analyze the real-life signal from a passive radar using the DAB signal as an illumination source. A 1 s-long segment of the recorded signal was selected for evaluation. The data were collected during the trials using a general-purpose software-defined receiver equipped with two directional antennas. The receiver was configured with a sampling rate of 2 MSa/s, an effective bandwidth of 1.5 MHz.

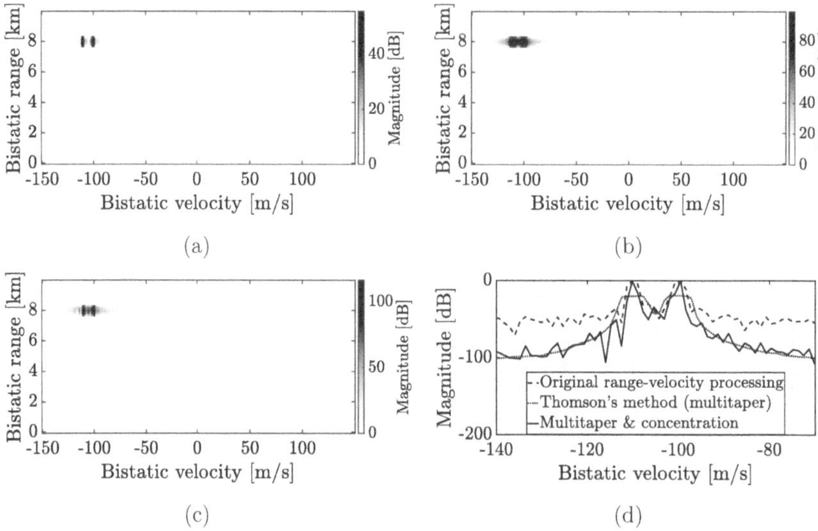

Figure 4.22 Passive radar simulation with two closely located targets and multitaper processing assuming $K = 6$. (a) Original range-velocity map; (b) range-velocity map for the multitaper (Thomson's) method; (c) range-velocity map for the multitaper technique with concentration; (d) cross-sections for the target echo.

A detailed description of the hardware setup, including the signal acquisition system's configuration and the measurement campaign's specific settings, can be found in [241]. Figure 4.23 delineates the outcomes.

Figure 4.23a shows the classical range-velocity map generated without enhancement techniques. A clear target echo, corresponding to a Boeing B789 Dreamliner, is observed at an approximate range of 8 km and radial velocity of 100 m/s. While the target is visible, the presence of tangible ambient noise increases the risk of false alarms due to the noise variance. To address this, Thomson's multitaper method was applied, in which multiple range-velocity maps, each computed with a different orthonormal window, are averaged. The result for $K = 8$ orthonormal tapers is shown in Figure 4.23b. As can be inferred, the noise level is greatly reduced, improving the robustness of target detection. However, this improvement comes at the expense of degraded velocity resolution, as the echo becomes smeared along the Doppler axis. This trade-off is characteristic of multitaper spectral estimation. To overcome this limitation, the multitaper method with concentration was applied. The resulting range-velocity map demonstrates a narrower, more focused target peak, facilitating improved detection accuracy. This enhancement is especially evident when analyzing cross-sections of the velocity axis at the target's range cell, where the benefits of echo sharpening and noise suppression are both clearly observed.

The presented technique has proven to be an effective tool for radar signal analysis. Moreover, as demonstrated in [242], it enhances the probability of target detection in low SNR conditions. In this section, we further analyzed the method's

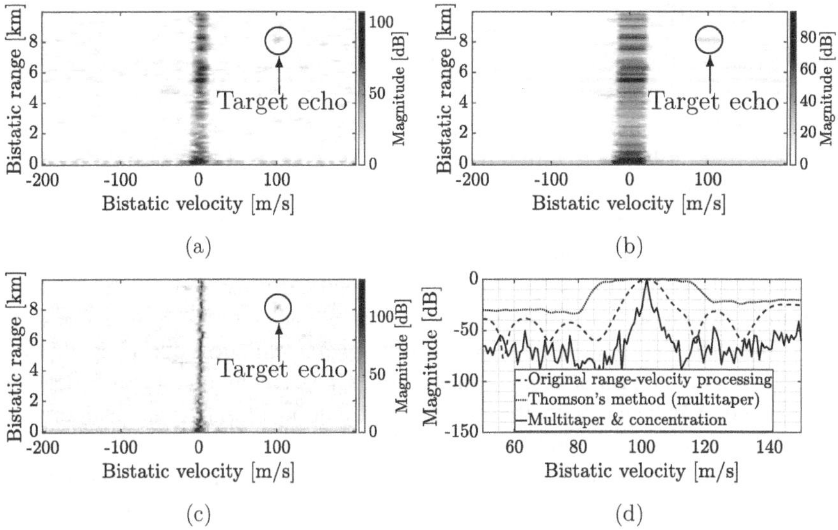

Figure 4.23 Real-life experiment results using DAB as an illuminator. (a) Original range-velocity map; (b) range-velocity map using Thomson's method for velocity estimation; (c) range-velocity map using the multitaper with concentration; (d) cross-sections through the target echo for all analyzed cases.

separability capabilities in scenarios involving two closely spaced targets located at the same range and moving with nearly identical velocities. The use of the multitaper approach not only flattens the noise surrounding the target echo but also concentrates its energy into a single, well-localized point corresponding to the target's velocity.

In the previous chapter, we presented similar considerations for frequency-modulated continuous-wave radar, where the method successfully enhanced a drone echo while simultaneously flattening the surrounding noise. In that case, the multitaper was applied in two dimensions due to the nature of the frequency-modulated continuous-wave radar systems. In this chapter, we focused solely on passive radar experiments, though the methodology can be applied to any system where velocity and/or range estimation relies on windowed Fourier transforms. By employing orthonormal Hermite functions as windowing functions, it is possible to suppress white noise. As a companion argument, spectrum concentration techniques enhance component separation in the frequency domain.

Nonetheless, it is important to note the parametric nature of the presented approach, which is primarily governed by the parameter K, representing the number of Hermite windows used and, consequently, the number of averaged range-velocity maps. Increasing K improves noise variance suppression but comes at the cost of reduced resolution. Conversely, choosing a value of K that is too small leads to minimal noise flattening while retaining sharp target echoes. Therefore, careful finetuning of K is crucial to balancing noise suppression and target resolution.

4.7 SUMMARY

We have reached the end of this chapter devoted to the estimation of target kinematic parameters in radar systems. Although this chapter does not exhaust all possible aspects of this vast and complex topic, it provides a solid foundation for learning, reviewing, and further exploration in the field of radar signal-processing.

Throughout this chapter, we have presented selected techniques for estimating the kinematic parameters of moving objects observed by radar sensors. Our focus was primarily on parameters such as velocity, acceleration, jerk, and snap, deliberately limiting the scope to the first few derivatives of motion, as this constraint is often sufficient for accurate target characterization in real-world scenarios. In particular, acceleration was given special attention due to its frequent occurrence and its pronounced influence on radar signal quality and detection performance. We began with the classical estimation of range and velocity, describing basic techniques and their limitations in the presence of target motion-induced migration effects. To address this, we introduced methods for compensating range and velocity migration, which are pivotal for accurate long-time coherent integration. Next, we explored several advanced strategies for acceleration estimation. The extended cross-ambiguity function, the generalized Radon-Fourier transform, and the Zak transform-based acceleration estimation were described as efficient approaches for modeling non-constant velocity motion. These techniques aim to compensate for acceleration effects to improve target localization and enhance the clarity of the range-velocity map. Simulation results and real radar data experiments validated the effectiveness of the discussed approaches. We then extended the discussion to high-order motion models by introducing jerk and snap as parameters for further refinement of the target motion characterization. Although higher-order derivatives can theoretically increase precision, our analysis suggests that only a few of the lowest-order terms typically suffice in practice. Finally, we discussed the multitaper spectrum estimation technique as a robust tool for alleviating fluctuations and enhancing target echo concentration. By leveraging orthonormal Hermite windows and combining multitaper averaging with spectral concentration, we were able to reduce the variance of background noise while preserving the sharpness of the target peak. This property of the joint multitaper method with concentration significantly improves detectability, particularly in scenarios with multiple closely spaced objects. Notably, these methods were successfully applied in both simulated and real-life passive radar signals.

In conclusion, this chapter has outlined key challenges and modern solutions in the domain of radar-based kinematic parameter estimation. By focusing on practically significant motion descriptors and combining classical radar processing with advanced signal enhancement techniques, we laid a framework for the accurate detection and classification of moving targets. The material presented here may serve as a principal outline of the current techniques and the material for further research.

4.8 CHAPTER IN A NUTSHELL

The critical content and findings presented in this chapter can be summarized as follows:

- In radar systems, moving targets are typically identified in the range-velocity domain by prominent peaks located on the corresponding two-dimensional map. Under ideal conditions, namely when the target maintains constant motion, these peaks can be sharpened and intensified by increasing the coherent integration time. However, when a target performs a maneuver or exhibits non-uniform motion, its echo tends to smear across the range-velocity plane, making it more challenging to detect and interpret accurately.
- For maneuvering targets, a common phenomenon is velocity migration, where the apparent Doppler frequency varies over time. Moreover, when the coherent processing interval is extended, range migration can also occur. This latter effect manifests as a displacement of the target echo from one range bin to another due to the target's continuous movement. Notably, range migration can affect even targets moving at constant velocity, depending on the radar signal bandwidth and the integration duration.
- The estimation of a target's kinematic parameters, such as velocity, acceleration, and higher-order motion descriptors, plays a pivotal role in improving detection performance under non-ideal motion conditions. These parameters can be used to perform signal compensation (also referred to as motion compensation or demodulation), which effectively mitigates the negative impact of motion-induced spreading, restoring the energy concentration of the echo.
- Manifold algorithmic approaches can be employed for kinematic parameter estimation. Broadly speaking, they fall into two main categories: iterative methods, which rely on successive refinement of the estimated parameters, and non-iterative (or gridless) techniques, which estimate the parameters directly based on analytical or transformation-based models. Each class of methods offers a trade-off between accuracy, robustness, and computational complexity.
- When high-order kinematic parameters are considered, such as acceleration (second-order derivative), jerk (third-order derivative), or snap (fourth-order derivative), the estimation and compensation process must be performed hierarchically. Specifically, one must begin by estimating and compensating for the highest-order term first, and then proceed sequentially to lower-order terms. This approach ensures proper demodulation and avoids error propagation across the compensation steps.
- A particularly effective approach for enhancing range-velocity map quality is the multitaper method for spectrum estimation. Originally designed for spectral analysis for variance reduction, this method has been successfully adapted to radar processing tasks. When applied to velocity

estimation, the multitaper technique not only flattens the surrounding noise floor but also improves the localization and sharpness of the target peak. Its combination with spectral concentration methods further enhances the suppression of background noise, resulting in better target-to-clutter contrast and improved detectability.

4.9 EXERCISES

1. Write a procedure that computes the correlation of two input signals by multiplying their spectra (use the FFT algorithm).
2. Write an implementation of the time-domain correlation. Compare the computation time of the time-domain correlation with your frequency-domain implementation from Exercise 1. Analyze signals of lengths from 2^4 to 2^{14} samples, increasing the exponent by one at each step. For each signal length, perform 100 iterations and take the minimum computation time as the result. Plot the computation times of the two correlation implementations as a function of signal length.
3. Assume that a monostatic radar observes a target moving with radial velocity of 100 m/s and that the observation time is 1 s. Determine the largest signal bandwidth B for which no range migration of the target echo will be observed. Also determine by how many range cells the target will shift if the bandwidth is 300 MHz.
4. Assume a passive radar that uses a signal of duration 100 ms sampled at 10 MSa/s. The signal contains an echo from a target whose amplitude is 30 dB below the clutter and inter-channel leakage. Determine the integration gain and the required noise level for which the target echo will exceed the detection threshold of 10 dB above the noise.
5. Using the idea from Figure 4.12a, write a procedure that computes the correlation of successive blocks of two signals.
6. Generate a complex baseband noise signal and apply low-pass filtering. You may assume a sampling rate of 1 MSa/s and a duration of 1 s. Create a second signal that is a delayed copy of the first with delay 100 μs. In this way, simulate a static target at a fixed range. Verify the procedure from Exercise 5 using the generated signals.
7. Introduce a small Doppler shift into the second signal, small enough so as not to cause range migration. Use a complex harmonic signal to implement the Doppler shift.
8. Using the idea from Figure 4.12b, write a procedure that computes the Fourier transform for each range cell and returns the range-velocity map. Analyze the obtained result. Test the procedure for various simulated ranges and velocities.
9. Compare the result with the one obtained in Exercise 8 of Chapter 1.
10. Examine how the range-velocity computation time changes when the correlation is performed using convolution in the time domain and when it is

performed using the Fourier transform. Analyze different signal lengths and observe the differences in processing time.

11. Generate an echo signal that contains two targets at the same range and the same velocity. Gradually increase the separation in range between the two targets and observe when you can reliably distinguish them. Repeat the experiments for signal bandwidths one half and two times the original bandwidth.
12. Generate an echo signal that contains two targets at the same range and the same velocity. Gradually increase the velocity of one of the targets and observe when you can reliably distinguish them. Repeat the experiments for signal durations that are two times longer and two times shorter than the original duration.
13. Extend Exercise 8 to include range migration. You may use a resampling technique that changes sample delays over time. Observe the effect on the range-velocity map.
14. To the simulation from Exercise 13, add target acceleration. Analyze the results on the range-velocity map.
15. Write a procedure that computes a range-velocity map with acceleration compensation as illustrated in Figure 4.6. Assume an analysis interval of accelerations around the value introduced in Exercise 14 and check whether acceleration compensation increases the echo magnitude. Repeat the calculations for a larger bank of Doppler filters and compare computation times.
16. Using the procedures developed in previous chapters for generating windows and their modifications, compute range-velocity maps for six windows as in Figure 4.13.
17. Write a procedure that estimates the target acceleration using the chosen estimator (4.89) or (4.91).
18. Derive estimators of instantaneous velocity, acceleration, jerk, and snap using derivatives of the window and of the signal with respect to time. Implement a procedure that computes these values.
19. Simulate an echo signal from a target that exhibits acceleration, jerk, and snap. Using the method of echo compensation from the highest-order polynomial coefficient to the lowest-order one, perform the decomposition of the echo signal. Compare results for different kinematic parameter values.
20. Observe what happens to the acceleration estimate if compensation for nonzero jerk and snap is omitted.
21. Extend the procedure from Exercise 15 to allow computation of a range-velocity map with compensation for acceleration, jerk, and snap. Simulate an echo signal containing acceleration, jerk, and snap. Estimate the kinematic parameters and compare estimation accuracy and computation time with the results from Exercise 19. Perform experiments for different sizes of the Doppler filter bank. Observe how computation density affects processing time.

22. Implement a procedure that realizes the generalized Radon-Fourier transform. Compare the estimation time obtained with this procedure for different grid densities. Determine the difference in precision and processing time between this procedure and the two previous methods.
23. Using Hermite windows, compute range-velocity maps averaged over two, four, and eight results. Observe how the target echo changes with the number of averages. Compare with the results obtained when averaging twenty distributions.
24. Write a procedure that computes a velocity-concentrated range-velocity map. Compare the echo magnitude before and after concentration.
25. Integrate the procedure from Exercise 23 with the procedure from Exercise 24 to obtain a concentrated range-velocity map with additionally flattened noise.

NOTES

1 The Fourier transform-based signal convolution becomes more efficient for the signal length of $N > 64$, albeit it is apparent for higher N.

2 In bistatic radars, we measure the target echo delay with respect to the reference signal reception, and this is called bistatic range – see Chapter 1 to recall fundamentals.

5 Conclusions, Challenges, Prospects

We have now reached the end of our journey. If you are reading this chapter, it means that you have either remained engaged throughout this book or are simply curious about how this story concludes. First and foremost, thank you for taking the time to explore this intellectual exercise. If you remember even a small fragment of what was written in any part of this book, its objective has been fulfilled. It is my hope that some readers will revisit this text from time to time to refresh their understanding of Fourier-based multidimensional and nonstationary signal-processing methods in the context of radar systems.

Although it has been emphasized several times that this book does not cover all aspects of the topic, its main goal was to provide a systematic overview of related techniques that help deepen the understanding of modern radar signal processing. The field is vast, and many areas remain unexplored or are described in scattered sources. This work aims to offer a coherent and accessible description of promising techniques for analyzing radar signals from a new perspective. If this book has helped you better understand even a single concept, then the effort invested in its creation was worthwhile. I encourage you to question, test, and build upon the ideas presented here, as the journey of learning never truly ends.

A MATLAB-based toolbox has been released alongside this book, allowing readers to apply the techniques described throughout the chapters to their own signals. The toolbox includes implementations of the key methods covered in the book, including time-frequency transformations and the W2D-FT-based two-dimensional signal analysis. It is designed to be modular, extendable, and easy to use. Whether for educational purposes, research, or prototyping, the toolbox offers a hands-on way to engage with the material and experiment with measured or simulated data. The toolbox is publicly available on GitHub under the following link:

`https://github.com/kabratkiewicz`

While this book outlines the current research landscape, several areas still deserve further exploration. The most promising directions for future research include:

- All the presented techniques have the potential to be implemented in real-world radar and spectrum monitoring systems. To achieve this, more sophisticated signal separation methods may be needed, particularly when multiple components coexist. This challenge applies to one-, two-, and multidimensional signals.
- Making the proposed techniques suitable for real-time applications remains difficult, especially for large-scale and high-dimensional data. This

DOI: 10.1201/9781003765783-5

goal demands both efficient algorithmic implementations and advancements in hardware, including specialized semiconductors.

- Multidimensional signal analysis can be further extended to multi-antenna systems, offering new and unconventional approaches to high-dimensional data processing.
- While one-dimensional signal decomposition is a well-established area of research with several mature and robust techniques, two-dimensional signal decomposition requires deeper insight into signal structure to enable effective component separation. The practical application of such techniques in radar systems may facilitate more efficient target imaging and detection in both active and passive radar configurations.
- Several methods described in this book rely on predefined parameters, such as the number of signal components or the initial window width. For autonomous systems, it is essential to develop adaptive algorithms that can automatically estimate or tune these parameters based on signal characteristics.
- This book adheres to the Fourier transform as the main tool for analyzing signals in the frequency domain. However, alternative transformations (those that overcome the limitations imposed by the Heisenberg-Gabor uncertainty principle) may offer even more flexibility and robustness. Incorporating such techniques will require additional theoretical groundwork and further mathematical exploration.
- In the context of target kinematic parameter estimation, there remains a fundamental trade-off between range and velocity migration versus integration time. Future work should address this issue, especially in scenarios with low signal-to-noise ratios and time constraints.

This book has been written for those, like myself, who were searching for clarity and structure in the world of time-frequency radar signal processing years ago. It is a compilation of insights, methods, and experiences gathered over the years, unified into a single narrative. I sincerely hope it serves as a helpful resource, a springboard for further learning, and perhaps even a spark for new discoveries. Thank you for joining me on this journey. The landscape of radar technology and signal analysis continues to evolve, and I hope this book becomes part of that ongoing transformation. May it guide, inspire, and support you as you pursue your own path in science and engineering.

References

1. L. Taycher. You Can Count the Number of Books in the World on 25,972,976 Hands, 2010.https://googleblog.blogspot.com/2010/08/you-can-count-number-of-books-in-world.html. Accessed: 2025-05-04.
2. M. I. Skolnik. Fifty Years of Radar. *Proceedings of the IEEE*, 73(2):182–197, 1985.
3. H. D. Griffiths. Radar Detection and Tracking of German V-2 Rocket Launches in WW2. *IEEE Aerospace and Electronic Systems Magazine*, 28(3):13–19, 2013.
4. V. S. Chernyak and I. Y. Immoreev. A Brief History of Radar. *IEEE Aerospace and Electronic Systems Magazine*, 24(9):B1–B32, 2009.
5. M. Guarnieri. The Early History of Radar [Historical]. *IEEE Industrial Electronics Magazine*, 4(3):36–42, 2010.
6. H. D. Griffiths, P. Knott, and W. Koch. Christian Hülsmeyer: Invention and Demonstration of Radar, 1904. *IEEE Aerospace and Electronic Systems Magazine*, 34(9):56–60, 2019.
7. N. J. Willis. *Bistatic Radar*. Artech House Radar Library. Institution of Engineering and Technology, 2005.
8. M. I. Skolnik. *Radar Handbook*, Third Edition. Electronics Electrical Engineering. McGraw-Hill Education, 2008.
9. M. Malanowski. *Signal Processing for Passive Bistatic Radar*. Artech House Radar Library. Artech House, 2019.
10. H. D. Griffiths and C. J. Baker. *An Introduction to Passive Radar*, Second Edition. Artech House Radar Library. Artech House, 2022.
11. E. Giaccari and G. Nucci. A Family of Air Traffic Control Radars. *IEEE Transactions on Aerospace and Electronic Systems*, AES-15(3):378–396, 1979.
12. P. Mattes. Systems of Systems: What, Exactly, Is an Integrated Air Defense System? In *The Mitchell Forum*, number 26, pp. 1–11, 2019. https://www.mitchellaerospacepower.org/app/uploads/2021/02/a2dd912f17e209f90f4aaab80b116e4d139eb4.pdf
13. E. D. Jansing. *Introduction to Synthetic Aperture Radar: Concepts and Practice*. McGraw-Hill Education, 2021.
14. F. Berizzi, M. Martorella, and E. Giusti. *Radar Imaging for Maritime Observation*. CRC Press, 2018.
15. K.-S. Chen. *Principles of Synthetic Aperture Radar Imaging: A System Simulation Approach*. CRC Press, 2016.
16. Z. Peng, C. Li, and F. Uysal. *Modern Radar for Automotive Applications*. The Institution of Engineering and Technology, 2022.
17. D. Burch. *Radar for Mariners*, Revised Edition. McGraw Hill LLC, 2013.
18. L. Zheng, M. Lops, Y. C. Eldar, and X. Wang. Radar and Communication Coexistence: An Overview: A Review of Recent Methods. *IEEE Signal Processing Magazine*, 36(5):85–99, 2019.
19. V. N. Bringi, K. V. Mishra, and M. Thurai. *Advances in Weather Radar*, Volume 1. The Institution of Engineering and Technology, 2023.
20. G. A. Fabrizio. *High Frequency Over-the-Horizon Radar: Fundamental Principles, Signal Processing, and Practical Applications*, 1st Edition. McGraw-Hill Education, 2013.

21. S. Trotta, D. Weber, R. W. Jungmaier, A. Baheti, J. Lien, D. Noppeney, M. Tabesh, C. Rumpler, M. Aichner, S. Albel, J. S. Bal, and I. Poupyrev. 2.3 SOLI: A Tiny Device for a New Human Machine Interface. In *2021 IEEE International Solid-State Circuits Conference (ISSCC)*, vol. 64, pp. 42–44, San Francisco, CA, USA, 2021.
22. W. L. Melvin, and J. A. Scheer. *Principles of Modern Radar: Volume 3: Radar Applications*. SciTech Publishing Inc, 2013.
23. J. Li, R. Hummel, P. Stoica, and E. G. Zelnio. *Radar Signal Processing and Its Applications*. Springer, 2013.
24. V. C. Chen. *The Micro-Doppler Effect in Radar*. Artech house, 2019.
25. M. Cherniakov. *Bistatic Radar: Principles and Practice*. John Wiley & Sons, 2008.
26. J. D. Taylor. *Advanced Ultrawideband Radar: Signals, Targets, and Applications*. CRC Press, 2016.
27. R. G. Lyons. *Understanding Digital Signal Processing*, First Edition. Addison-Wesley Longman Publishing Co., Inc., 1996.
28. B. R. Mahafza. *Radar Systems Analysis and Design Using MATLAB*. Chapman and Hall/CRC, 2005.
29. K. Abratkiewicz, A. Księżyk, M. Płotka, P. Samczyński, J. Wszołek, and T. P. Zieliński. SSB-Based Signal Processing for Passive Radar Using a 5G Network. *IEEE Journal of Selected Topics in Applied Earth Observations and Remote Sensing*, 16:3469–3484, 2023.
30. H. Rahman. *Fundamental Principles of Radar*. CRC Press, 2019.
31. B.G. Osgood. *Lectures on the Fourier Transform and Its Applications*, American Mathematical Society, 33:693, 2019.
32. J. V Stone. *The Fourier Transform: A Tutorial Introduction*. Sebtel Press, 2021.
33. I. N. Sneddon. *Fourier Transforms*. Courier Corporation, 1995.
34. E. W. Hansen. *Fourier Transforms: Principles and Applications*. John Wiley & Sons, 2014.
35. P. Dyke. *An Introduction to Laplace Transforms and Fourier Series*. Springer, 2001.
36. J. R. Hanna and J. H. Rowland. *Fourier Series, Transforms, and Boundary Value Problems*. Courier Corporation, 2008.
37. R. E. Edwards. *Fourier Series: A Modern Introduction Volume 2*, volume 85. Springer Science & Business Media, 2012.
38. J. F. James. *A Student's Guide to Fourier Transforms: With Applications in Physics and Engineering*, Third Edition. Student's Guides. Cambridge University Press, 2011.
39. R. N. Bracewell. *Solutions Manual to Accompany the Fourier Transform and Its Applications*. McGraw-Hill Electrical and Electronic Engineering Series. McGraw-Hill, 1978.
40. K. R. Rao, D. N. Kim, and J. J. Hwang. *Fast Fourier Transform – Algorithms and Applications*. Springer Science & Business Media, 2011.
41. J. B. J. Fourier and A. Freeman. *The Analytical Theory of Heat*. The University Press, 1878.
42. A. V. Oppenheim and R. W. Schafer. *Discrete-Time Signal Processing*. Prentice-Hall, 1989.
43. J. W. Cooley and J. W. Tukey. An Algorithm for the Machine Calculation of Complex Fourier Series. *Mathematics of Computation*, 19(90):297–301, 1965.
44. C. F. Gauss. Theoria Interpolationis Methodo Nova Tractata. In E. Schering, editor, *Werke, Band 3*, pp. 265–327. Königliche Gesellschaft der Wissenschaften zu Göttingen, 1866. Originally written in 1805, published posthumously in 1866.

45. A. Papoulis and S. U. Pillai. *Probability, Random Variables, and Stochastic Processes*, Fourth Edition. McGraw Hill, 2002.
46. P. Flandrin. *Explorations in Time-Frequency Analysis*. Cambridge University Press, 2018.
47. R. Silverman. Locally Stationary Random Processes. *IRE Transactions on Information Theory*, 3(3):182–187, 1957.
48. J. Xiao, P. Borgnat, P. Flandrin, and C. Richard. Testing Stationarity with Surrogates – A One-Class SVM Approach. In *2007 IEEE/SP 14th Workshop on Statistical Signal Processing*, pp. 720–724, Madison, WI, USA, 2007.
49. P. Borgnat, P. Flandrin, P. Honeine, C. Richard, and J. Xiao. Testing Stationarity with Surrogates: A Time-Frequency Approach. *IEEE Transactions on Signal Processing*, 58(7):3459–3470, 2010.
50. K. Kodera, C. De Villedary, and R. Gendrin. A New Method for the Numerical Analysis of Non-Stationary Signals. *Physics of the Earth and Planetary Interiors*, 12(2):142–150, 1976.
51. L. Cohen. *Time-Frequency Analysis: Theory and Applications*. Pnentice Hall, 1995.
52. S.-C. Pei and S.-G. Huang. STFT with Adaptive Window Width Based on the Chirp Rate. *IEEE Transactions on Signal Processing*, 60(8):4065–4080, 2012.
53. R. G. Baraniuk, P. Flandrin, A. J. E. M. Janssen, and O. J. J. Michel. Measuring Time-Frequency Information Content Using the Renyi Entropies. *IEEE Transactions on Information Theory*, 47(4):1391–1409, 2001.
54. S. Meignen, M. Colominas, and D.-H. Pham. On the Use of Renyi Entropy for Optimal Window Size Computation in the Short-Time Fourier Transform. In *ICASSP 2020–2020 IEEE International Conference on Acoustics, Speech and Signal Processing (ICASSP)*, pp. 5830–5834, Barcelona, Spain, 2020.
55. W. J. Williams, M. L. Brown, and A. O. Hero III. Uncertainty, Information, and Time-Frequency Distributions. In F. T. Luk, editor, *Advanced Signal Processing Algorithms, Architectures, and Implementations II*, vol. 1566, pp. 144– 156. International Society for Optics and Photonics, SPIE, 1991.
56. P. Flandrin, R. G. Baraniuk, and O. Michel. Time-Frequency Complexity and Information. In *Proceedings of ICASSP'94. IEEE International Conference on Acoustics, Speech and Signal Processing*, vol. iii, pp. III/329–III/332, Adelaide, SA, Australia, 1994.
57. W. J. Williams. Reduced Interference Distributions: Biological Applications and Interpretations. *Proceedings of the IEEE*, 84(9):1264–1280, 1996.
58. T.-H. Sang and W. J. Williams. Renyi Information and Signal-Dependent Optimal Kernel Design. In *1995 International Conference on Acoustics, Speech, and Signal Processing*, vol. 2, pp. 997–1000, Detroit, MI, USA, 1995.
59. S.-C. Pei and S.-G. Huang. Adaptive STFT with Chirp-Modulated Gaussian Window. In *2018 IEEE International Conference on Acoustics, Speech and Signal Processing (ICASSP)*, pp. 4354–4358, Calgary, AB, Canada, 2018.
60. K. Abratkiewicz and P. Samczyński. An Adaptive Spectrogram and Accelerogram Algorithm for Electronic Warfare Applications. In *2020 IEEE International Radar Conference (RADAR)*, pp. 530–535, Washington, DC, USA, 2020.
61. M. Zhu, X. Zhang, and Y. Qi. An Adaptive STFT Using Energy Concentration Optimization. In *2015 10th International Conference on Information, Communications and Signal Processing (ICICS)*, pp. 1–4, Singapore, 2015.

62. R. N. Czerwinski and D. L. Jones. Adaptive Short-Time Fourier Analysis. *IEEE Signal Processing Letters*, 4(2):42–45, 1997.
63. Y. Wei and G. Bi. Robust STFT with Adaptive Window Length and Rotation Direction. In *Fourth International Conference on Information, Communications and Signal Processing, 2003 and the Fourth Pacific Rim Conference on Multimedia. Proceedings of the 2003 Joint*, vol. 2, pp. 827–829, Singapore, 2003.
64. L. Li, H. Cai, H. Han, Q. Jiang, and H. Ji. Adaptive Short-Time Fourier Transform and Synchrosqueezing Transform for Non-Stationary Signal Separation. *Signal Processing*, 166:107231, 2020.
65. H. Xie, J. Lin, Y. Lei, and Y. Liao. Fast-Varying AM–FM Components Extraction Based on an Adaptive STFT. *Digital Signal Processing*, 22(4):664–670, 2012.
66. D. Gabor. Theory of Communication. *Journal of the Institution of Electrical Engineers*, 93(26):429–457, 1946.
67. W. Siebert. A Radar Detection Philosophy. *IRE Transactions on Information Theory*, 2(3):204–221, 1956.
68. A. Grossmann and T. Paul. Wave Functions on Subgroups of the Group of Affine Canonical Transformations. In *Resonances—Models and Phenomena: Proceedings of a Workshop Held at the Centre for Interdisciplinary Research Bielefeld University*, Bielefeld, Germany, pp. 128–138. Springer, 9–14, 1984.
69. A. Papoulis. *Signal Analysis*. McGraw-Hill, 1981.
70. D. Jones and T. Parks. Time-Frequency Window Leakage in the Short-Time Fourier Transform. *Circuits, Systems and Signal Processing*, 6(3):239–262, 1993.
71. S. Mann and S. Haykin. The Chirplet Transform: A Generalization of Gabor's Logon Transform. *Vision Interface*, 91:205–212, 1991.
72. D. Mihovilovic and R. N. Bracewell. Adaptive Chirplet Representation of Signals in the Time-Frequency Plane. *Electronics Letters*, 27(13):1159–1161, 1991.
73. S. Mann and S. Haykin. The Chirplet Transform: Physical Considerations. *IEEE Transactions on Signal Processing*, 43(11):2145–2160, 1995.
74. K. Czarnecki, D. Fourer, F. Auger, and M. Rojewski. A Fast Time-Frequency Multi-Window Analysis Using a Tuning Directional Kernel. *Signal Processing*, 147:110–119, 2018.
75. K. Abratkiewicz. Double-Adaptive Chirplet Transform for Radar Signature Extraction. *IET Radar, Sonar & Navigation*, 14(10):1463–1474, 2020.
76. A. Grossmann and J. Morlet. Decomposition of Hardy Functions into Square Integrable Wavelets of Constant Shape. *SIAM Journal on Mathematical Analysis*, 15(4):723–736, 1984.
77. X. Wang. Electronic Radar Signal Recognition Based on Wavelet Transform and Convolution Neural Network. *Alexandria Engineering Journal*, 61(5):3559–3569, 2022.
78. J. Torres, A. Marcano, and D. Andina. Radar Chirp Detection Through Wavelet Transform. In *2007 IEEE International Symposium on Industrial Electronics*, pp. 3420–3423, Harbin, China, 2007.
79. S. Zhang, J. Fan, L. Shou, and J. Dong. A Detection Method of Radar Signal by Wavelet Transforms. In *Fourth International Conference on Fuzzy Systems and Knowledge Discovery (FSKD 2007)*, vol. 2, pp. 710–714, Haikou, China, 2007.
80. R. N. Khushaba and A. J. Hill. Radar-Based Materials Classification Using Deep Wavelet Scattering Transform: A Comparison of Centimeter vs. Millimeter Wave Units. *IEEE Robotics and Automation Letters*, 7(2):2016–2022, 2022.

81. S. Mallat. *A Wavelet Tour of Signal Processing*. Academic Press, 1999.
82. D. B. Percival. *Wavelet Methods for Time Series Analysis*. Cambridge University Press, 2000.
83. M. Vetterli. *Wavelets and Subband Coding*. Prentice Hall PTR, 1995.
84. I. Daubechies. *Ten Lectures on Wavelets*. Society for Industrial and Applied Mathematics, 1992.
85. A. A. Petrosian and F. G. Meyer. *Wavelets in Signal and Image Analysis: From Theory to Practice*, vol. 19. Springer Science & Business Media, 2013.
86. K. Kodera, R. Gendrin, and C. Villedary. Analysis of Time-Varying Signals with Small BT Values. *IEEE Transactions on Acoustics, Speech, and Signal Processing*, 26(1):64–76, 1978.
87. F. Auger and P. Flandrin. Improving the Readability of Time-Frequency and Time-Scale Representations by the Reassignment Method. *IEEE Transactions on Signal Processing*, 43(5):1068–1089, 1995.
88. F. Auger, P. Flandrin, Y.-T. Lin, S. McLaughlin, S. Meignen, T. Oberlin, and H.-T Wu. Time-Frequency Reassignment and Synchrosqueezing: An Overview. *IEEE Signal Processing Magazine*, 30(6):32–41, 2013.
89. P. Flandrin, F. Auger, and E. Chassande-Mottin. Time-Frequency Reassignment: From Principles to Algorithms. In *Applications in Time-Frequency Signal Processing*, pp. 179–204. CRC Press, (Ed.) A. Papandreou-Suppappola, 2018.
90. R. Amiri and A. Shahzadi. Micro-Doppler Based Target Classification in Ground Surveillance Radar Systems. *Digital Signal Processing*, 101:102702, 2020.
91. B. Falęcki, K. Abratkiewicz, and P Samczyński. Measurement of Human Vital Signs Based on High-Resolution mmWave Micro-Doppler Radar Signatures. In *2024 International Radar Symposium (IRS)*, pp. 260–265, Wroclaw, Poland, 2024.
92. A. Izzo, L. Ausiello, C. Clemente, and J. J. Soraghan. Radar Micro-Doppler for Loudspeaker Analysis: An Industrial Process Application. In *International Conference on Radar Systems (Radar 2017)*, pp. 1–6. IET, Belfast, Ireland, 2017.
93. K. Abratkiewicz and J. Gambrych. Real-Time Variants of Vertical Synchrosqueezing: Application to Radar Remote Sensing. *IEEE Journal of Selected Topics in Applied Earth Observations and Remote Sensing*, 15:1760–1774, 2022.
94. D. White, M. Jahangir, A. Kumar Mishra, C. J. Baker, and M. Antoniou. Latent Variable and Classification Performance Analysis of Bird–Drone Spectrograms With Elementary Autoencoder. *IEEE Transactions on Radar Systems*, 3:115–123, 2025.
95. D. Belgiovane and C.-C. Chen. Micro-Doppler Characteristics of Pedestrians and Bicycles for Automotive Radar Sensors at 77 GHz. In *2017 11th European Conference on Antennas and Propagation (EUCAP)*, pp. 2912–2916, Paris, France, 2017.
96. K. Abratkiewicz, D. Gromek, K. Stasiak, and P. Samczyński. Time-Frequency Reassigned Micro-Doppler Signature Analysis Using the XY-DemoRad System. In *2019 Signal Processing Symposium (SPSympo)*, pp. 331–334, Krakow, Poland, 2019.
97. P. Samczynski, K. Stasiak, D. Gromek, K. Kulpa, and J. Misiurewicz. XY-DemoRad – Novel K- and mm-Band Radar Demo Kit for Educational and Commercial Applications. In *2019 20th International Radar Symposium (IRS)*, pp. 1–11, Ulm, Germany, 2019.
98. D. J. Thomson. Spectrum Estimation and Harmonic Analysis. *Proceedings of the IEEE*, 70(9):1055–1096, 1982.
99. D. B. Percival and A. T. Walden. *Spectral Analysis for Physical Applications: Multitaper and Conventional Univariate Techniques*. Cambridge University Press, 1993.

100. K. S. Riedel and A. Sidorenko. Minimum Bias Multiple Taper Spectral Estimation. *IEEE Transactions on Signal Processing*, 43(1):188–195, 1995.
101. S. Karnik, J. Romberg, and M. A. Davenport. Thomson's Multitaper Method Revisited. *IEEE Transactions on Information Theory*, 68(7):4864–4891, 2022.
102. I. Daubechies. Time-Frequency Localization Operators: A Geometric Phase Space Approach. *IEEE Transactions on Information Theory*, 34(4):605–612, 1988.
103. T. W. Parks and R. G. Shenoy. Time-Frequency Concentrated Basis Functions. In *International Conference on Acoustics, Speech, and Signal Processing*, vol. 5, pp. 2459–2462, Albuquerque, NM, USA, 1990.
104. P. Flandrin. Maximum Signal Energy Concentration in a Time-frequency Domain. In *ICASSP-88., International Conference on Acoustics, Speech, and Signal Processing*, vol. 4, pp. 2176–2179, New York, NY, USA, 1988.
105. M. Bayram and R. G. Baraniuk. Multiple window time-frequency analysis. In *Proceedings of Third International Symposium on Time-Frequency and Time-Scale Analysis (TFTS-96)*, pp. 173–176, Paris, France, 1996.
106. P. Flandrin. A Note on Reassigned Gabor Spectrograms of Hermite Functions. *Journal of Fourier Analysis and Applications*, 19:285–295, 2013.
107. G. Fraser and B. Boashash. Multiple Window Spectrogram and Time-Frequency Distributions. In *Proceedings of ICASSP'94. IEEE International Conference on Acoustics, Speech and Signal Processing*, vol. iv, pp. IV/293–IV/296, Adelaide, SA, Australia, 1994.
108. M. Bayram and R. G. Baraniuk. Multiple Window Time-Varying Spectrum Estimation. *Nonlinear and Nonstationary Signal Processing*, 49:292–316, 2000.
109. F. Cakrak and P. J. Loughlin. Multiple Window Time-Varying Spectral Analysis. *IEEE Transactions on Signal Processing*, 49(2):448–453, 2001.
110. M. J. Prerau, R. E. Brown, M. T. Bianchi, J. M. Ellenbogen, and P. L. Purdon. Sleep Neurophysiological Dynamics through the Lens of Multitaper Spectral Analysis. *Physiology*, 32(1):60–92, 2017.
111. J. Xiao and P. Flandrin. Multitaper Time-Frequency Reassignment. In *2006 14th European Signal Processing Conference*, pp. 1–5, Florence, Italy, 2006.
112. J. Xiao and P. Flandrin. Multitaper Time-Frequency Reassignment for Nonstationary Spectrum Estimation and Chirp Enhancement. *IEEE Transactions on Signal Processing*, 55(6):2851–2860, 2007.
113. K. Abratkiewicz and P. Samczyński. Multitaper Time-Frequency Reassigned Spectrogram in Micro-Doppler Radar Signal Analysis. In *2021 Signal Processing Symposium (SPSympo)*, pp. 1–5, Łódź, Poland, 2021.
114. M. Płotka, K. Abratkiewicz, and M. Malanowski. Improving FM-Based Passive Forward-Scattering Signature Readability Using Multitaper Reassignment. In *2023 24th International Radar Symposium (IRS)*, pp. 1–10, Berlin, Germany, 2023.
115. I. Daubechies, J. Lu, and H.-T. Wu. Synchrosqueezed Wavelet Transforms: An Empirical Mode Decomposition-Like Tool. *Applied and Computational Harmonic Analysis*, 30(2):243–261, 2011.
116. T. Oberlin, S. Meignen, and V. Perrier. The Fourier-Based Synchrosqueezing Transform. In *2014 IEEE International Conference on Acoustics, Speech and Signal Processing (ICASSP)*, pp. 315–319, Florence, Italy, 2014.
117. D. Fourer, F. Auger, K. Czarnecki, S. Meignen, and P. Flandrin. Chirp Rate and Instantaneous Frequency Estimation: Application to Recursive Vertical Synchrosqueezing. *IEEE Signal Processing Letters*, 24(11):1724–1728, 2017.

118. T. Oberlin, S. Meignen, and V. Perrier. Second-Order Synchrosqueezing Transform or Invertible Reassignment? Towards Ideal Time-Frequency Representations. *IEEE Transactions on Signal Processing*, 63(5):1335–1344, 2015.
119. Q. Legros and D. Fourer. A Novel Pseudo-Bayesian Approach for Robust Multi-Ridge Detection and Mode Retrieval. In *2021 29th European Signal Processing Conference (EUSIPCO)*, pp. 1925–1929. IEEE, Dublin, Ireland, 2021.
120. Q. Legros and D. Fourer. Pseudo-Bayesian Approach for Robust Mode Detection and Extraction Based on the STFT. *Sensors*, 23(1):85, 2022.
121. Q. Legros and D. Fourer. Time-Frequency Ridge Estimation of Multi-Component Signals Using Sparse Modeling of Signal Innovation. *arXiv preprint arXiv:2212.11343*, 2022.
122. J. M. Miramont, F. Auger, M. A. Colominas, N. Laurent, and S. Meignen. Unsupervised Classification of the Spectrogram Zeros with an Application to Signal Detection and Denoising. *Signal Processing*, 214:109250, 2024.
123. N. Laurent and S. Meignen. A Novel Ridge Detector for Nonstationary Multicomponent Signals: Development and Application to Robust Mode Retrieval. *IEEE Transactions on Signal Processing*, 69:3325–3336, 2021.
124. N. Laurent, M. A. Colominas, and S. Meignen. On Local Chirp Rate Estimation in Noisy Multicomponent Signals: With an Application to Mode Reconstruction. *IEEE Transactions on Signal Processing*, 70:3429–3440, 2022.
125. K. Abratkiewicz. Radar Detection-Inspired Signal Retrieval from the Short-Time Fourier Transform. *Sensors*, 22(16):5954, 2022.
126. D. Fourer, J. Harmouche, J. Schmitt, T. Oberlin, S. Meignen, F. Auger, and P. Flandrin. The ASTRES Toolbox for Mode Extraction of Non-Stationary Multicomponent Signals. In *2017 25th European Signal Processing Conference (EUSIPCO)*, pp. 1130–1134, Kos, Greece, 2017.
127. G. Thakur, E. Brevdo, N. S. Fučkar, and H. Wu. The Synchrosqueezing Algorithm for Time-Varying Spectral Analysis: Robustness Properties and New Paleoclimate Applications. *Signal Processing*, 93(5):1079–1094, 2013.
128. J. Harmouche, D. Fourer, F. Auger, P. Borgnat, and P. Flandrin. The Sliding Singular Spectrum Analysis: A Data-Driven Nonstationary Signal Decomposition Tool. *IEEE Transactions on Signal Processing*, 66(1):251–263, 2018.
129. M. Plotka, K. Abratkiewicz, M. Malanowski, K. Kulpa, and P. Samczyński. Passive Forward Scattering Signal Extraction Using Second-Order Vertical Synchrosqueezing. In *2021 21st International Radar Symposium (IRS)*, pp. 1–10, Berlin, Germany, 2021.
130. D.-H. Pham and S. Meignen. High-Order Synchrosqueezing Transform for Multicomponent Signals Analysis—With an Application to Gravitational-Wave Signal. *IEEE Transactions on Signal Processing*, 65(12):3168–3178, 2017.
131. K. Abratkiewicz. On the Instantaneous Angular Jerk Estimation in the Time-Frequency Domain. *IEEE Signal Processing Letters*, 28:798–802, 2021.
132. K. Abratkiewicz, J. Drozdowicz, and P. Samczyński. Vertical Synchrosqueezing for High-Resolution Radar Imaging. *IEEE Transactions on Geoscience and Remote Sensing*, 61:1–13, 2023.
133. D. Fourer and F. Auger. Second-Order Time-Reassigned Synchrosqueezing Transform: Application to Draupner Wave Analysis. In *2019 27th European Signal Processing Conference (EUSIPCO)*, pp. 1–5. IEEE, Coruna, Spain, 2019.

134. K. Abratkiewicz, J. Gambrych, K. Stasiak, and P. Samczyński. Estimation of Rotational Speed of Helicopter Rotor through Horizontal Synchrosqueezing. In *2021 Signal Processing Symposium (SPSympo)*, pp. 6–10, Łodz, Poland, 2021.
135. X. Tu, Q. Zhang, Z. He, Y. Hu, S. Abbas, and F. Li. Generalized Horizontal Synchrosqueezing Transform: Algorithm and Applications. *IEEE Transactions on Industrial Electronics*, 68(6):5293–5302, 2021.
136. L. Stanković, S. Stanković, and I. Djurović. An Architecture for Realization of the Cross-Terms Free Polynomial Wigner-Ville distribution. In *1997 IEEE International Conference on Acoustics, Speech, and Signal Processing*, vol. 3, pp. 2053–2056, Munich, Germany, 1997.
137. H. Ijima, A. Ohsumi, and I. Djurović. Parameter Estimation of FM Signals in Random Noise Using Wigner Distribution. In *SICE 2004 Annual Conference*, vol. 2, pp. 1417–1420, Sapporo, Japan, 2004.
138. I. Djurović and L. Stanković. STFT-Based Estimator of Polynomial Phase Signals. *Signal Processing*, 92(11):2769–2774, 2012.
139. V. Sucic, J. Lerga, and B. Boashash. Multicomponent Noisy Signal Adaptive Instantaneous Frequency Estimation Using Components Time Support Information. *IET Signal Processing*, 8(3):277–284, 2014.
140. M. Wang, A. K. Chan, and C. K. Chui. Linear Frequency-Modulated Signal Detection Using Radon-Ambiguity Transform. *IEEE Transactions on Signal Processing*, 46(3):571–586, March 1998.
141. S. Peleg and B. Friedlander. The Discrete Polynomial-Phase Transform. *IEEE Transactions on Signal Processing*, 43(8):1901–1914, 1995.
142. S. Kay and G. Boudreaux-Bartels. On the Optimality of the Wigner Distribution for Detection. In *ICASSP'85. IEEE International Conference on Acoustics, Speech, and Signal Processing*, vol. 10, pp. 1017–1020, Tampa, FL, USA, 1985.
143. J. Angeby. Estimating Signal Parameters Using the Nonlinear Instantaneous Least Squares Approach. *IEEE Transactions on Signal Processing*, 48(10):2721–2732, 2000.
144. R. Schmidt. Multiple Emitter Location and Signal Parameter Estimation. *IEEE Transactions on Antennas and Propagation*, 34(3):276–280, March 1986.
145. F. G. Geroleo, M. Brandt-Pearce, and C. L. Brown. Detection and Estimation of Multi-Pulse LFMCW Radar Signals. In *2010 IEEE Radar Conference*, pp. 1009–1013, Arlington, VA, USA, 2010.
146. I. Djurović. Covariance of Wigner Distribution-based Instantaneous Frequency Estimates. *Electronics Letters*, 54(6):391–393, 2018.
147. Y. Yao and X. Zhang. Parameter Estimation for HFM Signals Using Combined STFT and Iteratively Reweighted Least Squares Linear Fitting. *Signal Processing*, 92(11):2775–2783, 2014.
148. A. Serbes and O. Aldimashki. A Fast and Accurate Chirp Rate Estimation Algorithm Based on the Fractional Fourier Transform. In *2017 25th European Signal Processing Conference (EUSIPCO)*, pp. 1105–1109, Kos, Greece, 2017.
149. E. Świercz. Estimation of Polynomial Frequency Modulation Law for FM Signals Based on Modified Extended Generalized Chirp Transform. In *2017 Signal Processing Symposium (SPSympo)*, pp. 1–5, Jachranka, Poland, 2017.
150. P. Samczyński. Extended Generalized Chirp Transform for Signal Parameter Estimation in Bistatic Passive Pulse Radars. In *2013 14th International Radar Symposium (IRS)*, vol. 1, pp. 155–160, Dresden, Germany, 2013.

151. K. Abratkiewicz, P. Samczyński, and K. Czarnecki. Radar Signal Parameters Estimation Using Phase Accelerogram in the Time-Frequency Domain. *IEEE Sensors Journal*, 19(13):5078–5085, 2019.
152. K. Czarnecki and M. Moszyński. A Novel Method of Local Chirp-Rate Estimation of LFM Chirp Signals in the Time-Frequency Domain. In *2013 36th International Conference on Telecommunications and Signal Processing (TSP)*, pp. 704–708, Rome, Italy, 2013.
153. A. E. Barnes. The Calculation of Instantaneous Frequency and Instantaneous Bandwidth. *Geophysics*, 57(11):1520–1524, 1992.
154. D. J. Nelson. Instantaneous Higher Order Phase Derivatives. *Digital Signal Processing*, 12(2):416–428, 2002.
155. S. Fulop and K. Fitz. Algorithms for Computing the Time-Corrected Instantaneous Frequency (Reassigned) Spectrogram, With Applications. *The Journal of the Acoustical Society of America*, 119:360–71, 2006.
156. D. Nelson. Cross-Spectral Methods for Processing Speech. *The Journal of the Acoustical Society of America*, 110:2575–2592, 2001.
157. F. Auger, É. Chassande-Mottin, and P. Flandrin. On Phase-Magnitude Relationships in the Short-Time Fourier Transform. *IEEE Signal Processing Letters*, 19(5):267–270, 2012.
158. K. Abratkiewicz, P. Samczyński, R. Rytel-Andrianik, and Z. Gajo. Multipath Interference Removal in Receivers of Linear Frequency Modulated Radar Pulses. *IEEE Sensors Journal*, 21(17):19000–19012, 2021.
159. B. Boots, A. Okabe, and K. Sugihara. Spatial Tessellations: Concepts and Applications of Voronoi Diagrams. *Geographical Information Systems*, 1:503–526, 1999.
160. P. Flandrin. On Spectrogram Local Maxima. In *2017 IEEE International Conference on Acoustics, Speech and Signal Processing (ICASSP)*, pp. 3979–3983, New Orleans, LA, USA, 2017.
161. P. Flandrin. Time–Frequency Filtering Based on Spectrogram Zeros. *IEEE Signal Processing Letters*, 22(11):2137–2141, 2015.
162. R. Bardenet, J. Flamant, and P. Chainais. On the Zeros of the Spectrogram of White Noise. *Applied and Computational Harmonic Analysis*, 48(2):682–705, 2020.
163. N. Laurent, S. Meignen, M. A. Colominas, J. M. Miramont, and F. Auger. A Novel Approach Based on Voronoï Cells to Classify Spectrogram Zeros of Multicomponent Signals. In *ICASSP 2023–2023 IEEE International Conference on Acoustics, Speech and Signal Processing (ICASSP)*, pp. 1–5, Rhodes Island, Greece, 2023.
164. V. Bargmann. On a Hilbert Space of Analytic Functions and an Associated Integral Transform Part I. *Communications on Pure and Applied Mathematics*, 14(3):187–214, 1961.
165. K. Abratkiewicz. Radar Detection-Inspired Signal Retrieval from the Short-Time Fourier Transform. *Sensors*, 22(16):5954, 2022.
166. B. Fałęcki, K. Abratkiewicz, P. Gomez del Hoyo, and P. Samczyński. 3D Visualization of High-Resolution Radar-Measured Vital Sign Signatures Using Range-Time-Frequency Representation. In *21st International Conference on Smart Technologies (EUROCON)*, pp. 1–6. IEEE, Gdynia, Poland, 2025.
167. C. Özdemir. *Inverse Synthetic Aperture Radar Imaging with MATLAB Algorithms*. Wiley Series in Microwave and Optical Engineering. Wiley, 2012.
168. J. C. Curlander and R. N. McDonough. *Synthetic Aperture Radar: Systems and Signal Processing*. John Wiley & Sons, 1991.

169. I. G. Cumming and F. H. Wong. *Digital Processing of Synthetic Aperture Radar Data: Algorithms and Implementation*. Artech House, 2005.
170. V. C. Chen and H. Ling. *Time-Frequency Transforms for Radar Imaging and Signal Analysis*. Artech House, 2002.
171. M. A. Richards, et al. *Fundamentals of Radar Signal Processing*, volume 1. McGraw-Hill, 2005.
172. Y. Wang and Y. Jiang. Inverse Synthetic Aperture Radar Imaging of Maneuvering Target Based on the Product Generalized Cubic Phase Function. *IEEE Geoscience and Remote Sensing Letters*, 8(5):958–962, 2011.
173. F. Berizzi, E. D. Mese, M. Diani, and M. Martorella. High-Resolution ISAR Imaging of Maneuvering Targets by Means of the Range Instantaneous Doppler Technique: Modeling and Performance Analysis. *IEEE Transactions on Image Processing*, 10(12):1880–1890, 2001.
174. S. Brisken and J. G. Worms. ISAR Motion Parameter Estimation via Multilateration. In *2011 Microwaves, Radar and RemotE Sensing Symposium*, pp. 190–194, Kiev, Ukraine, 2011.
175. Q. Zhang, T. S. Yeo, G. Du, and S. Zhang. Estimation of Three-Dimensional Motion Parameters in Interferometric ISAR Imaging. *IEEE Transactions on Geoscience and Remote Sensing*, 42(2):292–300, 2004.
176. S. Zhang, Y. Liu, and X. Li. Pseudomatched-Filter-Based ISAR Imaging under Low SNR Condition. *IEEE Geoscience and Remote Sensing Letters*, 11(7):1240–1244, 2014.
177. S. Zhang, Y. Liu, and X. Li. Bayesian Bistatic ISAR Imaging for Targets with Complex Motion under Low SNR Condition. *IEEE Transactions on Image Processing*, 27(5):2447–2460, 2018.
178. F. Santi, D. Pastina, and M. Bucciarelli. Multi-Sensor ISAR Techniques for Motion Estimation of Pitching, Rolling and Yawing Targets. In *2013 IEEE Radar Conference (RadarCon13)*, pp. 1–6, Ottawa, ON, Canada, 2013.
179. J. Fu, M. Xing, M. Amin, and G. Sun. ISAR Translational Motion Compensation with Simultaneous Range Alignment and Phase Adjustment in Low SNR Environments. In *2021 IEEE Radar Conference (RadarConf21)*, pp. 1–6, Atlanta, GA, USA, 2021.
180. D. Li, M. Zhan, X. Zhang, Z. Fang, and H. Liu. ISAR Imaging of Nonuniformly Rotating Target Based on the Multicomponent CPS Model under Low SNR Environment. *IEEE Transactions on Aerospace and Electronic Systems*, 53(3):1119–1135, 2017.
181. D. Li, M. Zhan, H. Liu, Y. Liao, and G. Liao. A Robust Translational Motion Compensation Method for ISAR Imaging Based on Keystone Transform and Fractional Fourier Transform under Low SNR Environment. *IEEE Transactions on Aerospace and Electronic Systems*, 53(5):2140–2156, 2017.
182. L. Liu, F. Zhou, M. Tao, P. Sun, and Z. Zhang. Adaptive Translational Motion Compensation Method for ISAR Imaging under Low SNR Based on Particle Swarm Optimization. *IEEE Journal of Selected Topics in Applied Earth Observations and Remote Sensing*, 8(11):5146–5157, 2015.
183. Y. Li, J. Zhang, J. Zheng, J. Niu, Y. Zhou, and N. Meng. ISAR Imaging of Nonuniformly Rotating Targets with Low SNR Based on Third Order Autocorrelation Function. *IEEE Access*, 8:74707–74719, 2020.
184. J. Flamant, N. Le Bihan, and P. Chainais. Time-Frequency Analysis of Bivariate Signals. *Applied and Computational Harmonic Analysis*, 46(2):351–383, 2019.

185. J. Zak. Finite Translations in Solid-State Physics. *Physical Review Letters*, 19:1385–1387, 1967.
186. J. Zak. Dynamics of Electrons in Solids in External Fields. *Physical Review*, 168:686–695, Apr 1968.
187. A. K. Brodzik. Signal Extrapolation in the Real Zak Space. *IEEE Transactions on Signal Processing*, 50(8):1957–1964, 2002.
188. V. S. Bhat, G. Harshavardhan, and A. Chockalingam. Input-Output Relation and Performance of RIS-Aided OTFS with Fractional Delay-Doppler. *IEEE Communications Letters*, 27(1):337–341, 2023.
189. K. Abratkiewicz and Z. Gajo. Gridless High-Order Target Kinematic Parameter Estimation in Passive Radars—From Cross-Ambiguity Function to Zak Transform. *IEEE Journal of Selected Topics in Applied Earth Observations and Remote Sensing*, 17:15255–15268, 2024.
190. Q. Kemao. Two-Dimensional Windowed Fourier Transform for Fringe Pattern Analysis: Principles, Applications and Implementations. *Optics and Lasers in Engineering*, 45(2):304–317, 2007.
191. K. R. Fitz and S. A. Fulop. A Unified Theory of Time-Frequency Reassignment. *arXiv preprint arXiv:0903.3080*, 2009.
192. K. Abratkiewicz. Windowed Two-Dimensional Fourier Transform Concentration and Its Application to ISAR Imaging. *IEEE Transactions on Image Processing*, 32:6260–6273, 2023.
193. K. Stasiak, M. Ciesielski, J. Julczyk, W. Duda, G. Pietrzykowski, and P. Samczyński. Real-Time High Resolution Multichannel ISAR Imaging System. In *International Conference on Radar Systems (RADAR 2022)*, vol. 2022, pp. 389–394, Hybrid Conference, Edinburgh, UK, 2022.
194. M. Caris, S. Stanko, S. Palm, R. Sommer, A. Wahlen, and N. Pohl. 300 GHz Radar for High Resolution SAR and ISAR Applications. In *2015 16th International Radar Symposium (IRS)*, pp. 577–580, Dresden, Germany, 2015.
195. P. Knott and R. Perkuhn. Non-Destructive Permittivity Measurement of Thin Dielectric Sheets: Quality Conformance Testing for the Tracking and Imaging Radar TIRA. In *2015 German Microwave Conference*, pp. 25–28, Nuremberg, Germany, 2015.
196. X. Zhu, W. Hu, and B. Guo. A Review of Super-Resolution Inverse Synthetic Aperture Radar Imaging Algorithms. In *2017 7th IEEE International Conference on Electronics Information and Emergency Communication (ICEIEC)*, pp. 107–110, 2017.
197. L. Mingjiu, M. Lei, M. Jianchao, C. Wenfeng, Y. Jun, M. Xiaoyan, and C. Qi. Fast, Super-Resolution Sparse Inverse Synthetic Aperture Radar Imaging via Continuous Compressive Sensing. *IET Signal Processing*, 16(3):310–326, 2022.
198. T. Yang, H. Shi, J. Guo, and Z. Qiao. Orbital-Angular-Momentum-Based Super-Resolution ISAR Imaging for Maneuvering Targets: Modeling and Performance Analysis. *Digital Signal Processing*, 117:103197, 2021.
199. E. Giusti, D. Cataldo, A. Bacci, S. Tomei, and M. Martorella. ISAR Image Resolution Enhancement: Compressive Sensing Versus State-of-the-Art Super-Resolution Techniques. *IEEE Transactions on Aerospace and Electronic Systems*, 54(4):1983–1997, 2018.
200. J. H. G. Ender. On Compressive Sensing Applied to Radar. *Signal Processing*, 90(5):1402–1414, 2010. Special Section on Statistical Signal & Array Processing.

201. M. S. Kang and K. T. Kim. Compressive Sensing Approach for High-Resolution ISAR Image Reconstruction and Autofocus. *The Journal of Engineering*, 2019(20):7017–7020, 2019.
202. G. Zhao, Z. Wang, Q. Wang, G. Shi, and F. Shen. Robust ISAR Imaging Based on Compressive Sensing from Noisy Measurements. *Signal Processing*, 92(1):120–129, 2012.
203. Y. Y. Dou, Y. M. Wu, H. Q. Jin, Y. Q. Jin, J. Hu, and J. Cheng. The Efficient Norm Regularization Method Applying on the ISAR Image with Sparse Data. *IEEE Transactions on Geoscience and Remote Sensing*, 60:1–11, 2022.
204. M. S. Kang, S. J. Lee, S. H. Lee, and K. T. Kim. ISAR Imaging of High-Speed Maneuvering Target Using Gapped Stepped-Frequency Waveform and Compressive Sensing. *IEEE Transactions on Image Processing*, 26(10):5043–5056, 2017.
205. Q. Liu and Y. Wang. Spectral-Factorisation Root-MUSIC Algorithm for Super-Resolution ISAR Imaging. *The Journal of Engineering*, 2019:7125–7129(4), 2019.
206. R. Gong, L. Wang, and D. Zhu. High Resolution 3D InISAR Imaging of Space Targets Based on PFA Algorithm with Single Baseline. In *2023 24th International Radar Symposium (IRS)*, pp. 1–10, Berlin, Germany, 2023.
207. R. Deming, M. Best, and S. Farrell. Polar Format Algorithm for SAR Imaging with MATLAB. In *Algorithms for Synthetic Aperture Radar Imagery XXI*, vol. 9093, pp. 47–66. SPIE, Baltimore, MD, United States, 2014.
208. A. W. Doerry. Basics of Polar-Format Algorithm for Processing Synthetic Aperture Radar Images. Technical report, Sandia National Laboratories (SNL), Albuquerque, NM, and Livermore, CA, 2012.
209. K. Abratkiewicz. A Robust CLEAN-like Approach to ISAR Image Decomposition with Non-stationary Signal Components. *IEEE Transactions on Geoscience and Remote Sensing*, 2026. doi: 10.1109/TGRS.2026.3700848
210. R. A. Carmona, W.-L. Hwang, and B. Torresani. Characterization of Signals by the Ridges of Their Wavelet Transforms. *IEEE Transactions on Signal Processing*, 45(10):2586–2590, 1997.
211. A. Hanssen. Multidimensional Multitaper Spectral Estimation. *Signal Processing*, 58(3):327–332, 1997.
212. A. Wünsche. General Hermite and Laguerre Two-Dimensional Polynomials. *Journal of Physics A: Mathematical and General*, 33(8):1603, 2000.
213. A. Wünsche. Hermite and Laguerre 2D Polynomials. *Journal of Computational and Applied Mathematics*, 133(1):665–678, 2001.
214. S. Khan and M. A. Pathan. Lie-Theoretic Generating Relations of Hermite 2D Polynomials. *Journal of Computational and Applied Mathematics*, 160(1):139–146, 2003.
215. K. Abratkiewicz. Multitaper ISAR Noise Suppression. *IEEE Transactions on Geoscience and Remote Sensing*, 62:1–13, 2024.
216. V. Tuzlukov. *Signal Processing in Radar Systems*. CRC Press, 2017.
217. J. Palmer, S. Palumbo, A. Summers, D. Merrett, S. Searle, and S. Howard. An Overview of an Illuminator of Opportunity Passive Radar Research Project and Its Signal Processing Research Directions. *Digital Signal Processing*, 21(5):593–599, 2011.
218. D. Eustice, C. Baylis, and R. J. Marks. Woodward's Ambiguity Function: From Foundations to Applications. In *2015 Texas Symposium on Wireless and Microwave Circuits and Systems (WMCS)*, pp. 1–17, Waco, TX, USA, 2015.

219. P. M. Woodward. *Probability and Information Theory with Applications to Radar, A volume in International Series of Monographs on Electronics and Instrumentation*, volume 3. Elsevier, 1953.
220. K. S. Kulpa and J. Misiurewicz. Stretch Processing for Long Integration Time Passive Covert Radar. In *2006 CIE International Conference on Radar*, pp. 1–4, Shanghai, China, 2006.
221. Y. Li, T. Zeng, T. Long, and Z. Wang. Range Migration Compensation and Doppler Ambiguity Resolution by Keystone Transform. In *2006 CIE International Conference on Radar*, pp. 1–4. IEEE, Shanghai, China, 2006.
222. M. Malanowski. Detection and Parameter Estimation of Manoeuvring Targets with Passive Bistatic Radar. *IET Radar, Sonar & Navigation*, 6:739–745(6), 2012.
223. K. Abratkiewicz, M. Malanowski, and Z. Gajo. Target Acceleration Estimation in Active and Passive Radars. *IEEE Journal of Selected Topics in Applied Earth Observations and Remote Sensing*, 16:9193–9206, 2023.
224. X. Li, G. Cui, W. Yi, and L. Kong. Radar Maneuvering Target Detection and Motion Parameter Estimation Based on TRT-SGRFT. *Signal Processing*, 133:107–116, 2017.
225. P. You, Z. Ding, L. Qian, M. Li, X. Zhou, W. Liu, and S. Liu. A Motion Parameter Estimation Method for Radar Maneuvering Target in Gaussian Clutter. *IEEE Transactions on Signal Processing*, 67(20):5433–5446, 2019.
226. J. Tian, W. Cui, and S. Wu. A Novel Method for Parameter Estimation of Space Moving Targets. *IEEE Geoscience and Remote Sensing Letters*, 11(2):389–393, 2014.
227. Z. Solatzadeh and A. Zaimbashi. Accelerating Target Detection in Passive Radar Sensors: Delay-Doppler-Acceleration Estimation. *IEEE Sensors Journal*, 18(13):5445–5454, 2018.
228. X. Li, G. Cui, L. Kong, and W. Yi. Fast Non-Searching Method for Maneuvering Target Detection and Motion Parameters Estimation. *IEEE Transactions on Signal Processing*, 64(9):2232–2244, 2016.
229. X. Gao, H. Zhang, and T. Dang. Radar Detection and Fast Motion Parameter Estimation for Complex Manoeuvering Targets at High Speed and Acceleration. *IET Radar, Sonar & Navigation*, 16(12):1977–1996, 2022.
230. P. Huang, G. Liao, Z. Yang, X.-G. Xia, J.-T. Ma, and J. Ma. Long-Time Coherent Integration for Weak Maneuvering Target Detection and High-Order Motion Parameter Estimation Based on Keystone Transform. *IEEE Transactions on Signal Processing*, 64(15):4013–4026, 2016.
231. Q. Liu, J. Guo, Z. Liang, and T. Long. Motion Parameter Estimation and HRRP Construction for High-Speed Weak Targets Based on Modified GRFT for Synthetic-Wideband Radar with PRF Jittering. *IEEE Sensors Journal*, 21(20):23234–23244, 2021.
232. M. Żywek. Rocket Launch Detection Using VHF DVB-T Passive Bistatic Radar. In *2024 International Radar Symposium (IRS)*, pp. 293–297, Wroclaw, Poland, 2024.
233. J. Xu, J. Yu, Y.-N. Peng, and X.-G. Xia. Radon-Fourier Transform for Radar Target Detection, I: Generalized Doppler Filter Bank. *IEEE Transactions on Aerospace and Electronic Systems*, 47(2):1186–1202, 2011.
234. J. Xu, J. Yu, Y.-N. Peng, and X.-G. Xia. Radon-Fourier Transform for Radar Target Detection (II): Blind Speed Sidelobe Suppression. *IEEE Transactions on Aerospace and Electronic Systems*, 47(4):2473–2489, 2011.
235. J. Yu, J. Xu, Y.-N. Peng, and X.-G. Xia. Radon-Fourier Transform for Radar Target Detection (III): Optimality and Fast Implementations. *IEEE Transactions on Aerospace and Electronic Systems*, 48(2):991–1004, 2012.

236. J. Xu, X.-G. Xia, S.-B. Peng, J. Yu, Y.-N. Peng, and L.-C. Qian. Radar Maneuvering Target Motion Estimation Based on Generalized Radon-Fourier Transform. *IEEE Transactions on Signal Processing*, 60(12):6190–6201, 2012.
237. X. Zhou, P. Wang, J. Chen, Z. Men, W. Liu, and H. Zeng. A Modified Radon Fourier Transform for GNSS-Based Bistatic Radar Target Detection. *IEEE Geoscience and Remote Sensing Letters*, 19:1–5, 2022.
238. I. Gladkova. Zak Transform and a New Approach to Waveform Design. *IEEE Transactions on Aerospace and Electronic Systems*, 37(4):1458–1464, 2001.
239. I. Gladkova. Design of Frequency Modulated Waveforms via the Zak Transform. *IEEE Transactions on Aerospace and Electronic Systems*, 40(1):355–359, 2004.
240. C. Moscardini, D. Petri, A. Capria, M. Conti, M. Martorella, and F. Berizzi. Batches Algorithm for Passive Radar: A Theoretical Analysis. *IEEE Transactions on Aerospace and Electronic Systems*, 51(2):1475–1487, 2015.
241. G. Mazurek. DAB Signal Preprocessing for Passive Coherent Location. *Sensors*, 22(1):378, 2022.
242. K. Abratkiewicz and G. Mazurek. Passive Radar Noise Flattening Using Multitaper Spectrum Estimation. In *2025 Signal Processing Symposium (SPSympo)*, pp. 1–6, Warsaw, Poland,n2025.

Index

A

Acceleration estimation, 231–252
extended cross-ambiguity function, 232–235
generalized Radon-Fourier transform, 235–241
Zak transform-based acceleration estimation, 241–252
Active radar, 4
monostatic configuration, 4
Adaptive short-time Fourier transform (adaptive STFT), 46–49
Adaptive STFT. *See* Short-time Fourier transform (STFT)
Adaptive windowed two-dimensional chirplet transform, 159–166
chirp-rate adaptation, 159–166
Adaptive windowed two-dimensional Fourier transform (adaptive W2D-FT), 153–159
window size selection, 153–159
Adaptive W2D-FT. *See* Adaptive windowed two-dimensional Fourier transform (adaptive W2D-FT)
Ambiguity function, 12–15
auto-ambiguity function, 6–7, 13–15
cross-ambiguity function, 13
integrated sidelobe level, 13
peak-to-sidelobe ratio, 13–14
peak width, 13
sidelobes, 13–15
waveform assessment, 13–15
Autocorrelation, 6, 11

B

Bandwidth, 6, 9, 41–46
radar waveform, 6
range resolution, 9
STFT window-width choice, 41–46
ultrawideband radar, 6
Barker code, 14–15
Bistatic radar, 4–10
bistatic angle, 10
geometry, 4–5, 8–9
passive radar geometry, 4–5, 8–9
range equation, 7–8
range estimation, 8–9
range resolution, 9
velocity estimation, 10
velocity resolution, 10
Bivariate spectrum, 141–148
construction from one-dimensional signal, 141–148
W2D-FT, 148–153

C

Carrier frequency, 2, 6, 9–10
Chirp rate, 43–48, 53–58, 80–81, 96–106, 159–166, 241
estimation, 53–58, 97–106
W2D-CT, 159–166
Chirp signal, 14–15, 49–58
Coherent integration, 235
long-time coherent integration, 235–241
Complex exponential representation, 20–21
Continuous Fourier transform (CFT), 20–21, 132–133
Convolution theorem, 30–31
Fourier transform property, 30–31
Correlation, 11–13
autocorrelation, 11
cross-correlation, 11, 19
delay estimation, 11–12
matched filtering relation, 12, 19

Cross-ambiguity function, 13, 232–235
  extended cross-ambiguity function, 232–235
Cross-correlation. *See* Correlation

D

DFT. *See* Discrete Fourier transform (DFT)
Dimension-selective reassignment, 173–190, 198–209
  ISAR image noise flattening, 198–209
  multitaper reassignment, 198–209
  W2D-FT, 173–190
Discrete Fourier transform (DFT), 22–26
  computational complexity, 25
  inverse DFT, 22
  matrix representation, 23–24
  relation to DTFT, 22
  spectrum magnitude, 22–24
  two-dimensional, 133
Discrete-time Fourier transform (DTFT), 21–24
  continuous-frequency representation, 22–24
  inverse DTFT, 22
  relation to DFT, 22
  sampled signals, 21–22
  two-dimensional, 133
Doppler frequency, 2, 10, 13–15
  bistatic, 10
  target velocity estimation, 2, 10
Doppler shift, 6–7, 13–15, 18–19
  micro-Doppler signatures, 18–19
DTFT. *See* Discrete-time Fourier transform (DTFT)

E

ECAF. *See* Extended cross-ambiguity function (ECAF)
Echo sharpening, 261–268
Energy concentration, 44–46, 57–58, 68, 153–215
Extended cross-ambiguity function (ECAF), 232–235
  acceleration estimation, 232–235

F

Fast Fourier transform (FFT), 24–26
  computational complexity, 25
  radix-2 algorithm, 25–26
  spectrogram computation, 39–40
  two-dimensional implementation, 133
FFT. *See* Fast Fourier transform (FFT)
Four-dimensional reassignment, 166–171, 209–215
  range-velocity map case study, 209–215
  W2D-FT concentration, 166–171
Fourier transform, 19–33, 131–140
  complex exponential representation, 20–21
  continuous, 20–21, 132–133
  discrete, 22–26, 133
  discrete-time, 21–24, 133
  fast, 24–26, 133
  one-dimensional, 19–33
  properties, 26–33, 134–140
  two-dimensional, 131–140
Fourier transform properties, 26–33, 134–140
  convolution theorem, 30–31
  differentiation in time, 31–32
  duality, 29
  frequency shifting, 28
  linearity, 26–27
  multiplication in time domain, 32–33
  Parseval's theorem, 30
  scaling in time, 28–29
  time reversal, 32
  time shifting, 27–28

Frequency-domain representation, 20–24, 39, 133
Frequency shifting, 28
Fourier transform property, 28

G

Gaussian window, 37–39, 43–49, 148–150, 154–160, 190
STFT window function, 37–38
W2D-FT, 148–150, 154–160
Generalized Radon-Fourier transform (GRFT), 235–241
acceleration estimation, 235–241
long-time coherent integration, 235–241
target motion compensation, 235–241
GRFT. *See* Generalized Radon-Fourier transform (GRFT)

H

Hamming window, 37–39
Heisenberg uncertainty principle, 41–42, 58, 78, 171, 192
Hermite function, 71–74, 79, 198–210, 261–268
Hermite window, 198–210, 262–268
multitaper spectrum estimation, 261–268
multitaper W2D-FT, 198–210
High-order target kinematic parameter estimation, 252–261
Horizontal synchrosqueezing, 92–97

I

Illuminators of opportunity, 14
Instantaneous frequency, 58–67, 76–97, 166–190, 202, 249
estimation, 58–67, 76–97
W2D-FT reassignment, 166–190
Inverse discrete Fourier transform (IDFT). *See* Discrete Fourier transform (DFT)
Inverse discrete-time Fourier transform (IDTFT). *See* Discrete-time Fourier transform (DTFT)
Inverse short-time Fourier transform (ISTFT), 38–39
signal reconstruction, 38
Invertible W2D-FT, 190–192
signal reconstruction, 191–192
ISAR image, 124–131, 174–190, 196–209
decomposition, 196–198
filtering, 124–131
noise flattening, 198–209
W2D-FT reassignment, 174–190
ISTFT. *See* Inverse short-time Fourier transform (ISTFT)

J

Jamming signals, 19

L

Linear frequency-modulated (FM) signal, 14–15, 18
Linearity, 26–27
Fourier transform property, 26–27
Local stationarity, 35–36
Long-time coherent integration, 235–241
generalized Radon-Fourier transform, 235–241

M

Matched filter, 12, 19
correlation-based processing, 12, 19
impulse response, 12
signal-to-noise ratio maximization, 12
Micro-Doppler phenomenon, 5
Micro-Doppler signatures, 18–19, 64–66, 117–124

spatial micro-Doppler signature, 117–124
target feature extraction, 117–124
Monostatic radar, 4–10
geometry, 4–5
range equation, 7
range estimation, 8
range resolution, 9
velocity estimation, 10
velocity resolution, 10
Multicomponent signal, 19, 36, 40–46, 58, 63, 79, 93, 97, 103, 106, 194
Multidimensional signal analysis, 115–215
bivariate spectrum, 141–148
W2D-FT, 148–215
Multitaper methods, 68–76, 198–215, 261–268
Hermite functions, 71–74, 198–210, 261–268
ISAR image noise flattening, 198–209
orthonormal windows, 69–75, 198–209, 261–268
reassignment, 68–76, 198–215
Slepian sequences, 68–70
spectrum estimation, 261–268
Multitaper reassignment, 68–76, 198–215
dimension-selective W2D-FT reassignment, 198–209
four-dimensional W2D-FT reassignment, 209–215
Multistatic radar, 4–5, 9
Multitaper spectrum estimation, 261–268
echo sharpening, 261–268
noise flattening, 261–268

N

Noise flattening, 68–76, 198–209, 261–268
ISAR image, 198–209
multitaper spectrum estimation, 261–268
multitaper STFT, 68–76
Noise suppression, 198–215, 261–268
Nonstationary signal, 35–36, 49–58, 76–97, 115, 190–198
multicomponent signal, 36, 40–46
STFT, 36
W2D-FT signal decomposition, 190–198

O

One-dimensional Fourier transform, 19–33
One-dimensional signal, 18–19
jamming signals, 19
micro-Doppler signatures, 18–19
radar pulses, 18
range-compressed signal, 19
Orthonormal windows, 69–75, 198–209, 261–268
Hermite functions, 71–74, 198–210, 261–268
Slepian sequences, 68–70

P

Parseval's theorem, 30
Fourier transform property, 30
Passive radar, 4–5, 8–9, 14
bistatic geometry, 4–5, 8–9
illuminators of opportunity, 14
signals of opportunity, 4
waveform control limitation, 4
Peak-to-sidelobe ratio. *See* Ambiguity function
Pulse radar, 2, 6–7

Q

Quasi-monostatic radar, 4–5, 8–10

R

Radar architecture, 1–2
Radar classification, 3–6
- frequency bands, 6
- geometry, 4–5
- principle of operation, 4
- waveform types, 6–7

Radar cross-section (RCS), 7
Radar equation, 7–8
- bistatic, 7–8
- monostatic, 7

Radar pulses, 18, 33–36, 40
- Fourier spectrum, 33–34
- nonstationarity, 33–36
- spectrogram, 40

Radar waveform, 1, 6–7, 13–15
- auto-ambiguity function, 6–7, 13–15
- bandwidth, 6, 9, 14–15
- continuous waveform, 6
- pulse waveform, 6–7
- sidelobes, 13–15

Range-compressed signal, 19
Range-Doppler processing, 115–116, 141–144, 227
Range estimation, 2, 8–9, 221–225
- bistatic radar, 8–9
- correlation-based, 11–12
- monostatic radar, 8
- target kinematic parameter estimation, 221–225

Range migration, 225–231
- compensation, 225–231

Range resolution, 9
- bandwidth dependence, 9
- bistatic radar, 9
- monostatic radar, 9

Range-velocity map, 145–148, 162–166, 209–215, 221–268
- acceleration estimation, 241–252
- multitaper spectrum estimation, 261–268
- W2D-FT reassignment, 209–215

RCS. *See* Radar cross-section (RCS)
Reassignment, 58–76, 166–190, 198–215
- dimension-selective, 173–190, 198–209
- four-dimensional, 166–171, 209–215
- frequency reassignment, 58–68
- multitaper, 68–76, 198–215
- time-frequency, 58–76
- W2D-FT, 166–190, 198–215

Rectangular window, 37–39
spectral leakage, 37–38
Rényi entropy, 43–46, 53–55, 67, 89, 156–170, 184, 212

S

Scaling in time, 28–29
- Fourier transform property, 28–29

Short-time chirplet transform (STCT), 49–58
- chirp-rate adaptation, 49–58
- instantaneous bandwidth, 49–58
- instantaneous frequency, 49–58

Short-time Fourier transform (STFT), 36–49
- adaptive STFT, 46–49
- inverse STFT, 38–39
- multitaper STFT, 68–76
- nonstationary signals, 36
- spectrogram, 39–46
- window function, 36–38
- window-width choice, 41–46

Signal decomposition, 24, 190–198
- W2D-FT, 190–198

Signal parameter estimation, 97–106
- chirp-rate estimation, 97–106
- frequency estimation, 97–106
- phase derivative, 97–106
- spectrogram morphology, 106–111

Signal reconstruction, 38–39, 77–92, 190–198

inverse STFT, 38–39
vertical synchrosqueezing, 77–92
W2D-FT, 190–198
Signal retrieval, 190–198
component extraction, 190–198
concentrated W2D-FT, 190–198
Signals of opportunity, 4
Slepian sequences, 68–70
Spatial frequency, 132–190, 202–215
W2D-FT, 148–190
W2D-FT reassignment, 166–190
Spatial micro-Doppler signature, 117–124
target feature extraction, 117–124
Spectral leakage, 37–38, 68–76, 198
Spectrogram, 39–46, 106–111
energy distribution, 39–46
micro-Doppler signatures, 64–66, 117–124
morphology, 106–111
Rényi entropy, 43–46
synchrosqueezing, 77–97
Stationarity, 35–36
local stationarity, 35–36
observation scale, 35–36
STCT. *See* Short-time chirplet transform (STCT)
STFT. *See* Short-time Fourier transform (STFT)
Synchrosqueezing, 76–97
horizontal, 92–97
signal reconstruction, 77–92
vertical, 77–92

T

Target detection, 1–2, 5, 11–15, 19
ambiguity function, 13–15
correlation, 11
jamming signals, 19
Target kinematic parameter estimation, 220–268
acceleration estimation, 231–252
high-order target kinematic parameter estimation, 252–261
multitaper spectrum estimation, 261–268
range and velocity estimation, 221–225
range and velocity migration, 225–231
Zak transform-based acceleration estimation, 241–252
Target range, 2, 7–9
Target velocity, 2, 10, 221–225
Taylor expansion, 126, 194
Time-frequency analysis, 19, 35–111
nonstationary signals, 35–36
parameter estimation, 97–106
reassignment, 58–76
short-time chirplet transform, 49–58
short-time Fourier transform, 36–49
spectrogram, 39–46
synchrosqueezing, 76–97
Time-frequency distribution, 39–40, 58–76, 106–111
Time-frequency reassignment. *See* Reassignment
Time-frequency representation, 39–46, 58–97
Time-frequency resolution trade-off, 41–46
Heisenberg uncertainty principle, 41–42
window-width choice, 41–46
Time-range domain, 191–198
W2D-FT signal reconstruction, 191–198
Two-dimensional discrete Fourier transform. *See* Two-dimensional Fourier transform
Two-dimensional Fourier transform, 131–140

continuous, 132–133
discrete, 133
discrete-time, 133
fast, 133
properties, 134–140
Two-dimensional signal, 131–153, 190–198
bivariate spectrum, 141–148
signal decomposition, 190–198
W2D-FT, 148–153

U

Ultrawideband radar, 6
Uncertainty principle. *See* Heisenberg uncertainty principle

V

Velocity estimation, 2, 10, 221–225
bistatic radar, 10
Doppler-frequency relation, 2, 10
monostatic radar, 10
target kinematic parameter estimation, 221–225
Velocity migration, 225–231
compensation, 225–231
Velocity resolution, 10
Vertical synchrosqueezing, 77–92
signal reconstruction, 77–92

W

W2D-CT. *See* Windowed two-dimensional chirplet transform (W2D-CT)
W2D-FT. *See* Windowed two-dimensional Fourier transform (W2D-FT)
Wavelength, 2, 7, 10
Wideband radar, 6
Window function, 36–49, 148–159, 190–198
Gaussian window, 37–39, 148–150
Hamming window, 37–39
Hermite function, 71–74, 198–210, 261–268
rectangular window, 37–39
STFT, 36–38
W2D-FT, 148–159
Windowed two-dimensional chirplet transform (W2D-CT), 159–166
adaptive W2D-CT, 159–166
chirp-rate adaptation, 159–166
Windowed two-dimensional Fourier transform (W2D-FT), 148–215
adaptive W2D-FT, 153–159
bivariate spectrum, 148–153
concentration, 166–190
dimension-selective reassignment, 173–190
four-dimensional reassignment, 166–171
multitaper reassignment, 198–215
range-velocity map case study, 209–215
readability improvement, 153–166
signal retrieval, 190–198
spatial frequency, 148–190
Window-width choice, 41–46
Heisenberg uncertainty principle, 41–42
STFT, 41–46

Z

Zak transform, 241–252
acceleration estimation, 241–252
range-velocity map, 241–252
Zak transform-based acceleration estimation, 241–252

For Product Safety Concerns and Information please contact our EU representative GPSR@taylorandfrancis.com Taylor & Francis Verlag GmbH, Kaufingerstraße 24, 80331 München, Germany